Imaging, Modeling and Assimilation in Seismology

Imaging, Modeling and Assimilation in Seismology

Edited by
Yong-Gang Li

De Gruyter 高等教育出版社 HIGHER EDUCATION PRESS

Physics and Astronomy Classification 2010: 91.30.Ab, 91.30.Bi, 91.30.Jk, 91.30.pd, 93.85.Rt.

ISBN 978-3-11-025902-5
e-ISBN 978-3-11-025903-2

Library of Congress Cataloging-in-Publication Data

Imaging, modeling, and assimilation in seismology / edited by Yong-Gang Li.
p. cm.
Includes bibliographical references and index.
1. Seismology – Data processing. 2. Seismology – Mathematical models. I. Li, Yong-Gang.
QE539.2.D36I46 2011
551.2201'13–dc23
2011035607

Bibliographic information published by the Deutsche Nationalbibliothek

The Deutsche Nationalbibliothek lists this publication in the Deutsche Nationalbibliografie; detailed bibliographic data are available in the Internet at http://dnb.d-nb.de.

Printing and binding: Hubert & Co. GmbH & Co. KG, Göttingen
♾ Printed on acid-free paper

Printed in Germany

www.degruyter.com

Preface

This book is a monograph of the earth science specializing in observational, computational, and interpretational seismology, containing the full-3D waveform tomography method and one-return propagator method in modeling and imaging of subsurface structure, observations and 3-D finite-difference simulations of fault-zone trapped wave for high-resolution characterization of fault internal structure and physical properties, co-seismic rock damage and post-mainshock heal, spontaneous dynamic source and strong ground motion simulations, and earthquake hazard assessment, discrete element method for earthquake mechanics and fracture modeling, and stress load and unload measurements for earthquake prediction and hazard assessment. The editor approaches this as a broad interdisciplinary effort, with well balanced observational, modeling and applied aspects. Linked with these topics, the book highlights the importance for imaging the detailed subsurface structures and fault zones at seismic depths that are closely related to earthquake occurrence and rupture dynamics.

Researchers and graduate students in solid earth sciences will broaden their horizons about observational, computational and applied geophysics, seismology and earthquake sciences. This book covers multi-disciplinary topics to allow readers to grasp the new methods and techniques used in data acquisition, assimilation, analysis and numerical modeling for structural, physical and mechanical interpretation of earthquake phenomena, and to strengthen the understanding of earthquake processes and hazards, thus helping readers to evaluate potential earthquake risk in seismogenic regions globally. Readers of this book can make full use of the present knowledge and techniques to reduce earthquake disasters.

Contents

Imaging, Modeling and Assimilation in Seismology: An Overview

Yong-Gang Li

This book presents recent findings, methods and techniques in observational, computational and analytical seismology for earthquake sciences. Authors from global institutions present multi-disciplinary topics with case studies to illuminate high-resolution imaging of complex crustal structures and active faults by full-3D waveform tomography, one-return propagator and fault zone trapped waves, as well as earthquake physics and hazard assessment by dynamic rupture, discrete element modeling and stress load/unload analysis.

In order to relate present-day crustal stresses and fault motions to the geological structures formed by previous ruptures, we must understand the evolution of fault systems on many spatial and temporal scales in the complex earth crust. Extensive researches in the field, in laboratories, and with numerical simulations have illuminated that the fault zone undergoes high, fluctuating stress and pervasive cracking during an earthquake (e.g., Aki, 1984; Mooney and Ginzburg, 1986; Scholz, 1990; Rice, 1992; Kanamori, 1994). Rupture models that involve variations in fault-zone fluid pressure over the earthquake cycle have been proposed (e.g., Dieterich, 1979; Blanpied et al., 1992; Olsen et al., 1998). Structural fault variations and rheological fault variations (e.g., Sibson et al., 1975; Angevine et al., 1982; Chester, 1994; Hickman, 1995; Taira et al., 2008) as well as variations in strength and stress may affect the earthquake rupture (e.g., Vidale et al., 1994; Beroza et al., 1995; Marone et al., 1995; Massonnet et al., 1996; Schaff and Beroza, 2004; Rubinstein and Beroza, 2005). Actually, earthquake-related fault-zone damage and healing *in-situ* have been documented (e.g. Karageorgi et al., 1997; Li et al., 1998, 2006, 2007; Korneev et al., 2000, Vidale and Li, 2003; Yasuhara et al., 2004). Since the spatial extent of fault weakness, and the loss and recovery of strength across the earthquake cycle are critical ingredients in understanding of fault mechanics, knowledge of the fine-scale internal structures in the earth crust, such as active faults and plate boundaries, holds the key to understanding the physics of earthquakes and estimation of earthquake risk.

Seismic studies show that the active faults, like the San Andreas fault in California, undergo strong dynamic stresses and pervasive cracking during major ruptures, on which a distinct low-velocity zone (LVZ) has been imaged with high-resolution (e.g., Thurber et al., 2004; Korneev et al., 2003; Li and Malin, 2008, 2010; Wu et al., 2010). We interpreted this LVZ as a remnant of damage zone (process zone in dynamic rupture) that accumulated damage from historical earthquakes. The zone co-seismically weakens during the earthquake and subsequently heals (partially) during the interseismic period. The damage structure of fault zones is of great interest because the factors that control the initiation, propagation, and termination of rupture are not well understood. Since the fault plane is thought to be a weakness plane in the earth crust, it facilitates slip to occur under the prevailing stress orientation. As suggested by laboratory experiments, shear faulting is highly resisted in brittle material and proceeds as re-activated faults along surfaces which have already encountered considerable damage (e.g., Dieterich, 1997; Marone, 1998). Field evidences show that the rupture plane of slip on a mature fault occurs at a more restricted position, the edge of damage zone at the plane of contact with the intact wall rock (Chester et al., 1993; Chester and Chester, 1998). Assuming this to be an actual picture of rupture preparation on the major fault such as the San Andreas Fault in California, US (Li et al., 2011, see Chapter 3 in this book) and the Longmen-Shan Fault in Sichuan, China (Li et al., 2011, see Chapter 4 in this book), monitoring microseismic events that occur close to or on the principal rupture plane would be crucial for earthquake prediction. The slip of these events in series with the main fault is most likely to load the principal slip plane to a point of a major through-going rupture. In contrast, the slip on minor faults parallel to the main fault, but laterally offset some distance (hundreds of meters to several kilometers), is likely to unload than to load the main fault. In these circumstance, where the principal fault plane accompanied with damage zone at depth is a challenging problem to seismologists and geologists.

Clues to the role of fault zone strength in the earthquake may come from many directions, but are not yet clearly in focus. While we know slip is localized on faults because of their lower strength than the surrounding bedrock, there are certain critical parameters are practically unknown. For example, one of the long-standing problems in geodesy and crustal dynamics is an apparent difference between the dynamic elastic modules of the crustal rocks determined from seismic velocities and the static elastic modules appropriate for the secular tectonic deformation. Another question is whether the geodetically observed co-seismic strain across the fault zones is primarily due to static or dynamic stress changes. In the latter case, the permanent co-seismic strain might result from dynamic reductions in rigidity such that the dynamic strains are remnant after shaking stops (e.g., Fialko et al., 2001; Fialko, 2004; Duan, 2010). If the fault strength is proportional to the effective shear modulus of rocks within the fault zone, observations of the “soft” fault zones may be a direct evidence for weakness (in both relative and absolute sense) of large seismogenic faults. Testing of these hypotheses requires an accurate description of the complexity of mechanical properties of fault zones along strike and with depth as well as the crustal heterogeneity.

Detailing the fault structure and local variations in seismic velocities has also implications for near-fault hazards and expected ground shaking. Greater amplitude shaking is expected near faults due to both proximity to the fault and due to localized amplification in damaged material. Structurally, major crustal faults are often marked by zones of lowered velocity with a width of a few hundred meters to a few kilometers (e.g., Michelini and McEvilly, 1991; Li et al, 2000, 2002, 2004; Li et al., 2007; Wu et al., 2010). The low-velocity zone can be caused by intense fracturing during earthquakes, brecciation, liquid-saturation, and possibly high pore-fluid pressure near the fault. Recent results have suggested damage zones can extend up to a kilometer from the main slip trace (e.g., Fialko et al., 2002; Cochran et al., 2009) with implications for rupture mechanics and seismic hazard estimates. Examining the geometry and damage structure of a fault will help us understand the origin of spatial and temporal variations in rock damage and the evolution of heterogeneities in stress and strain on a well-developed fault zone, like the San Andreas fault. The fine structure of fault zones is of great interest because it controls to the initiation, propagation, and termination of rupture. Observations suggest that fault zone complexity may segment fault zones (Aki, 1984; Ellsworth, 1992) or control the timing of moment release in earthquakes (e.g., Harris and Day, 1997). Other studies predict that a larger portion of rupture energy is expended in cracking and damaging rocks in less developed fault zones (Mooney and Ginzburg, 1986), or increased roughness of the fault zone (Sagy et al., 2007).

Moreover, quantitative studies of earthquakes based on the fault model in the past decades took kinematic and dynamic approaches. The kinematic approach is basically solving an inverse problem in which we determine the fault slip function in space and time from observed seismic records by means of the elasto-dynamic representation theorem; kinematic model parameters are also interpreted in terms of fracture mechanics. On the other hand, the dynamic approach attempts to predict the fault slip function based on a given distribution of rupture strength over the fault plane and the loading stress condition by means of the principle of rupture mechanics. The two approaches are now combined to produce the distribution of stress drop and that of fracture strength over the fault plane. For all these models, knowledge of spatial and temporal variations in fault geometry and physical properties will help predict the behavior of future earthquakes, and such knowledge will help evaluate the models as well. It would be crucial to know what they are physically, especially when we need to predict ground motion for a future earthquake on a given active fault.

In this book, we introduce the new methodology and technology used in data assimilation for defining subsurface complexity, seismically imaging the multi-scale crustal heterogeneity and fault zone geometry, characterizing fault damage magnitude and heal progression, and its physical properties with high-resolution. We also introduce a sophisticated finite-element method for ground-motion simulation with dynamic rupture and source characterization, the discrete element method with fracture mechanics for earthquake simulation, and the stress load-unload response ratio evaluation for earthquake prediction. This book includes seven chapters.

The Chapters:

Chapter 1: "Full-Wave Seismic Data Assimilation: A Unified Methodology for Seismic Waveform Inversion" by Po Chen

Po Chen (2011) presents theoretical background and recent advances of full-physics seismic data assimilation in Chapter 1. First, he reviews some recent applications of the full-physics methodology in seismic tomography and source parameter inversion (Chen et al., 2007a, b), and then discusses some challenging issues related to the computational implementation and the effective exploitation of seismic waveform data as case study in this chapter. Using the data assimilation methodology, he has formulated the seismological inverse problem for estimating seismic source and earth structure parameters in the form of weak-constraint generalized inverse, in which the seismic wave equation and the associated initial and boundary conditions are allowed to contain errors. The resulting Euler-Lagrange equations are closely related to the adjoint method and the scattering-integral method, which have been successfully applied in full-3D, full-physics seismic tomography and earthquake source parameter inversions (Chen et al., 2010a, b). Capabilities of this unified methodology are demonstrated using examples and several challenging issues facing today's seismologists are discussed at the end of this chapter.

The study of the solid Earth system by seismological methods involving with the proper use of simulation-based predictions to make valid scientific inferences requires a continually iterated cycle of model formulation verification, simulation-based predictions, validation against observations, and data assimilation to improve the model and reinitiate the cycle at a higher level where the model is deficient. Recent advances in parallel computing technology and numerical methods have made large-scale, three-dimensional numerical simulations of seismic wave-fields much more affordable, and they have opened up the possibility of full-physics seismic data assimilation, in which seismological observations of ground motions are combined with the underlying dynamic principles described by seismic wave equations to produce realistic estimations of the properties of the seismic sources that generate seismic waves and the geological structures through which seismic waves propagate. In this chapter, Chen formulates the seismic data assimilation problem in the generalized inverse framework (Evensen, 2009). The full physics of the underlying dynamic principles is accounted for using the three-dimensional seismic wave equation. The starting point of the derivation is the Bayes' theorem, which defines the posterior probability density functions of the poorly known seismic source and structure parameters conditioned on a set of seismological observations. The adjoint method, which was adopted to solve seismic imaging problems in Tarantola (1984) and later extended to solve both seismic source and structure inverse problems in Tromp et al. (2005), can be considered as a simplification of the generalized inverse, in which the dynamic model is assumed to have zero model error. A slight alteration in the derivation of the generalized inverse naturally leads to the scattering integral method used for full-3D waveform tomography in Chen et al. (2007a, b) and for earthquake moment tensor inversion in Zhao et al. (2006).

Chapter 2: "One-Return Propagators and the Applications in Modeling and Imaging" by Ru-Shan Wu, Xiao-Bi Xie and Shengwen Jin

Ru-Shan Wu, Xiao-Bi Xie and Shengwen Jin (2011) present the elastic one-return propagator and its application in modeling and imaging of crustal complex in Chapter 2. The earth has been revealed to have hierarchical, multi-scaled heterogeneities everywhere, and keeping change all the time. The place is more heterogeneous, the more interesting processes (subduction, collision, accretion) can be found around that place. Seismology played and is still playing an important role in the movement. The introduction of broadband and multi-component seismometers, installation of various seismic networks and arrays (from global, regional and local networks, to small aperture arrays), digital recording and networking, etc. have tremendously increased the data amount and quality available to the seismological community. Combined with even faster progress of super-computing and parallel processing, the opportunity of discoveries and contributions in front of seismologists are ever greater than before. Wu (2003) has initiated advance wave propagation and scattering theory and high-resolution geophysical imaging methods (diffraction and scattering tomography). Wu et al. (2001) are taking full advantage of the new development in data acquisition and computing capacity, to join the adventure of penetrating the Earth and discovering new features. They have studied the small-scale heterogeneities in the crust and upper mantle and their statistical characteristics (modeled as random media) directly (from well-logging data) and indirectly (from seismic wave scattering). Influences of diffraction/scattering and random scattering by small-scale heterogeneities to tomography of large-scale structures are also in their research topics.

Based on the multiple forward-scattering and single backscattering (MFSB) approximation (Wu et al., 1995, Wu and Xie, 2006), the authors derived a method which can calculate forward propagations and primary reflections for acoustic and elastic waves in complex velocity models shown in this chapter. This method can be implemented using an iterative marching algorithm shuttling between the space and wave-number domain. Single backscattering is calculated for each marching step (a thin-slab) to handle primary reflections. Therefore, it is also called the one-return method. For models where reverberation and resonance scattering can be neglected, this method provides an accurate and highly efficient algorithm. Two versions of the one-return method, the thin slab method and screen method are given in this chapter. From the screen approximation, which involves a small-angle approximation for the wave-medium interaction, the forward scattered or converted waves are mainly controlled by velocity perturbations; while the backscattered or reflected waves are mainly controlled by impedance perturbations. This method is suitable for simulating wave propagation in multiple-scale velocity models and at high-frequencies. Numerical examples are presented to demonstrate its wide applications in seismic modeling, imaging and inversion of complicated subsurface crustal structure.

Chapter 3: "Fault-Zone Trapped Waves: High-Resolution Characterization of the Damage Zone on the Parkfield San Andreas Fault at Depth" by Yong-Gang Li, Peter E. Malin and Elizabeth S. Cochran

Yong-Gang Li, Peter Malin and Elizabeth Cochran (2011) present their recent results from fault-zone trapped waves to characterize the damage zone on the Parkfield San Andreas Fault (SAF) at seismogenic depths with high resolution in Chapter 3. The internal structure of major faults holds the key to understanding how earthquakes come about and how plate boundaries are lubricated. This chapter provides fundamental information on the nature and source of the fault damage and healing in situ, and therefore enables us to better understand the relationship between the fault zone structure and dynamic rupture as well as the physical basis of the earthquake cycle related to the build-up and release of stresses in large earthquakes.

The fault-zone trapped waves (FZTWs) were first discovered at the Oroville and San Andreas fault zones in California (Li et al., 1990; Li and Leary, 1990). Because FZTWs arise from constructive interference of multiple reflections at the boundaries between the low-velocity fault zone and high-velocity surrounding rocks, their dispersive features, including amplitudes and frequency contents, are strongly dependent on the fault-zone geometry and physical properties. These waves have been shown to be ability to reveal detailed information of the fine structure at the heart of fault zones and its variations laterally and with depth. Therefore, we can use FZTWs to resolve fine internal damage structure in the range of tens to several hundreds of meters. The previous results from the FZTWs recorded at ruptures of the 1992 *M*7.4 Landers and 1999 *M*7.1 Hector Mine earthquakes in California indicated that the active faults undergo strong dynamic stresses and pervasive cracking resulted from these earthquakes, on which we have delineated a distinct low-velocity zone (LVZ) using FZTWs (Li et al., 1994; 2002). They interpret this LVZ as a remnant of damage zone (break-down zone in dynamic rupture) that accumulates damages from historical earthquakes. The zone co-seismically weakens during the earthquake and subsequently heals (partially) during the interseismic period (Li et al., 1998; Vidale and Li, 2003). The damage zone on these ruptures is approximately 200-m wide with the velocity reduction of 20%-50% from wall-rock velocities and likely extends across the seismogenic depths. The distinct LVZ naturally forms a waveguide to trap seismic waves as a source is located within or close to it.

In this chapter, Li et al. (2011) present the new result from a combination of FZTWs generated by earthquakes and explosions and recorded at a cross-fault surface array and borehole seismographs at the San Andreas Fault Observatory at Depth (SAFOD) site to document the details of fault zone geometry, subsurface fine structure and material properties. Observations and 3-D finite-difference simulations of these FZTWs at dominant frequencies of 2-10 Hz show the downward tapering SAF characterized by a 30-40-m wide fault core with the maximum velocity reduction up to ∼50% embedded in a 100-200-m wide zone with velocities reduced by 25%-40% in average from wall-rock velocities. The width and velocity reduction of the damage zone at 3 km depth delineated by FZTWs are verified by the direct measurements in SAFOD drilling and logging studies at this depth (Hickman et al., 2007). The results indicate the localization of severe rock damage on the SAF likely reflecting pervasive cracking caused by historical earthquakes on it. The magnitude of damage varies with depth and along the fault strike due to rupture distributions and stress variations over multiple length and time scale, and is also asymmetric across the main slip plane but extends farther on the southwest side of the main fault trace. Based on the depths of earthquakes generating prominent FZTWs, we estimate that the low-velocity damage zone along the strike-slipping SAF at Parkfield extends at least to depths of ∼7-8 km.

Chapter 4: "Fault-Zone Trapped Waves at a Dip Fault: Documentation of Rock Damage on the Thrusting Longmen-Shan Fault Ruptured in the 2008 *M*8 Wenchuan Earthquake" by Yong-Gang Li, Jin-Rong Su, and Tian-Chang Chen

In Chapter 4, Yong-Gang Li, Jin-Rong Su, and Tian-Chang Chen (2011) present observations and 3-D finite-difference simulations of fault-zone trapped waves (FZTWs) recorded at the south Longmen-Shan fault (LSF) with varying dip angles, which ruptured in the 2008 *M*8 Wenchuan earthquake in Sichuan, China. Li et al. examined rock damage and heal on the LSF using the data recorded at Sichuan Seismic Network and portable stations in the source region where the LSF is characterized by reverse thrusting. The dominating FZTWs generated by on-fault aftershocks illustrate the coherent interference phenomenon of wave propagation in a highly fractured low-velocity fault zone bounded by high-velocity crustal intact rocks. 3-D finite-difference simulations of these FZTWs show a distinct low-velocity wave-guide (LVWG) a few hundred-meters wide on the south LSF, in which the maximum velocity reduction is ∼50% or more at shallow depth, likely extends across seismogenic depths at varying dip angles with depth. Because of the sensitivity of trapped wave excitation to the source location from the LVGW, it allows to depict the principal slip plane along with the LSF in the Wenchuan *M*8 mainshock at depth inferred by precise locations of aftershocks generating prominent FZTWS. The width, velocity and shape of the LSF at shallow depth delineated by FZTWS are generally consistent with the results from geological mapping and fault-zone drilling at the southern LSF (Xu et al., 2008a, b, 2009).

Li et al. (2011) interpret this LVGW as a damage zone in dynamic rupture that accumulated damage from historical earthquakes, particularly from the 2008 *M*8 Wenchuan earthquake. The fault zone co-seismically weakening during the major earthquake and subsequently healing (partially) on the LSF have been studied using similar earthquakes occurring before and after the Wenchuan earthquake. They examined the changes in amplitude and dispersion of FZTWS recorded at the same seismic station for earthquakes at the same places between 2006 and 2009. Results suggest that seismic velocities within the south LSF zone could be co-seismically reduced by ~10%-15% due to the rock damage caused by the *M*8 mainshock on May 12, 2008. The moving-window cross-correlations of waveforms for body waves and FZTWs from repeated aftershocks show that seismic velocities near the LSF increased by ~5% or even more in the first year after the Wenchuan earthquake, with an approximately logarithmic healing rate with time and the largest healing in the earliest stage, indicating the post-main shock healing with rigidity recovery of fault-zone rocks damaged in the *M*8 quake. The magnitude of rock damage and heal observed at the Longman-Shan fault is larger than those observed at the San Andreas fault ruptured in the 2004 *M*6 Parkfield earthquake and also on ruptures of the 1992 *M*7.4 Landers and 1999 *M*7.1 Hector Mine earthquakes. This difference is probably related to the different earthquake magnitude, faulting mechanism and stress drop.

The significance of this investigation is to illuminate the subsurface rock damage along a dip fault at seismogenic depths associated with the 2008 *M*8 Wenchuan earthquake. The data from the 2008 Wenchuan earthquake add the information into previous results of the spatio-temporal variations of fault zone properties associated with large earthquakes (Li et al., 1994, 1998, 2002, 2003, 2004, 2006; Vidale and Li, 2003). The spatial extent of fault-zone damage and the loss and recovery of strength across the earthquake cycle are critical ingredients in understanding of fault mechanics and physics. A comparison of the results from a reverse-thrusting dip faults, like the Longmen-Shan fault, with those at the strike-slipping faults, like the San Andreas fault at Parkfield in California, is helpful in examination if the magnitude of fault rock damage is a function of earthquake size and to evaluate potential earthquake risk in seismogenic regions globally. These results also provide the useful information of subsurface rock damage caused by the 2008 Wenchuan earthquake in site selection for re-construction in the earthquake hazardous areas.

The application of fault-zone trapped waves at the Parkfield San Andreas Fault and south Longmen-Shan fault described in Chapter 3 and Chapter 4 can be taken as case studies of the fault-zone trapped wave method and technique used in seismic imaging, modeling and characterization of crustal fault rock damage magnitude and lateral extension at seismogenic depth.

Chapter 5: "Ground-Motion Simulations with Dynamic Source Characterization and Parallel Computing" by Benchun Duan

Benchun Duan (2011) presents dynamic source characterization and parallel computing technique for ground-motion simulations in Chapter 5. An emerging trend in simulation of ground-motion caused by earthquakes is to use spontaneous rupture models to characterize dynamic earthquake sources and to use rapid developing high performance computing resources with evolving, sophisticated computer algorithms to predict time histories of ground motion in seismically active regions. Here, a sophisticated finite element method (FEM) algorithm EQdyna is introduced and two examples of ground-motion related applications are presented. Duan first reviews the basics of spontaneous rupture models (e.g., Andrews, 1976; Dieterich, 1979; Day, 1982; Ruina, 1983) and recent development of the FEM algorithm in parallelization using a hybrid MPI/OpenMP approach, and then provides its application to the convergence test of a benchmark problem in this chapter.

Synthetic seismograms for possible future moderate and large earthquakes are one of the key products that seismologists can provide to engineers. These seismograms are used by engineers in the design of earthquake-resistant structures. Thus, ground-motion simulations are a vital bridge between earthquake seismology and earthquake engineering. In particular, for close-in distances near active faults, the ground motion recordings from real earthquakes are sparse. Synthetic seismograms can fill this gap. In addition, with increasing usage of nonlinear analyses in the seismic design of structure, time histories of ground motion becomes more important for completely determining structure response and damage estimation from future significant earthquakes. Synthetic seismograms also allow engineers to examine the variability in the structure response to different earthquake scenarios with different rupture directivity and slip distributions, which would not be available in recorded data. With rapid development of modern high performance computing systems, particularly cluster systems with CMPs (Chip MultiProcessors), parallel computing has been becoming increasingly important and popular in numerical simulations of rupture dynamics and ground motion. Parallel computing allows seismologists to explore small-scale rupture complexities observed in large earthquakes and to augment high frequency limits of deterministically simulated ground motions.

Chapter 6: "Load-Unload Response Ratio and Its New Progress" by Xiang-Chu Yin, Yue Liu, Lang-Ping Zhang, and Shuai Yuan

In Chapter 6, Xiang-Chu Yin, Yue Liu, Lang-Ping Zhang, and Shuai Yuan introduce the stress load-unload response ratio (LURR) for earthquake prediction and its new progress in detail. The motivation, basic ideas, fundamental problems of the LURR and assessment of earthquake prediction status using LURR are discussed in this chapter. From the viewpoint of mechanics, the physical essence of earthquake is an abrupt shear rupture in seismic source region accompanying with sudden release of strain energy and a damage process of the focal media leading to the abrupt shear rupture (Meakin, 1991). In other words, the earthquake genesis process is damage evolution. From the microscopic viewpoint, there are a large number of disordered defects (cracks, fissures, joints, faults, caves etc.) with different sizes, shapes and orientations in rock. The damage process involves the nucleation and extension of micro-damages, coalescence between micro-damages and the formation of a main crack that leads to the eventual fracture (e.g., Wei et al., 2000;

Xie et al., 2002). It is an irreversible, far-from-equilibrium, nonlinear, multi-scale and multi-physics phenomenon, which has been intensively studied for decades but a series of fundamental questions are still open (e.g., Yin et al., 1995, 2004). From the macroscopic viewpoint, the constitutive relation between stress and strain is a comprehensive description of the mechanical behaviors of any materials. If the load acting on the material increases monotonously, the material will experience the regimes of elastic, damage and failure or destabilization. The most essential characteristic of the elastic regime is its reversibility, i.e., the positive process and the contrary process are reversible. In other words, the loading modulus and the unloading one are equal to each other. Contrary to the elastic regime, the damage one is irreversible, hence the loading response is different from the unloading one, or the loading modulus is different from the unloading one. This difference indicates the deterioration of material due to damage. Yin formulated the load-unload response ratio in rock block for measuring the damage degree and proximity to failure, which acts as a precursor for earthquake prediction or forecasting.

Yin et al. (2011) present the evaluation law in LURR to be applied before large earthquakes and the dimensional method to evaluate the applicable range of the LURR for earthquake prediction. The results from laboratory experiment, numerical simulation, analytical and the real seismic data show that the value of LURR fluctuate around unit 1 in the early stage of interseismic period, then the value rise swiftly and to its peak point (abbreviated PP) before the earthquake occurs. The catastrophic events do not happen at the time of peak point, but after it, namely the catastrophic events lag behind the PP. The evolution law of LURR is important for an actual earthquake prediction, by which we may predict the occurrence time of the earthquake quantitatively (by scale of months) if the time of the PP can be verified. Above all, monitoring the variations of LURR allows us to track the seismogenic process in the earthquake cycle eventually achieve prediction of the forthcoming earthquake. Yin et al. (2011) provides the maps of the LURR anomaly regions in the mainland of China calculated in the end of 2003, 2004, 2005 and 2006 and also the epicenters distribution of earthquakes with magnitude $M_L \geqslant 5$ occurred in the next year (2004, 2005, 2006 and 2007). The variation of LURR observed around the $M8$ Wenchuan earthquake occurring on May 12, 2008 is also taken as a case study in this chapter.

Chapter 7: "Discrete Element Method and Its Applications in Earthquake and Rock Fracture Modeling" by Yucang Wang, Sheng Xue, and Jun Xie

In Chapter 7, Yucang Wang, Sheng Xue, and Jun Xie (2011) present the discrete element method (DEM), a powerful numerical tool in many scientific and engineering applications. Here, the DEM is used to model fault mechanics and brittle rock fracture associated with earthquakes. One advantage of DEM is that highly complex systems can be modeled using basic methodologies without any assumptions on the constitutive behaviors of the materials and any predisposition about where and how cracks may occur and propagate. Owing to its discrete nature, DEM is extremely suitable to model large deformation and dynamics phenomena (Wang and Mora, 2008). While different types of DEMs have appeared in the past decades, the Esys_Particle code has been developed most recently (Wang and Mora, 2009). Wang et al. (2011) outline the recent developments of the Esys_Particle code, including incorporation of single particle rotation, new contact law, and parameter calibration, parallel algorithm, coupling of thermal and hydrodynamic effects. They discuss the major differences of their model from other DEMs (e.g., Mora et al., 1994; 1998; Wang et al., 2000; Hazzard et al., 2000; Hentz et al., 2004). The applicability of the Esys_Particle is illustrated through several numerical simulations in this chapter. Most qualitative features of rock fracture observed in laboratory tests are well reproduced (e.g., Place and Mora, 1999, 2000), these include, fracture of brittle materials under compression, wing crack extension both in 2-D and 3-D, crushing of aggregates and fracture caused by dynamic impact.

Compared with the other DEMs, the Esys_Particle shows advantageous in the following aspects. Firstly the Esys_Particle has a unique and explicit representation of 3-D particle orientation using unit quaternion. In other DEMs, three angular velocities around three orthogonal axes are used to implicitly represent orientations that cannot extract the exact orientation for each particle by simply integrating from the three angular velocities in 3-D case. Secondly, when dealing with particle rotations, the Esys_Particle is numerically more stable than other DEMs, because in other DEMs the incremental method is adopted to update particle angular velocities, in which errors increase much faster than the algorithm used in the Esys_Particle when the time step increases (Wang, 2009). Thirdly, in most existing DEMs, shear forces and torques are computed in an incremental fashion (Potyondy and Cundall, 2004). It should be pointed out that when torsion and rolling exist at the same time, rolling changes the axis of torsion, which is not decoupled in the incremental method. Using the new decomposition technique in the Esys_Particle, rolling and torsion are completely decoupled and uniquely determined, and so are forces and torques between particles. In summary, in dealing with particle rotations and calculating torques, the finite deformation scheme used in the Esys_Particle respects the physical principle and is numerically stable, while the incremental method used in other DEMs ignores the certain physical principle and is only accurate when time step and rotation are very small.

As a first step towards studying the underlying physical mechanism for the LURR observations (Yin et al., 2011) (see Chapter 6 in this book), the Esys_Particle model has been used to simulate a 2-D elastic-brittle system in which a sinusoidal stress perturbation is added to the gradual compressional loading to simulate loading and unloading cycles. In each case, fractures develop and seismic energy is radiated within the model as the system is compressed until the sample fails catastrophically, and LURR is calculated. The results show that LURR values become high and then drop prior to the main event, and remain low thereafter, similar to those that has often been observed in earthquake prediction practice. Statistical tests of LURR values in shearing fault model show the similar results (Wang et al., 2004). The results suggest that LURR method developed by Yin et al. (2004, 2011) provides a good predictor for catastrophic failure in elastic-brittle systems, and provide encouragement for the prospects of earthquake prediction using LURR and the use of advanced numerical models to probe the physics of earthquakes. The potential applications of the Esys_Particle may include: Hydraulic fracturing in mining engineering and geothermal energy extraction; Induced seismicity by reservoir and mining activities; Tsunami source generation; Effect of underground fluid on the earthquake generation; Dynamic triggering of earthquakes; Landslide and slope stability.

The purpose of this book is to introduce the new approaches in solid-earth geophysics research in general. The following new methods and results presented in this book will be of particular interest to the readers:

(i) The full-3D waveform tomography method, and one-return propagator method in modeling and imaging of subsurface structure.
(ii) Observations and 3-D finite-difference simulations of fault-zone trapped wave for high-resolution delineation of fault internal structure and physical properties.
(iii) Rock damage and heal in major earthquakes.
(iv) Spontaneous dynamic source and strong ground motion simulations.
(v) Discrete element method for earthquake and fracture modeling, and earthquake hazard assessment.
(vi) Observations of precursory swarm occurrence, stress loading and unloading, seismicity pattern restrospection, and simulation-based, statistical physics methods for earthquake prediction.

This book is a self-contained volume that starts with an overview of the subject then explores each topic with in depth detail. Extensive reference lists and cross references with other volumes facilitate further research. Full-color figures and tables support the text and aid in understanding. Content suited for both the senior researchers and graduate students in geosciences who will broaden their horizons about observational, computational and applied seismology and earthquake sciences. This book covers multi-disciplinary topics to allow readers to grasp the new methods and techniques used in data analysis and numerical modeling for structural, physical and mechanical interpretation of earthquake phenomena, to aid the understanding of earthquake processes and hazards, and thus help readers to evaluate potential earthquake risk in seismogenic regions globally.

References

Aki, K. (1984). Asperities, barriers, characteristic earthquakes, and strong motion prediction. Journal of Geophysical Research, 89, 5867-5872.

Andrews, D. J. (1996). Rupture velocity of plane strain shear cracks. Journal of Geophysical Research, 81, 5679-5687.

Angevine, C. L., D. L., Turcotte, and M. D. Furnish (1982). Pressure solution lithification as a mechanism for the stick-slip behavior of faults. Tectonics, 1, 151-160.

Beroza, G. C., A. T. Cole, and W. L. Ellsworth (1995). Stability of coda wave attenuation during the Loma Prieta, California, earthquake sequence. Journal of Geophysical Research, 100, 3977-3987.

Blanpied, M. L., D. A. Lockner, and J. D. Byerlee (1992). An earthquake mechanism based on rapid sealing of faults. Nature, 359, 574-576.

Chen, P. (2011). Full-wave seismic data assimilation: a unified methodology for seismic waveform inversion, in Imaging, Modeling and Assimilation in Seismology, edited by Y. G. Li, China High Education Press, Beijing, De Gruyter, Boston, USA.

Chen, P., L. Zhao, and T. J. Jordan (2007b). Full 3D tomography for the crustal structure of the Los Angeles region. Bulletin of the Seismological Society of America, 97(4): 1094-1120.

Chen, P., T. H. Jordan, and E. -J. Lee (2010a). Perturbation kernels for generalized seismological data functionals (GSDF). Geophysical Journal International, 183 (2): 869-883.
Chen, P., T. H. Jordan, and L. Zhao (2010b). Resolving fault plane ambiguity for small earthquakes. Geophysical Journal International, 181 (1): 493-501.
Chen, P., T. J. Jordan, T. H., and L. Zhao (2007a). Full three-dimensional tomography: a comparison between the scattering-integral and adjoint-wavefield methods. Geophysical Journal International, 170 (1): 175-181.
Chester, F. M. (1994). Rheological model crust faults: Influence of fluids on strength and stability, 487-500, USGS Proceedings of Workshop LXIII, The Mechanical Involvement of Fluids in Faulting, Open-File Report 94-228.
Chester, F. M., and J. S. Chester (1998). Ultracataclasite structure and friction processes of the San Andreas fault. Tectonophysics, 295, 199-221.
Cochran, E. S., Y. G. Li, P. M. Shearer, S. Barbot, Y. Fialko, J. E. Vidale (2009). Seismic and geodetic evidence for extensive, long-lived fault damage zones. Geology, 37(4), 315-318.
Day, S. M. (1982). Three-dimensional simulation of spontaneous rupture: the effect of nonuniform prestress. Bull. Seism. Soc. Am., 72, 1881-1902.
Dieterich, J. H. (1997). Modeling of rock friction 1. Experimental results and constitutive equations. Journal of Geophysical Research, 84, 2169-2175.
Dieterich, J. H. (1997). Modeling of rock friction 1. Experimental results and constitutive equations. Journal of Geophysical Research, 84, 2161-2168.
Duan, B. C. (2010). Inelastic response of compliant fault zones to nearby earthquakes. Geophys. Res. Lett., L16303, doi: 10.1029/2010GL044150.
Duan, B. C. (2011). Ground-motion simulations with dynamic source characterization and parallel computing, in Imaging, Modeling and Assimilation in Seismology, edited by Y. G. Li, China High Education Press, Beijing, De Gruyter, Boston, USA.
Ellsworth, W. L, A. T. Cole, G. C. Beroza, and M. C. Verwoerd (1992). Changes in crustal wave velocity association with the 1989 Loma Prieta, California earthquake. EOS, Transactions, 73, 360.
Evensen, G. (2009). Data Assimilation: The Ensemble Kalman Filter. Springer.
Fialko, Y. (2004). Probing the mechanical properties of seismically active crust with space geodesy: study of the co-seismic deformation due to the 1992 *M*w 7.3 Landers (southern California) earthquake. Journal of Geophysical Research, 109, B03307, doi: 10.1029/2003JB002756.
Fialko, Y., D. Sandwell, D. Agnew, M. Simons, P. Shearer, and B. Minster (2002). Deformation on nearby faults induced by the 1999 Hector Mine earthquake. Science, 297, 1858-1862.
Fialko, Y., M. Simons, and D. Agnew (2001). The complete (3-D) surface displacement field in the epicentral area of the 1999 *M*w7.1 Hector Mine earthquake, California, from space geodetic observations. Geophys. Res. Lett., 28, 3063-3066.
Fichtner, A., B. Kennett, H. Igel, and H. Bunge (2009). Full seismic waveform tomography for upper-mantle structure in the Australasian region using adjoint methods. Geophysical Journal International, 179 (3): 1703-1725.
Harris, R. A. and S. M. Day (1997). Effects of a low-velocity zone on a dynamic rupture. Bull. Seism. Soc. Am., 87, 1267-1280.

Hazzard, J. F., D. S. Collins, W. S. Pettitt, and R. P. Young (2000). Simulation of unstable fault slip in granite using a bonded-particle model. Pure and Applied Geophysics, 159: 221–245.

Hentz S., F. V. Donze, L. Daudeville (2004). Discrete element modeling of concrete submitted to dynamics loading at high strain rates. Comput. Struct., 82: 2509-2524.

Hickman, S., M. D. Zoback, W. Ellsworth, N. Boness, P. Malin, S. Roecker and C. Thurber (2007). Structure and properties of the San Andreas Fault in central California: recent results from the SAFOD experiment, in H. Ito, J. Behrmann, S. Hickman, H. Tobin and G. Kimura (eds.). Report from IODP/ICDP Workshop on Fault Zone Drilling, Miyasaki, Japan. Scientific Drilling, Special Issue, 1, 29-32.

Hickman, S., R. Sibson, and R. Bruhn (1995). Introduction to special section: mechanical involvement of fluids in faulting. Journal of Geophysical Research, 100, 12831-12840.

Kanamori, H. and C. R. Allen (1986). Earthquake repeat time and average stress drop, in Earthquake Source Mechanics, AGU Geophys. Mono. 37, eds. S. Das et al., 227-236.

Karageorgi, E. D., T. V. McEvilly, and R. W. Clymer (1997). Seismological studies at Parkfield IV: variations in controlled-source waveform parameters and their correlation with seismicity, 1987 to 1995. Bull. Seism. Soc. Am., 87, 39-49.

Korneev, V. A., T. V. McEvilly, and E. D. Karageorgi (2000). Seismological studies at Parkfield VIII: modeling the observed travel-time changes. Bull. Seism. Soc. Am., 90, 702-708.

Li, H., L. Zhu, and H. Yang (2007). High-resolution structure of the Landers fault zone inferred from aftershock waveform data. Geophysical Journal International, 171: 1295-1307, doi: 10.1111/j.1365-246X.2007.03608.x.

Li, Y. G. and P. C. Leary (1990). Fault zone trapped seismic waves. Bull. Seism. Soc. Am., 80, 1245-1271.

Li, Y. G. and P. E. Malin (2008). San Andreas Fault damage at SAFOD viewed with fault-guided waves. Geophys. Res. Lett., 35, L08304, doi: 10.1029/2007GL032924.

Li, Y. G., J. E. Vidale, K. Aki, and F. Xu (2000). Depth-dependent structure of the Landers fault zone from trapped waves generated by aftershocks. Journal of Geophysical Research, 105, 6237-6254.

Li, Y. G., J. E. Vidale, K. Aki, F. Xu, T. Burdette (1998). Evidence of shallow fault zone strengthening after the 1992 M7.5 Landers, California, earthquake, Science, 279, 217-219.

Li, Y. G., J. E. Vidale, S. M. Day and D. Oglesby (2002). Study of the M7.1 Hector Mine, California, earthquake fault plane by fault-zone trapped waves. Hector Mine Earthquake Special Issue, Bull. Seism. Soc. Am., 92, 1318-1332.

Li, Y. G., J. E., Vidale, and S. E. Cochran (2004). Low-velocity damaged structure of the San Andreas fault at Parkfield from fault-zone trapped waves. Geophy. Res. Lett., 31, L12S06.

Li, Y. G., J. Y. Su, and T. C. Chen (2011). Fault-zone trapped waves at a dip fault: documentation of rock damage on the thrusting Longmen-Shan fault ruptured in the 2008 *M*8 Wenchuan earthquake, in Imaging, Modeling and Assimilation in Seismology, edited by Y. G. Li, China High Education Press, Beijing, De Gruyter, Boston, USA.

Li, Y. G., K. Aki, D. Adams, A. Hasemi, and W. H. K. Lee (1994). Seismic guided waves trapped in the fault zone of the Landers, California, earthquake of 1992. Journal of Geophysical Research, 99, 11705-11722.

Li, Y. G., P. C. Leary, K. Aki, and P. E. Malin (1990). Seismic trapped modes in the Oroville and San Andreas fault zones. Science, 249, 763-766.

Li, Y. G., P. Chen, E. S. Cochran, and J. E. Vidale (2007). Seismic velocity variations on the San Andreas Fault caused by the 2004 *M*6 Parkfield earthquake and their implications. EPS, 59, 21-31.

Li, Y. G., P. Chen, E. S. Cochran, J. E., Vidale, and T. Burdette (2006). Seismic evidence for rock damage and healing on the San Andreas fault associated with the 2004 *M*6 Parkfield earthquake, Special issue for Parkfield *M*6 earthquake, Bull. Seism. Soc. Am., 96 (4): S1-15, doi: 10.1785/0120050803.

Li, Y. G., P. Malin, and E. Cochran (2011). Fault-zone trapped waves: high-resolution characterization of the damage zone on the Parkfield San Andreas Fault at depth, in Imaging, Modeling and Assimilation in Seismology, edited by Y. G. Li, China High Education Press, Beijing , De Gruyter, Boston, USA.

Marone, C. (1998). Laboratory-derived friction laws and their application to seismic faulting. Annu. Rev. Earth Planet. Sci., 26, 643-696.

Marone, C., J. E. Vidale, and W. L. Ellsworth (1995). Fault healing inferred from time dependent variations in source properties of repeating earthquakes. Geophys. Res. Lett. 22, 3095-3098.

Massonnet, D, W. Thatcher, and H. Vadon, (1996). Detection of postseismic fault-zone collapse following the Landers earthquake. Nature, 382, 612-616.

Meakin, P. (1991). Model for material failure and deformation. Science, 252, 226-234.

Michelini, A. and T. V. McEvilly (1991). Seismological studies at Parkfield I. Simultaneous inversion for velocity structure and hypocenters using cubic B-splines parameterization. Bull. Seism. Soc. Am., 81, 524-552.

Mooney, W. D., and A. Ginzburg (1986). Seismic measurements of the internal properties of fault zones. Pure and Applied Geophysics, 124, 141-157.

Mora, P., and D. Place (1994). Simulation of the frictional stick-slip instability. Pure and Applied Geophysics, 143, 61-87.

Mora, P., D. Place (1998). Numerical simulation of earthquake faults with gouge: towards a comprehensive explanation for the heat flow paradox. Journal of Geophysical Research, 103, 21067-21089.

Olsen, M., C. H. Scholz, and A. Leger (1999). Healing and sealing of a simulated fault gouge under hydrothermal conditions for fault healing. Journal of Geophysical Research, 103, 7421-7430.

Place, D., and P. Mora (2000). Numerical simulation of localization phenomena in a fault zone. Pure and Applied Geophysics, 157: 1821-1845.

Place, D., Mora, P. (1999). A lattice solid model to simulate the physics of rocks and earthquakes: incorporation of friction. J. Comp. Phys., 150: 332-372.

Rice, J. R. (1992). Fault stress states, pore pressure distributions, and the weakness of the San Andreas fault, in Fault Mechanics and Transport Properties of Rocks, edited by B. Evans and T.-F. Wong, 475-503, Academic, San Diego, California.

Rubinstein, J. L., and G. C. Beroza (2005). Depth constraints on nonlinear strong ground motion from the 2004 Parkfield earthquake. Seism. Res. Lett., 32, L14313,

doi: 10.1029/2005GL023189.
Ruina, A. (1983). Slip instability and state variable friction laws. Journal of Geophysical Research, 88, 10359-10370.
Sagy, A., E. Brodsky, and G. Axen (2007). Evolution of fault-surface roughness with slip. Geology, 35, 283-286.
Schaff, D. P., and G. C. Beroza (2004). Coseismic and postseismic velocity changes measured by repeating earthquakes. Journal of Geophysical Research, 109, B103002, doi: 10.1029/2004JB003011.
Scholz, C. H. (1990). Wear and gouge formation in brittle faulting. Geology, 15, 493-495.
Sibson, R. H., J. M. Moore, and A. H. Rankin (1975). Seismic pumping—a hydrothermal fluid transport mechanism. Geological Society of London Journal, 131, 653-659.
Taira, T., P. G. Silver, F. Niu, and R. M. Nadeau (2008). Detecting seismogenic stress evolution and constraining fault-zone rheology in San Andreas Fault following the 2004 Parkfield earthquake. Journal of Geophysical Research, 113, B03303, doi: 0.1029/2007JB00515.
Tarantola, A. (1984). Inversion of seismic reflection data in the acoustic approximation. Geophysics, 49 (8): 1259-1266.
Thurber, C., S. Roecker, H. Zhang, S. Baher and W. Ellsworth (2004). Fine-scale structure of the San Andreas fault zone and location of the SAFOD target earthquakes. Geophys. Res. Letter, 31, L12S02, doi:10.1029/2003GL019398.
Tromp, J., Tape, C., and Q. Liu (2005). Seismic tomography, adjoint methods, time reversal and banana-doughnut kernels. Geophysical Journal International, 160 (1): 195-216.
Vidale, J. E. and Y. G. Li (2003). Damage to the shallow Landers fault from the nearby Hector Mine earthquake. Nature, 421, 524-526.
Vidale, J. E., W. L. Ellsworth, A. Cole, and C. Marone (1994). Rupture variation with recurrence interval in eighteen cycles of a small earthquake. Nature, 368, 624-626.
Wang, Y. C., S. Xue, and J. Xie (2011). Discrete element method and its applications in earthquake and rock fracture modeling, in Imaging, Modeling and Assimilation in Seismology, edited by Y. G. Li, China High Education Press, Beijing, De Gruyter, Boston, USA.
Wang, Y.C., and P. Mora (2008). Modeling wing crack extension: implications to the ingredients of discrete element model. Pure and Applied Geophysics, 165, 609-620.
Wang, Y.C., and P. Mora (2009). Esys_Particle: a new 3-D discrete element model with single particle rotation, in Advances in Geocomputing, edited by H. L. Xing, Springer.
Wang, Y.C., X. C. Yin, F. J. Ke, M. F. Xia, and K. Y. Peng. (2000). Numerical Simulation of rock failure and earthquake process on mesoscopic scale. Pure and Applied Geophysics, 157, 1905-1928.
Wei, Y. J., M. F. Xia, F. J. Ke, X. C. Yin and Y. L. Bai (2000). Evolution induced catastrophe and its predictability. PAGEOPH, 157, 1929-1943.
Wu, J., J. A. Hole, and J. A. Snoke (2010). Fault-zone structure at depth from differential dispersion of seismic guided waves: evidence for a deep waveguide on the San Andreas fault. Geophysical Journal International, 182, 343-354, doi: 10.111.j.1365-246X.2010.04612.x.
Wu, R. S. (2003). Wave propagation, scattering and imaging using dual-domain one-way and one-return propagators. Pure and Applied Geophysics, 160, 509-539.

Wu, R. S., and X. Jia (2006). Accuracy improvement for super-wide angle one-way waves by wavefront reconstruction. Expanded Abstracts, SEG 76th Annual Meeting, 2976-2980.

Wu, R. S., L. J. Huang, and X. B. Xie (1995). Backscattered wave calculation using the De Wolf approximation and a phase-screen propagator. Expanded Abstracts, SEG 65th Annual Meeting, 1293-1296.

Wu, R. S., X. B. Xie, and S. W. Jin (2011). One-return propagators and the applications in modeling and imaging, in Imaging, Modeling and Assimilation in Seismology, edited by Y. G. Li, China High Education Press, Beijing, De Gruyter, Boston, USA.

Xia, M. F., Y. J. Wei, F. J. Ke, Y. L. Bai (2002). Critical sensitivity and trans-scale fluctuations in catastrophic rupture. Pure and Applied Geophysics, 159, 2491-2509.

Xu Z. et al. (2008a). Uplift of the Longmen Shan range and the Wenchuan earthquake. Episodes, 31(3): 291-301.

Xu Z. et al. (2008b). Wenchuan earthquake and scientific drilling. Acta Geologica Sinica (Chinese Edition), 82 (12): 1613-1622.

Xu, X., X. Wen, G. Yu, G. Chen, Y. Klinger, J. Hubbard, and J. Shaw (2009). Coseismic reverse- and oblique-slip, surface faulting generated by the 2008 *M*w 7.9 Wenchuan earthquake, China. Geology, 37 (6): 515-518.

Yasuhara, H., C. Marone, and D., Elsworth (2005). Fault zone restrengthening and frictional healing: the role of pressure solution. Journal of Geophysical Research, 110, B06310, doi: 10.1029/2004JB003327.

Yin, X. C., H. Z. Yu, V. Kukshenko, Z. Y. Xu, Z. S. Wu, M. Li, K. Y. Peng, S. Elizarov and Q. Li (2004). Load-unload response ratio (LURR), accelerating energy release (AER) and state vector evolution as precursors to failure of rock specimens. Pure and Applied Geophysics, 161 (11-12): 2405-2416.

Yin, X. C., X. Z. Chen, Z. P. Song, and C. Yin (1005). A new approach to earthquake prediction: the load/unload response ratio (LURR) theory. Pure and Applied Geophysics, 145, 701-715.

Yin, X. C., Y. Liu, L. P., Zhang, and S. Yuan (2011) Load-unload response ratio and its new progress, in Imaging, Modeling and Assimilation in Seismology, edited by Y. G. Li, China High Education Press, Beijing, De Gruyter, Boston, USA.

Zhao, L., P. Chen, and T. H. Jordan (2006). Strain green's tensors, reciprocity, and their applications to seismic source and structure studies. Bull. Seism. Soc. Am., 96 (5): 1753-1763.

Keywords: Data assimilation, Seismic tomography and source parameter, Full-waveform inversion, One-return approximation, Seismic imaging, Fault-zone trapped waves, Rock damage and heal, Dynamic rupture models, Ground motion prediction, Load-unload response ratio, Discrete element method, Earthquake simulation and prediction

Author Information

Yong-Gang Li

Research Professor at University of Southern California, Los Angeles, CA 90089, USA.
E-mail: ygli@usc.edu

Chapter 1
Full-Wave Seismic Data Assimilation: A Unified Methodology for Seismic Waveform Inversion

Po Chen

This chapter presents a unified methodology for seismic waveform inversion based on the data assimilation theory. The melding of observations with dynamic models to provide estimations of the state variables and/or model parameters is known as "data assimilation", which is making rapid progress in meteorology and oceanography. Using the data assimilation methodology, I will formulate the full-wave seismological inversion for seismic source and the earth structure parameters in the form of the weakly constrained generalized inverse. In such a weakly constrained formulation, the dynamic model (i.e., the seismic wave equation and its associated initial and boundary conditions) as well as the measurements are allowed to contain errors and the solution of the inverse problem does not have to satisfy the dynamics exactly. The strongly constrained inverse can be obtained as a limiting case where the errors in the dynamic model approach zero. Deficiencies in dynamic models are usually difficult, if possible at all, to eliminate, but the weakly constrained generalized inverse provides a systematic means to accommodate model errors in the data assimilation process. The Euler-Lagrange equations of the generalized inverse are closely related to the adjoint method and the scattering-integral (SI) method, which have been successfully applied in full-wave seismic tomography. I will review some recent applications of the full-wave methodology and discuss some challenging issues related to the computational implementation and the effective exploitation of the seismic waveform data.

Keywords: Data assimilation, Seismic tomography and source parameter, Full-waveform inversion, Adjoint and scattering-integral methods

1.1 Introduction

The seismic inverse problem deals with the inversion of seismological observations for properties of the seismic source that generates seismic waves and/or the geological medium through which seismic waves propagate. This inverse problem has much in common with the data assimilation problem frequently found in studies of the earth's ocean, atmosphere and other natural systems. The term "data assimilation" often refers to the melding of observations with dynamic models to produce realistic estimations of the state variables and/or model parameters and it is driving rapid advances in areas such as oceanography (e.g., Bennett, 1992; Malanotte-Rizzoli, 1996; Wunsch, 1996) and meteorology (e.g., Bengtsson et al., 1981; Daley, 1993; Kalnay, 2003). Recent advances in parallel computing technology and numerical methods (e.g., Graves, 1996; Akcelik et al., 2002; Komatitsch and Tromp, 2002; Komatitsch et al., 2004) have made large-scale, three-dimensional numerical simulations of seismic wave-fields much more affordable and they have opened up the possibility of full-wave seismic data assimilation, in which seismological observations are combined with the seismic wave equation to produce realistic estimations of the properties of the seismic source and the earth structure. I use the term "full-wave" to indicate that the dynamic model used in this chapter is the viscoelastic wave equation and its associated initial and boundary conditions, rather than approximations to the viscoelastic wave equation (e.g., the eikonal equation, the parabolic wave equation), and the methodology allows us to fully utilize the complete waveform data, as well as other types of seismological observations derived from the waveform data, such as cross-correlation travel-time measurements of body waves, dispersion measurements of surface waves, etc.

Schemes for solving data assimilation problems can be related either to estimation theory (e.g., Simon, 2006; Evensen, 2009), which can be formulated within a Bayesian statistical framework (Wikle and Berliner, 2007), or to control theory that solves optimization problems constrained by differential equations (e.g., Schulz, 2006). In this chapter, I formulate the full-wave seismic data assimilation problem in the form of the weakly constrained generalized inverse (Evensen, 2009). The starting point of the derivation is the Bayes' theorem, which defines the posterior probability density functions of the poorly known seismic source and the earth structure parameters conditioned on a set of seismological observations. Unlike conventional formulations of the seismic inverse problem, the weakly constrained generalized inverse allows the dynamic model (e.g., the wave equation and its initial and boundary conditions) as well as the observations to contain uncertainties and the solution of the inverse problem does not have to satisfy the dynamic model exactly. The strongly constrained inverse can be obtained as a limiting case where the uncertainties in the dynamic model approach zero.

The uncertainties in the seismic wave equation and their initial and boundary conditions are modeled using additive stochastic noise processes. The motivation for introducing stochastic noises into the dynamic model is two-fold. First, our deterministic model is not perfect and it is difficult, if possible at all, to fully eliminate all its deficiencies. Second, the impact of the uncertainties in the dynamic model, particularly, the impact on the estimation of model parameters, needs to be evaluated. The origins of the uncertainties are sometimes difficult to classify or explain. Some types of uncertainties depend on chance (i.e., aleatory or statistical) and others are due to the lack of knowledge (i.e., epistemic or systematic). Because of such unknown or unexplained origins, we often work toward reducing them to stochastic processes and try to quantify their statistical properties through direct or indirect methods. In this chapter the statistical properties of the noises are assumed to be Gaussian. The Gaussian assumption brings much convenience into the derivation and a direct benefit in terms of readability is that the derived equations can be easily related to conventional formulations (e.g., Tarantola, 1988). For probability densities that are non-Gaussian, the formulation still has applicability if a Gaussian distribution provides a sufficiently good approximation to the actual distribution within the range of consideration. An optimal Gaussian approximation can be found from the first and second moments of the actual distribution.

The adjoint method, which was adopted to solve seismic imaging problems in Tarantola (1984, 1988) and later extended to solve both seismic source and structural inverse problems in Tromp et al. (2005), can be considered as a limiting case of the generalized inverse, in which the dynamic model is assumed to have zero model uncertainties. A slight alteration of the derivation naturally leads to the scattering-integral method (Zhao et al., 2005; Chen et al., 2007a) used for full-3D waveform tomography in Chen et al. (2007b) and for earthquake centroid moment tensor (CMT) inversion (Zhao et al., 2006) and finite moment tensor (FMT) inversion (Chen et al., 2010b). Through the derivation of the generalized inverse, it is possible to interpret different seismological inversion methods in a unified context and to understand the assumptions and approximations they rely on.

In this article, I will also analyze the computational issues involved in full-wave seismic data assimilation in terms of compute cycles and data storage. Capabilities of the full-wave methodology will be demonstrated using examples and several challenging issues will be discussed.

1.2 Generalized Inverse

We consider elastic waves propagating inside a three-dimensional domain V with boundary ∂V. Points inside V or on ∂V are denoted as $\mathbf{x}$ and time is denoted as t. The medium is described by the mass density $\rho(\mathbf{x})$ and a fourth-order rate-of-relaxation tensor $\boldsymbol{\psi}(\mathbf{x},t)$ with its components given by $\psi^{ijkl}(\mathbf{x},t)$. The forcing of the system is described by a body-force density $\mathbf{f}_s(\mathbf{x},t)$ with its components given by $f_s^i(\mathbf{x},t)$ and a second-order moment tensor density $\mathbf{M}_s(\mathbf{x},t)$ with its components given by $M_s^{ij}(\mathbf{x},t)$. Here we have introduced s to index the seismic sources used in the inversion. The seismic wave-field is described by the source-specific displacement field $\mathbf{u}_s(\mathbf{x},t)$ with its components given by $u_s^i(\mathbf{x},t)$ and the second-order stress tensor field $\boldsymbol{\sigma}_s(\mathbf{x},t)$ with its components given by $\sigma_s^{ij}(\mathbf{x},t)$. We assume that the dynamic model described by the seismic wave equation has an additive model residual,

$$\rho(\mathbf{x})\partial_t^2 u_s^i(\mathbf{x},t) - \sum_j \partial_j \left[\int_{-\infty}^{\infty} d\tau \sum_{kl} \psi^{ijkl}(\mathbf{x},t-\tau)\partial_l u_s^k(\mathbf{x},\tau) + M_s^{ij}(\mathbf{x},t) \right] \\ - f_s^i(\mathbf{x},t) = q_s^i(\mathbf{x},t), \quad (1.1)$$

where we have assumed a constitutive relation of the form

$$\sigma_s^{ij}(\mathbf{x},t) = \int_{-\infty}^{\infty} d\tau \sum_{kl} \psi^{ijkl}(\mathbf{x},t-\tau)\partial_l u_s^k(\mathbf{x},\tau) + M_s^{ij}(\mathbf{x},t), \quad (1.2)$$

and ∂_t represents the partial derivative with respect to time t, and ∂_j represents the partial derivative with respect to the j-th component of $\mathbf{x}$, $\partial/\partial x^j$. Here we have introduced $q_s^i(\mathbf{x},t)$, which is a component of an additive stochastic noise process $\mathbf{q}_s(\mathbf{x},t)$. We assume that $q_s^i(\mathbf{x},t)$ has Gaussian statistics and its mean (i.e., modeling bias) is given by $\overline{q}_s^i(\mathbf{x},t)$. In this chapter, we express summations over indices explicitly and do not use repeated indices to imply the Einstein summation convention. We consider the quiescent-past initial condition and the free-surface boundary condition, also with additive residuals,

$$u_s^i(\mathbf{x},0) = a_s^i(\mathbf{x}), \quad \partial_t u_s^i(\mathbf{x},0) = b_s^i(\mathbf{x}) \tag{1.3}$$

$$\sum_j \hat{n}^j(\mathbf{x}) \int_{-\infty}^{\infty} d\tau \sum_{kl} \psi^{ijkl}(\mathbf{x},t-\tau)\partial_l u_s^k(\mathbf{x},\tau) = c_s^i(\mathbf{x},t), \quad \mathbf{x} \in \partial V, \tag{1.4}$$

where $a_s^i(\mathbf{x})$, $b_s^i(\mathbf{x})$, and $c_s^i(\mathbf{x},t)$ are components of additive stochastic noise processes $\mathbf{a}_s(\mathbf{x})$, $\mathbf{b}_s(\mathbf{x})$ and $\mathbf{c}_s(\mathbf{x},t)$. We assume they have Gaussian statistics with mean (i.e., modeling bias) given by $\overline{a}_s^i(\mathbf{x})$, $\overline{b}_s^i(\mathbf{x})$, and $\overline{c}_s^i(\mathbf{x},t)$, respectively. The vector $\hat{\mathbf{n}}(\mathbf{x})$, with its components given by $\hat{n}^j(\mathbf{x})$, denotes the normal directions on ∂V. The extension to other types of initial and/or boundary conditions is straightforward.

Some common origins of model uncertainties (the uncertainties in the model parameters are considered separately below) include but are not limited to the errors in the mathematical model, the numerical method used for solving the mathematical model, etc. Possible reasons that can cause deviations from the traction-free boundary condition may include lithosphere-atmosphere coupling, deviation from the continuum model for materials in the near-surface environment, numerical errors caused by, for instance, errors in the numerical representation of the actual topography, etc. For applications involving absorbing boundary conditions, spurious reflections may still exist at corners or for grazing incident waves and can contaminate the rest of the wave field. The earth is constantly in motion. The quiescent-past initial condition for one seismic event can be violated in practice if we consider motions caused by, for instance, other seismic events, the earth's ambient noise field and the constant hum of the earth caused by atmosphere-ocean-seafloor coupling (Rhie and Romanowicz, 2004).

For each seismogram, we consider a finite set of data functionals, indexed by n, that measure the misfit between the observed seismogram $\overline{u}_s^i(\mathbf{x}_r,t)$ and the model-predicted (i.e. synthetic) seismogram $u_s^i(\mathbf{x}_r,t)$,

$$d_{srn}^i = D_n\left[\overline{u}_s^i(\mathbf{x}_r,t), u_s^i(\mathbf{x}_r,t)\right], \tag{1.5}$$

where r is an index to reference different receivers. We assume D_n is constructed to satisfy

$$D_n\left[u_s^i(\mathbf{x}_r,t), u_s^i(\mathbf{x}_r,t)\right] = 0. \tag{1.6}$$

The measurement process generally involves nonlinear operations on both observed seismograms and the corresponding synthetics.

1.2.1 Prior Probability Densities

Assume that it is available to have a prior estimate for mass density ${}_0\rho(\mathbf{x})$. Furthermore, we assume that the poorly known mass density $\rho(\mathbf{x})$ has distributed errors with Gaussian statistics and an error covariance $C_{\rho\rho}(\mathbf{x},\mathbf{x}')$. We define the inverse of the error covariance as $W_{\rho\rho}(\mathbf{x},\mathbf{x}')$, which satisfies

$$\int_V dV(\mathbf{x}'')C_{\rho\rho}(\mathbf{x},\mathbf{x}'')W_{\rho\rho}(\mathbf{x}'',\mathbf{x}') = \delta(\mathbf{x}-\mathbf{x}'), \tag{1.7}$$

where $\delta(\mathbf{x}-\mathbf{x}')$ is the Dirac delta function. The prior probability density function for mass density $\boldsymbol{\rho}(\mathbf{x})$ can then be expressed as

$$p(\rho) \propto \exp\left\{-\frac{1}{2}\int_V dV(\mathbf{x})\int_V dV(\mathbf{x}')\left[\rho(\mathbf{x})-{}_0\rho(\mathbf{x})\right]\mathbf{W}_{\rho\rho}(\mathbf{x},\mathbf{x}')\left[\rho(\mathbf{x}')-{}_0\rho(\mathbf{x}')\right]\right\}. \tag{1.8}$$

We have not included a denominator that normalizes the right-hand-side, thereby using, proportional to, $\propto$, rather than equal to, $=$.

Assume the component of the fourth-order rate-of-relaxation tensor, $\boldsymbol{\psi}^{ijkl}(\mathbf{x},t)$, follows Gaussian statistics with a prior estimate ${}_0\boldsymbol{\psi}^{ijkl}(\mathbf{x},t)$ and error covariance $C_{\psi^{ijkl}\psi^{ijkl}}(\mathbf{x},t;\mathbf{x}',t')$. Similar to (1.7), we define the inverse of the error covariance as $W_{\psi^{ijkl}\psi^{ijkl}}(\mathbf{x},t;\mathbf{x}',t')$, which satisfies

$$\int_V dV(\mathbf{x}'')\int_0^T dt''C_{\psi^{ijkl}\psi^{ijkl}}(\mathbf{x},t;\mathbf{x}'',t'')W_{\psi^{ijkl}\psi^{ijkl}}(\mathbf{x}'',t'';\mathbf{x}',t') = \delta(\mathbf{x}-\mathbf{x}')\delta(t-t') \tag{1.9}$$

where $[0,T]$ is the time interval of our analysis. The prior probability density for $\psi^{ijkl}(\mathbf{x},t)$ can then be expressed as

$$p(\psi^{ijkl}) \propto \exp\left\{-\frac{1}{2}\int_V dV(\mathbf{x})\int_V dV(\mathbf{x}')\int_0^T dt\int_0^T dt'\left[\psi^{ijkl}(\mathbf{x},t)-{}_0\psi^{ijkl}(\mathbf{x},t)\right]\right.$$
$$\left. W_{\psi^{ijkl}\psi^{ijkl}}(\mathbf{x},t;\mathbf{x}',t')\left[\psi^{ijkl}(\mathbf{x}',t')-{}_0\psi^{ijkl}(\mathbf{x}',t')\right]\right\}. \tag{1.10}$$

For notational convenience in the following derivation, we express the prior probability density for $\boldsymbol{\psi}(\mathbf{x},t)$ as a multiplication of the probability density for each component

$$p(\boldsymbol{\psi}) \propto \exp\left\{-\frac{1}{2}\int_V dV(\mathbf{x})\int_V dV(\mathbf{x}')\int_0^T dt\int_0^T dt'\sum_{ijkl}\left[\psi^{ijkl}(\mathbf{x},t)-{}_0\psi^{ijkl}(\mathbf{x},t)\right]\right.$$
$$\left. W_{\psi^{ijkl}\psi^{ijkl}}(\mathbf{x},t;\mathbf{x}',t')\left[\psi^{ijkl}(\mathbf{x}',t')-{}_0\psi^{ijkl}(\mathbf{x}',t')\right]\right\}. \tag{1.11}$$

Because of the symmetry of the rate-of-relaxation tensor, out of the 81 components, only 21 of them are independent. To incorporate equality constraints among elastic parameters, the delta distribution can be introduced to represent the corresponding conditional probabilities. The delta distribution can be considered as the Gaussian distribution with its variance approaching zero. We note that by modifying (1.11) and the structure of $W_{\psi^{ijkl}\psi^{ijkl}}(\mathbf{x},t;\mathbf{x}',t')$ to account for cross-dependences among components of $\boldsymbol{\psi}(\mathbf{x},t)$, the derivations in the following can be extended to account for cases where the errors in the components of $\boldsymbol{\psi}(\mathbf{x},t)$ are not independent. The rate-of-relaxation tensor also needs to satisfy certain stability requirements, which can result in positivity constraints. In this case, the Gaussian assumption is valid only locally when the reference rate-of-relaxation tensor satisfies the stability requirements and the variances are not too large.

We also define probability density functions for the body-force density $\mathbf{f}_s(\mathbf{x},t)$ and the moment tensor density $\mathbf{M}_s(\mathbf{x},t)$ using the assumption of Gaussian statistics and we obtain

$$p(\mathbf{f}_s) \propto \exp\left\{ -\frac{1}{2}\int_V dV(\mathbf{x}) \int_V dV(\mathbf{x}') \int_0^T dt \int_0^T dt' \right.$$
$$\left. \sum_i \left[f_s^i(\mathbf{x},t) - {}_0f_s^i(\mathbf{x},t)\right] W_{f_s^i f_s^i}(\mathbf{x},t;\mathbf{x}',t') \left[f_s^i(\mathbf{x},t) - {}_0f_s^i(\mathbf{x},t)\right] \right\}, \tag{1.12}$$

$$p(\mathbf{M}_s) \propto \exp\left\{ -\frac{1}{2}\int_V dV(\mathbf{x}) \int_V dV(\mathbf{x}') \int_0^T dt \int_0^T dt' \right.$$
$$\left. \sum_{ij} \left[M_s^{ij}(\mathbf{x},t) - {}_0M_s^{ij}(\mathbf{x},t)\right] W_{M_s^{ij} M_s^{ij}}(\mathbf{x},t;\mathbf{x}',t') \left[M_s^{ij}(\mathbf{x},t) - {}_0M_s^{ij}(\mathbf{x},t)\right] \right\}, \tag{1.13}$$

where ${}_0f_s^i(\mathbf{x},t)$ and ${}_0M_s^{ij}(\mathbf{x},t)$ are components of ${}_0\mathbf{f}_s(\mathbf{x},t)$ and ${}_0\mathbf{M}_s(\mathbf{x},t)$, which are prior estimates of $\mathbf{f}_s(\mathbf{x},t)$ and $\mathbf{M}_s(\mathbf{x},t)$, $W_{f_s^i f_s^i}(\mathbf{x},t;\mathbf{x}',t')$ and $W_{M_s^{ij} M_s^{ij}}(\mathbf{x},t;\mathbf{x}',t')$ are inverses of the corresponding error covariance functions with definitions similar to (1.9).

For the model residual introduced in (1.1), as well as residuals in the initial and boundary conditions in (1.3) and (1.4), we also define Gaussian probability density functions

$$p(\mathbf{q}_s) \propto \exp\left\{ -\frac{1}{2}\int_V dV(\mathbf{x}) \int_V dV(\mathbf{x}') \int_0^T dt \int_0^T dt' \right.$$
$$\left. \sum_i \left[q_s^i(\mathbf{x},t) - \bar{q}_s^i(\mathbf{x},t)\right] W_{q_s^i q_s^i}(\mathbf{x},t;\mathbf{x}',t') \left[q_s^i(\mathbf{x}',t') - \bar{q}_s^i(\mathbf{x}',t')\right] \right\}, \tag{1.14}$$

$$p(\mathbf{a}_s) \propto \exp\left\{ -\frac{1}{2}\int_V dV(\mathbf{x}) \int_V dV(\mathbf{x}') \right.$$
$$\left. \sum_i \left[a_s^i(\mathbf{x}) - \bar{a}_s^i(\mathbf{x})\right] W_{a_s^i a_s^i}(\mathbf{x},\mathbf{x}') \left[a_s^i(\mathbf{x}') - \bar{a}_s^i(\mathbf{x}')\right] \right\}, \tag{1.15}$$

$$p(\mathbf{b}_s) \propto \exp\left\{ -\frac{1}{2}\int_V dV(\mathbf{x}) \int_V dV(\mathbf{x}') \right.$$
$$\left. \sum_i \left[b_s^i(\mathbf{x}) - \bar{b}_s^i(\mathbf{x})\right] W_{b_s^i b_s^i}(\mathbf{x},\mathbf{x}') \left[b_s^i(\mathbf{x}') - \bar{b}_s^i(\mathbf{x}')\right] \right\}, \tag{1.16}$$

$$p(\mathbf{c}_s) \propto \exp\left\{ -\frac{1}{2}\int_{\partial V} dS(\mathbf{x}) \int_{\partial V} dS(\mathbf{x}') \int_0^T dt \int_0^T dt' \right.$$
$$\left. \sum_i \left[c_s^i(\mathbf{x},t) - \bar{c}_s^i(\mathbf{x},t)\right] W_{c_s^i c_s^i}(\mathbf{x},t;\mathbf{x}',t') \left[c_s^i(\mathbf{x}',t') - \bar{c}_s^i(\mathbf{x}',t')\right] \right\}. \tag{1.17}$$

Here we have introduced $W_{q_s^l q_s^l}$, $W_{a_s^l a_s^l}$, $W_{b_s^l b_s^l}$ and $W_{c_s^l c_s^l}$, which are inverses of their corresponding covariance functions.

1.2.2 Bayes' Theorem

We assume that the errors in the misfit measurements follow Gaussian statistics, considering (1.6), we can express the conditional probability as

$$p\left(d_{srn}^i|\rho,\boldsymbol{\psi},\mathbf{f}_s,\mathbf{M}_s,\mathbf{q}_s,\mathbf{a}_s,\mathbf{b}_s,\mathbf{c}_s\right) \propto \exp\left\{-\frac{W_{d_{srn}^i d_{srn}^i}}{2}\left(d_{srn}^i\right)^2\right\}, \tag{1.18}$$

where $W_{d_{srn}^i d_{srn}^i}$ is the inverse of the variance for the errors in the misfit measurements. We introduce a measurement vector $\mathbf{d}_s$ with its components given by all the misfit measurements for source index s. Assuming the errors in the misfit measurements are uncorrelated, we can express the conditional probability for $\mathbf{d}_s$ as

$$p\left(\mathbf{d}_s|\rho,\boldsymbol{\psi},\mathbf{f}_s,\mathbf{M}_s,\mathbf{q}_s,\mathbf{a}_s,\mathbf{b}_s,\mathbf{c}_s\right) \propto \exp\left\{-\frac{\mathbf{d}_s^T\mathbf{W}_{\mathbf{d}_s\mathbf{d}_s}\mathbf{d}_s}{2}\right\}, \tag{1.19}$$

where $\mathbf{W}_{\mathbf{d}_s\mathbf{d}_s}$ is a diagonal matrix with its diagonal elements given by $W_{d_{srn}^i d_{srn}^i}$. We note that by modifying the structure of the matrix $\mathbf{W}_{\mathbf{d}_s\mathbf{d}_s}$, (1.19) can be generalized to account for correlated measurement errors as well. For seismic data assimilation problems involving multiple seismic sources, the conditional probability density can be expressed as

$$\begin{aligned}&p\left(\mathbf{d}_1,\cdots,\mathbf{d}_{N_s}|\rho,\boldsymbol{\psi},\mathbf{f}_1,\mathbf{M}_1,\mathbf{q}_1,\mathbf{a}_1,\mathbf{b}_1,\mathbf{c}_1,\cdots,\mathbf{f}_{N_s},\mathbf{M}_{N_s},\mathbf{q}_{N_s},\mathbf{a}_{N_s},\mathbf{b}_{N_s},\mathbf{c}_{N_s}\right)\\ &\propto \exp\left\{-\frac{1}{2}\sum_{s=1}^{N_s}\mathbf{d}_s^T\mathbf{W}_{\mathbf{d}_s\mathbf{d}_s}\mathbf{d}_s\right\},\end{aligned} \tag{1.20}$$

where N_s is the total number of seismic sources used in the inversion. The likelihood function can then be expressed as

$$\begin{aligned}&l\left(\rho,\boldsymbol{\psi},\mathbf{f}_1,\mathbf{M}_1,\mathbf{q}_1,\mathbf{a}_1,\mathbf{b}_1,\mathbf{c}_1,\cdots,\mathbf{f}_{N_s},\mathbf{M}_{N_s},\mathbf{q}_{N_s},\mathbf{a}_{N_s},\mathbf{b}_{N_s},\mathbf{c}_{N_s}|\mathbf{d}_1,\cdots,\mathbf{d}_{N_s}\right)\\ &\propto \exp\left\{-\frac{1}{2}\sum_{s=1}^{N_s}\mathbf{d}_s^T\mathbf{W}_{\mathbf{d}_s\mathbf{d}_s}\mathbf{d}_s\right\}.\end{aligned} \tag{1.21}$$

Following Bayes' theorem, we obtain the posterior probability density as follows:

$$\begin{aligned}&p\left(\rho,\boldsymbol{\psi},\mathbf{f}_1,\mathbf{M}_1,\mathbf{q}_1,\mathbf{a}_1,\mathbf{b}_1,\mathbf{c}_1,\cdots,\mathbf{f}_{N_s},\mathbf{M}_{N_s},\mathbf{q}_{N_s},\mathbf{a}_{N_s},\mathbf{b}_{N_s},\mathbf{c}_{N_s}|\mathbf{d}_1,\cdots,\mathbf{d}_{N_s}\right) \propto\\ &l\left(\rho,\boldsymbol{\psi},\mathbf{f}_1,\mathbf{M}_1,\mathbf{q}_1,\mathbf{a}_1,\mathbf{b}_1,\mathbf{c}_1,\cdots,\mathbf{f}_{N_s},\mathbf{M}_{N_s},\mathbf{q}_{N_s},\mathbf{a}_{N_s},\mathbf{b}_{N_s},\mathbf{c}_{N_s}|\mathbf{d}_1,\cdots,\mathbf{d}_{N_s}\right)\\ &\times p(\rho)p(\boldsymbol{\psi})\prod_{s=1}^{N_s}p(\mathbf{f}_s)p(\mathbf{M}_s)p(\mathbf{q}_s)p(\mathbf{a}_s)p(\mathbf{b}_s)p(\mathbf{c}_s) = \exp\left\{-\frac{1}{2}\mathbf{Y}\right\}.\end{aligned} \tag{1.22}$$

Here we have defined an objective function $\mathbf{Y}$, which has the expression

$$\mathbf{Y} = \mathbf{Y}_1+\mathbf{Y}_2+\mathbf{Y}_3+\mathbf{Y}_4, \tag{1.23}$$

$$\mathbf{Y}_1 = \sum_{s=1}^{N_s}\mathbf{d}_s^T\mathbf{W}_{\mathbf{d}_s\mathbf{d}_s}\mathbf{d}_s, \tag{1.24}$$

$$\begin{aligned}\mathbf{Y}_2 = &\sum_{s=1}^{N_s}\iint_V dV(\mathbf{x})dV(\mathbf{x}')\int_0^T dt\int_0^T dt'\sum_i\left[q_s^i(\mathbf{x},t)-\bar{q}_s^i(\mathbf{x},t)\right]\\ &W_{q_s^iq_s^i}(\mathbf{x},t;\mathbf{x}',t')\left[q_s^i(\mathbf{x}',t')-\bar{q}_s^i(\mathbf{x}',t')\right]\\ &+\sum_{s=1}^{N_s}\iint_V dV(\mathbf{x})dV(\mathbf{x}')\sum_i\left[a_s^i(\mathbf{x})-\bar{a}_s^i(\mathbf{x})\right]W_{a_s^ia_s^i}(\mathbf{x},\mathbf{x}')\left[a_s^i(\mathbf{x}')-\bar{a}_s^i(\mathbf{x}')\right]\\ &+\sum_{s=1}^{N_s}\iint_V dV(\mathbf{x})dV(\mathbf{x}')\sum_i\left[b_s^i(\mathbf{x})-\bar{b}_s^i(\mathbf{x})\right]W_{b_s^ib_s^i}(\mathbf{x},\mathbf{x}')\left[b_s^i(\mathbf{x}')-\bar{b}_s^i(\mathbf{x}')\right]\\ &+\sum_{s=1}^{N_s}\iint_{\partial V} dS(\mathbf{x})dS(\mathbf{x}')\int_0^T dt\int_0^T dt'\sum_i\left[c_s^i(\mathbf{x},t)-\bar{c}_s^i(\mathbf{x},t)\right]\\ &W_{c_s^ic_s^i}(\mathbf{x},t;\mathbf{x}',t')\left[c_s^i(\mathbf{x}',t')-\bar{c}_s^i(\mathbf{x}',t')\right],\end{aligned}\tag{1.25}$$

$$\begin{aligned}\mathbf{Y}_3 = &\sum_{s=1}^{N_s}\iint_V dV(\mathbf{x})dV(\mathbf{x}')\int_0^T dt\int_0^T dt'\sum_i\left[f_s^i(\mathbf{x},t)-{}_0f_s^i(\mathbf{x},t)\right]\\ &W_{f_s^if_s^i}(\mathbf{x},t;\mathbf{x}',t')\left[f_s^i(\mathbf{x}',t')-{}_0f_s^i(\mathbf{x}',t')\right]\\ &+\sum_{s=1}^{N_s}\iint_V dV(\mathbf{x})dV(\mathbf{x}')\int_0^T dt\int_0^T dt'\sum_{ij}\left[M_s^{ij}(\mathbf{x},t)-{}_0M_s^{ij}(\mathbf{x},t)\right]\\ &W_{M_s^{ij}M_s^{ij}}(\mathbf{x},t;\mathbf{x}',t')\left[M_s^{ij}(\mathbf{x}',t')-{}_0M_s^{ij}(\mathbf{x}',t')\right],\end{aligned}\tag{1.26}$$

$$\begin{aligned}\mathbf{Y}_4 = &\iint_V dV(\mathbf{x})dV(\mathbf{x}')\left[\rho(\mathbf{x})-{}_0\rho(\mathbf{x})\right]W_{\rho\rho}(\mathbf{x},\mathbf{x}')\left[\rho(\mathbf{x}')-{}_0\rho(\mathbf{x}')\right]\\ &+\iint_V dV(\mathbf{x})dV(\mathbf{x}')\int_0^T dt\int_0^T dt'\sum_{ijkl}\left[\psi^{ijkl}(\mathbf{x},t)-{}_0\psi^{ijkl}(\mathbf{x},t)\right]\\ &W_{\psi^{ijkl}\psi^{ijkl}}(\mathbf{x},t;\mathbf{x}',t')\left[\psi^{ijkl}(\mathbf{x}',t')-{}_0\psi^{ijkl}(\mathbf{x}',t')\right].\end{aligned}\tag{1.27}$$

Thus, for Gaussian priors and likelihood, maximization of the posterior probability density in (1.22) is equivalent to minimization of $\mathbf{Y}$ as defined in (1.23)-(1.27). The objective function $\mathbf{Y}$ will have a global minimum, but it may not be unique. It may also possess a number of local minima and there is a risk of converging into one of those local minima using iterative optimization algorithms such as steepest-descent, conjugate-gradient and Gauss-Newton methods.

1.2.3 Euler-Lagrange Equations

To minimize the objective function $\mathbf{Y}$ in (1.23), we can invoke the Hamilton's principle (Arnold, 1989) (i.e., the principle of stationary action) by calculating the variational derivatives of $\mathbf{Y}$ and requiring that they approach zeros while the arbitrary perturbations of the generalized coordinates go to zero. The resulting equations are called Euler-Lagrange equations.

We first define an "adjoint" vector field ${}^{\dagger}\mathbf{u}_s(\mathbf{x},t)$ with its components given by ${}^{\dagger}u_s^i(\mathbf{x},t)$ satisfying

$$ {}^{\dagger}u_s^i(\mathbf{x},T-t) = \int_V dV(\mathbf{x}') \int_0^T dt' W_{q_s^i q_s^i}(\mathbf{x},t;\mathbf{x}',t') \left[q_s^i(\mathbf{x}',t') - \bar{q}_s^i(\mathbf{x}',t')\right]. \tag{1.28} $$

If we multiply the covariance function $C_{q_s^i q_s^i}(\mathbf{x},t;\mathbf{x}'',t'')$ and carry out the spatial integration on $\mathbf{x}$ and the temporal integration on t on both sides of (1.28), we obtain

$$ \begin{aligned} &\rho(\mathbf{x})\partial_t^2 u_s^i(\mathbf{x},t) - \sum_j \partial_j \left[\int_{-\infty}^{\infty} d\tau \sum_{kl} \psi^{ijkl}(\mathbf{x},t-\tau)\partial_l u_s^k(\mathbf{x},\tau) + M_s^{ij}(\mathbf{x},t)\right] - f_s^i(\mathbf{x},t) \\ &= \bar{q}_s^i(\mathbf{x},t) + \int_V dV(\mathbf{x}') \int_0^T dt' C_{q_s^i q_s^i}(\mathbf{x},t;\mathbf{x}',t')\, {}^{\dagger}u_s^i(\mathbf{x}',T-t'), \end{aligned} \tag{1.29} $$

which is the original seismic wave equation (1.1) with a representation of the model residual involving the modeling bias $\bar{q}_s^i(\mathbf{x},t)$ and a product between the model residual covariance and the adjoint wave field.

The variational derivative of $\mathbf{Y}$ with respect to the displacement field $\mathbf{u}_s(\mathbf{x},t)$ involves $\mathbf{Y}_1$ in equation (1.24) and the first term of $\mathbf{Y}_2$ in equation (1.25). Here we introduce the "seismogram perturbation kernel", $J_{srn}^i(t)$, which is defined as the Fréchet kernel of the misfit measurement d_{srn}^i with respect to the synthetic seismogram

$$ \delta d_{srn}^i = \int_0^T dt J_{srn}^i(t)\,\delta u_s^i(\mathbf{x}_r,t). \tag{1.30} $$

A perturbation of $\mathbf{Y}_1$ can now be expressed using

$$ \sum_{s=1}^{N_s} \delta\mathbf{d}_s^T \mathbf{W}_{\mathbf{d}_s\mathbf{d}_s}\mathbf{d}_s = \sum_{s=1}^{N_s}\sum_i \int_V dV(\mathbf{x}) \int_0^T dt \sum_{rn} \delta(\mathbf{x}-\mathbf{x}_r) J_{srn}^i(t) W_{d_{srn}^i d_{srn}^i} d_{srn}^i \delta u_s^i(\mathbf{x},t). \tag{1.31} $$

Using the adjoint wave field as defined in (1.28) and considering the wave equation in (1.1), a perturbation of the first term on the right-hand-side of (1.25) can be expressed in terms of a perturbation of the displacement field as

$$ \begin{aligned} &\sum_{s=1}^{N_s} \int_V dV(\mathbf{x}) \int_0^T dt \sum_i \delta q_s^i(\mathbf{x},t)\, {}^{\dagger}u_s^i(\mathbf{x},T-t) \\ &= \sum_{s=1}^{N_s} \int_V dV(\mathbf{x}) \int_0^T dt \sum_i \rho(\mathbf{x})\, {}^{\dagger}u_s^i(\mathbf{x},T-t)\, \partial_t^2 \delta u_s^i(\mathbf{x},t) \\ &\quad - \sum_i\sum_j \partial_j \left[\int_{-\infty}^{\infty} d\tau \sum_{kl} \psi^{ijkl}(\mathbf{x},t-\tau)\partial_l \delta u_s^k(\mathbf{x},\tau)\right] {}^{\dagger}u_s^i(\mathbf{x},T-t). \end{aligned} \tag{1.32} $$

Transferring the second-order temporal derivative on $\delta u_s^i(\mathbf{x},t)$ in (1.32) to the adjoint wave field through integration-by-parts, we obtain

$$\begin{aligned}
&\sum_{s=1}^{N_s}\int_V dV(\mathbf{x})\rho(\mathbf{x})\sum_i\int_0^T dt\,{}^\dagger u_s^i(\mathbf{x},T-t)\partial_t^2\delta u_s^i(\mathbf{x},t)\\
&=\sum_{s=1}^{N_s}\sum_i\int_V dV(\mathbf{x})\rho(\mathbf{x})\left[\delta\dot{u}_s^i(\mathbf{x},t)^\dagger u_s^i(\mathbf{x},T-t)|_0^T+\delta u_s^i(\mathbf{x},t)^\dagger\dot{u}_s^i(\mathbf{x},T-t)|_0^T\right]\\
&+\int_V dV(\mathbf{x})\int_0^T dt\rho(\mathbf{x})^\dagger\ddot{u}_s^i(\mathbf{x},T-t)\delta u_s^i(\mathbf{x},t).
\end{aligned}\tag{1.33}$$

The spatial derivatives with respect to $\delta u_s^i(\mathbf{x},t)$ in (1.32) can also be transferred to the adjoint wave field by applying Green's identity. After some algebraic manipulations, the second term on the right-hand-side of (1.32) can be expressed as

$$\begin{aligned}
&\sum_{s=1}^{N_s}\sum_i\int_V dV(\mathbf{x})\int_0^T dt\sum_j\partial_j\left[\int_{-\infty}^{\infty}d\tau\sum_{kl}\psi^{ijkl}(\mathbf{x},t-\tau)\partial_l\delta u_s^k(\mathbf{x},\tau)\right]\dagger\, u_s^i(\mathbf{x},T-t)\\
&=\sum_{s=1}^{N_s}\sum_i\int_{\partial V}dS(\mathbf{x})\int_0^T dt\left\{{}^\dagger u_s^i(\mathbf{x},T-t)\sum_j\hat{n}^j(\mathbf{x})\left[\int_{-\infty}^{\infty}d\tau\sum_{kl}\psi^{ijkl}(\mathbf{x},t-\tau)\partial_l\delta u_s^k(\mathbf{x},\tau)\right]\right.\\
&\left.-\sum_j\hat{n}^j(\mathbf{x})\left[\int_{-\infty}^{\infty}d\tau\sum_{kl}\psi^{ijkl}(\mathbf{x},T-t-\tau)\partial_l{}^\dagger u_s^k(\mathbf{x},\tau)\right]\delta u_s^i(\mathbf{x},t)\right\}\\
&+\int_V dV(\mathbf{x})\int_0^T dt\sum_j\partial_j\left[\int_{-\infty}^{\infty}d\tau\sum_{kl}\psi^{ijkl}(\mathbf{x},T-t-\tau)\partial_l{}^\dagger u_s^k(\mathbf{x},\tau)\right]\delta u_s^i(\mathbf{x},t).
\end{aligned}\tag{1.34}$$

Collecting terms containing $\delta u_s^i(\mathbf{x},t)$ in (1.31)-(1.34), we obtain the variational derivative of $\mathbf{Y}$ with respect to $\mathbf{u}_s(\mathbf{x},t)$ as follows:

$$\begin{aligned}
\frac{\delta\mathbf{Y}}{2\delta u_s^i(\mathbf{x},t)}&=\rho(\mathbf{x})^\dagger\ddot{u}_s^i(\mathbf{x},T-t)-\sum_j\partial_j\left[\int_{-\infty}^{\infty}d\tau\sum_{kl}\psi^{ijkl}(\mathbf{x},T-t-\tau)\partial_l{}^\dagger u_s^k(\mathbf{x},\tau)\right]\\
&+\sum_{rn}\delta(\mathbf{x}-\mathbf{x}_r)J_{srn}^i(t)W_{d_{srn}^i d_{srn}^i}d_{srn}^i.
\end{aligned}\tag{1.35}$$

Requiring $\delta\mathbf{Y}\to 0$ for arbitrary $\delta u_s^i(\mathbf{x},t)\to 0$, we obtain

$$\begin{aligned}
&\rho(\mathbf{x})(\partial_t)^{2\,\dagger}u_s^i(\mathbf{x},t)-\sum_j\partial_j\left[\int_{-\infty}^{\infty}d\tau\sum_{kl}\psi^{ijkl}(\mathbf{x},t-\tau)\partial_l^\dagger u_s^k(\mathbf{x},\tau)\right]\\
&=-\sum_{rn}\delta(\mathbf{x}-\mathbf{x}_r)J_{srn}^i(T-t)W_{d_{srn}^i d_{srn}^i}d_{srn}^i.
\end{aligned}\tag{1.36}$$

Here we have reversed the sign of time by making the substitution $T-t\to t$. Equation (1.36) is the adjoint wave equation, which is forced by misfit measurements located at receiver locations and weighted by the corresponding time-reversed seismogram perturbation kernels and the inverse of the data covariance.

The variational derivatives of $\mathbf{Y}$ with respect to the initial conditions $\mathbf{u}_s(\mathbf{x},0)$ and $\dot{\mathbf{u}}_s(\mathbf{x},0)$ can be obtained by collecting terms containing $\delta u_s^i(\mathbf{x},0)$ and $\delta \dot{u}_s^i(\mathbf{x},0)$ from (1.33) and considering the second and third terms of $\mathbf{Y}_2$ on the right-hand-side of (1.25). The variational derivatives with respect to the initial conditions are

$$\frac{\delta \mathbf{Y}}{2\delta u_s^i(\mathbf{x},0)} = \int_V dV(\mathbf{x}') W_{a_s^i a_s^i}(\mathbf{x},\mathbf{x}') \left[u_s^i(\mathbf{x}',0) - \vec{a}_s^i(\mathbf{x}')\right] - {}^{\dagger}\dot{u}_s^i(\mathbf{x},T), \tag{1.37}$$

$$\frac{\delta \mathbf{Y}}{2\delta \dot{u}_s^i(\mathbf{x},0)} = \int_V dV(\mathbf{x}') W_{b_s^i b_s^i}(\mathbf{x},\mathbf{x}') \left[\dot{u}_s^i(\mathbf{x}',0) - \vec{b}_s^i(\mathbf{x}')\right] - {}^{\dagger}u_s^i(\mathbf{x},T). \tag{1.38}$$

Requiring $\delta \mathbf{Y} \to 0$ for arbitrary $\delta u_s^i(\mathbf{x},0) \to 0$ and $\delta \dot{u}_s^i(\mathbf{x},0) \to 0$, we obtain

$$u_s^i(\mathbf{x},0) = \vec{a}_s^i(\mathbf{x}) + \int_V dV(\mathbf{x}') C_{a_s^i a_s^i}(\mathbf{x},\mathbf{x}')^{\dagger}\dot{u}_s^i(\mathbf{x}',T), \tag{1.39}$$

$$\dot{u}_s^i(\mathbf{x},0) = \vec{b}_s^i(\mathbf{x}) + \int_V dV(\mathbf{x}') C_{b_s^i b_s^i}(\mathbf{x},\mathbf{x}')^{\dagger}u_s^i(\mathbf{x}',T). \tag{1.40}$$

Considering (1.33), the variational derivatives of $\mathbf{Y}$ with respect to $\mathbf{u}_s(\mathbf{x},T)$ and $\dot{\mathbf{u}}_s(\mathbf{x},T)$ provide us a set of initial conditions for the adjoint wave field,

$${}^{\dagger}u_s^i(\mathbf{x},0) = 0, \quad {}^{\dagger}\dot{u}(\mathbf{x},0) = 0. \tag{1.41}$$

The variational derivative of $\mathbf{Y}$ with respect to the boundary condition involves the first surface integral on the right-hand-side of (1.34) and the fourth term on the right-hand-side of (1.25).

$$\sum_j \hat{n}^j(\mathbf{x}) \left[\int_{-\infty}^{\infty} d\tau \sum_{kl} \psi^{ijkl}(\mathbf{x},t-\tau)\partial_l u_s^k(\mathbf{x},\tau)\right]$$
$$= \vec{c}_s^i(\mathbf{x},t) + \int_{\partial V} dS(\mathbf{x}') \int_0^T dt' C_{c_s^i c_s^i}(\mathbf{x},t;\mathbf{x}',t')^{\dagger}u_s^i(\mathbf{x}',T-t'). \tag{1.42}$$

The variational derivative of $\mathbf{Y}$ with respect to $u_s^i(\mathbf{x},t)$ for $\mathbf{x} \in \partial V$ is provided by the second surface integral on the right-hand-side of (1.34). Requiring this variational derivative to be zero, we obtain the boundary condition for the adjoint wave equation,

$$\sum_j \hat{n}^j(\mathbf{x}) \left[\int_{-\infty}^{\infty} d\tau \sum_{kl} \psi^{ijkl}(\mathbf{x},T-t-\tau)\partial_l {}^{\dagger}u_s^k(\mathbf{x},\tau)\right] = 0. \tag{1.43}$$

Considering the wave equation in (1.1), the variational derivative of $\mathbf{Y}$ with respect to the body-force density $\mathbf{f}_s$ involves the first term of $\mathbf{Y}_2$ on the right-hand-side of (1.25) and the first term of $\mathbf{Y}_3$ on the right-hand-side of (1.26). Requiring this variational derivative to be zero, we obtain

$$f_s^i(\mathbf{x},t) = {}_0f_s^i(\mathbf{x},t) + \int_V dV(\mathbf{x}') \int_0^T dt' C_{f_s^i f_s^i}(\mathbf{x},t;\mathbf{x}',t')^{\dagger}u_s^i(\mathbf{x}',T-t'). \tag{1.44}$$

The variational derivative of $\mathbf{Y}$ with respect to the moment tensor density $\mathbf{M}_s$ also involves the first term of $\mathbf{Y}_2$ on the right-hand-side of (1.25) and the second term of $\mathbf{Y}_3$ on the right-hand-side of (1.26). Requiring this variational derivative to be zero and applying the divergence theorem, we obtain

$$M_s^{ij}(\mathbf{x},t) = {}_0M_s^{ij}(\mathbf{x},t) - \int_V dV(\mathbf{x}') \int_0^T dt' C_{M_s^{ij} M_s^{ij}}(\mathbf{x},t;\mathbf{x}',t') \partial_j {}^\dagger u_s^i(\mathbf{x}',T-t'). \quad (1.45)$$

The variational derivative of $\mathbf{Y}$ with respect to the mass density $\rho(\mathbf{x})$ involves the first term of $\mathbf{Y}_2$ on the right-hand-side of (1.25) and the first term of $\mathbf{Y}_4$ on the right-hand-side of (1.27). Collecting terms containing $\delta\rho(\mathbf{x})$, we obtain

$$\rho(\mathbf{x}) = {}_0\rho(\mathbf{x}) - \int_V dV(\mathbf{x}') C_{\rho\rho}(\mathbf{x},\mathbf{x}') \sum_{s=1}^{N_s} \int_0^T dt \sum_i {}^\dagger u_s^i(\mathbf{x}',T-t) \partial_t^2 u_s^i(\mathbf{x}',t). \quad (1.46)$$

The variational derivative of $\mathbf{Y}$ with respect to the rate-of-relaxation tensor $\boldsymbol{\psi}(\mathbf{x},t)$ involves the first term of $\mathbf{Y}_2$ on the right-hand-side of (1.25) and the second term of $\mathbf{Y}_4$ on the right-hand-side of (1.27). Collecting terms containing $\delta\boldsymbol{\psi}^{ijkl}(\mathbf{x},t)$ and applying the divergence theorem, we obtain

$$\begin{aligned} \psi^{ijkl}(\mathbf{x},t) = {}_0\psi^{ijkl}(\mathbf{x},t) - \int_V dV(\mathbf{x}') \int_0^T dt' \Big\{ & C_{\psi^{ijkl}\psi^{ijkl}}(\mathbf{x},t;\mathbf{x}',t') \\ & \sum_{s=1}^{N_s} \int_0^T d\tau \left[\partial_j {}^\dagger u_s^i(\mathbf{x}',T-t'-\tau) \partial_l u_s^k(\mathbf{x}',\tau) \right] \Big\}. \end{aligned} \quad (1.47)$$

By now, we have completed our derivation of the Euler-Lagrange equations for the full-wave seismic data assimilation problem. The equations can be solved iteratively, i.e., the structure and source parameters that maximize the posterior probability for one iteration can be used as the prior estimates for the next iteration. We summarize the Euler-Lagrange equations in the following and use index γ to indicate the iteration number. The Euler-Lagrange equations consist of the adjoint wave equation, together with the adjoint initial and boundary conditions that correspond to the initial and boundary conditions of the forward problem (1.3) and (1.4),

$$\begin{aligned} & {}_\gamma\rho(\mathbf{x})(\partial_t)^2 {}^\dagger u_s^i(\mathbf{x},t) - \sum_j \partial_j \left[\int_{-\infty}^{\infty} d\tau \sum_{kl} {}_\gamma\psi^{ijkl}(\mathbf{x},t-\tau) \partial_l {}^\dagger u_s^k(\mathbf{x},\tau) \right] \\ & = -\sum_{rn} \delta(\mathbf{x}-\mathbf{x}_r) J_{srn}^i(T-t) W_{d_{srn}^i d_{srn}^i} d_{srn}^i, \end{aligned} \quad (1.48)$$

$${}^\dagger u_s^i(\mathbf{x},0) = 0, \quad {}^\dagger \dot{u}(\mathbf{x},0) = 0, \quad (1.49)$$

$$\sum_j \hat{n}^j(\mathbf{x}) \left[\int_{-\infty}^{\infty} d\tau \sum_{kl} {}_\gamma\psi^{ijkl}(\mathbf{x},T-t-\tau) \partial_l {}^\dagger u_s^k(\mathbf{x},\tau) \right] = 0, \quad (1.50)$$

a set of equations for updating the source and structure parameters,

$${}_{\gamma+1}f_s^i(\mathbf{x},t) = {}_\gamma f_s^i(\mathbf{x},t) + \int_V dV(\mathbf{x}') \int_0^T dt' C_{f_s^i f_s^i}(\mathbf{x},t;\mathbf{x}',t') {}^\dagger u_s^i(\mathbf{x}',T-t'), \quad (1.51)$$

$${}_{\gamma+1}M_s^{ij}(\mathbf{x},t) = {}_\gamma M_s^{ij}(\mathbf{x},t) - \int_V dV(\mathbf{x}') \int_0^T dt' C_{M_s^{ij} M_s^{ij}}(\mathbf{x},t;\mathbf{x}',t') \partial_j {}^\dagger u_s^i(\mathbf{x}',T-t'), \quad (1.52)$$

$$_{\gamma+1}\rho(\mathbf{x}) = {}_{\gamma}\rho(\mathbf{x}) - \int_V dV(\mathbf{x}')C_{\rho\rho}(\mathbf{x},\mathbf{x}')\sum_{s=1}^{N_s}\int_0^T dt\sum_i {}^{\dagger}u_s^i(\mathbf{x}',T-t)\partial_t^2 u_s^i(\mathbf{x}',t), \quad (1.53)$$

$$_{\gamma+1}\psi^{ijkl}(\mathbf{x},t) = {}_{\gamma}\psi^{ijkl}(\mathbf{x},t) - \int_V dV(\mathbf{x}')\int_0^T dt'\Bigg\{C_{\psi^{ijkl}\psi^{ijkl}}(\mathbf{x},t;\mathbf{x}',t')$$
$$\sum_{s=1}^{N_s}\int_0^T d\tau\left[\partial_j {}^{\dagger}u_s^i(\mathbf{x}',T-t'-\tau)\partial_l u_s^k(\mathbf{x}',\tau)\right]\Bigg\}, \quad (1.54)$$

and the forward wave equation, together with the initial and boundary conditions,

$$_{\gamma+1}\rho(\mathbf{x})\partial_t^2 u_s^i(\mathbf{x},t) - \sum_j \partial_j\left[\int_{-\infty}^{\infty} d\tau\sum_{kl} {}_{\gamma+1}\psi^{ijkl}(\mathbf{x},t-\tau)\partial_l u_s^k(\mathbf{x},\tau) + {}_{\gamma+1}M_s^{ij}(\mathbf{x},t)\right]$$
$$- {}_{\gamma+1}f_s^i(\mathbf{x},t) = \overline{q}_s^i(\mathbf{x},t) + \int_V dV(\mathbf{x}')\int_0^T dt' C_{q_s^i q_s^i}(\mathbf{x},t;\mathbf{x}',t')^{\dagger}u_s^i(\mathbf{x}',T-t'), \quad (1.55)$$

$$u_s^i(\mathbf{x},0) = \overline{a}_s^i(\mathbf{x}) + \int_V dV(\mathbf{x}')C_{a_s^i a_s^i}(\mathbf{x},\mathbf{x}')^{\dagger}\dot{u}_s^i(\mathbf{x}',T), \quad (1.56)$$

$$\dot{u}_s^i(\mathbf{x},0) = \overline{b}_s^i(\mathbf{x}) + \int_V dV(\mathbf{x}')C_{b_s^i b_s^i}(\mathbf{x},\mathbf{x}')^{\dagger}u_s^i(\mathbf{x}',T), \quad (1.57)$$

$$\sum_j \hat{n}^j(\mathbf{x})\left[\int_{-\infty}^{\infty} d\tau\sum_{kl} {}_{\gamma+1}\psi^{ijkl}(\mathbf{x},t-\tau)\partial_l u_s^k(\mathbf{x},\tau)\right]$$
$$= \overline{c}_s^i(\mathbf{x},t) + \int_{\partial V} dS(\mathbf{x}')\int_0^T dt' C_{c_s^i c_s^i}(\mathbf{x},t;\mathbf{x}',t')^{\dagger}u_s^i(\mathbf{x}',T-t'). \quad (1.58)$$

These equations define the Euler-Lagrange equations for the weakly constrained full-wave seismic data assimilation problem. The forward wave equation is forced by terms representing model residuals and the adjoint wave equation is forced by misfit measurements that quantify the discrepancies between the observations and their corresponding model predictions. Nonlinearity of the inverse problem can be accounted for by solving (1.48)-(1.58) iteratively for a number of times. Euler-Lagrange equations for other types of seismic wave equations and/or initial/boundary conditions can be derived by following the same procedure as demonstrated in this section. It will be shown in Sections **1.4** and **1.5** that the adjoint method and the scattering-integral method, which have been successfully applied in full-3D, full-wave tomography studies, are different variations of the generalized inverse and can both be derived from equations (1.48)-(1.58).

1.3 Data Functionals

The generalized inverse formulated in the previous section provides us a unified methodology for assimilating seismological observations into dynamic ground-motion models. The Euler-Lagrange equations (1.48)-(1.58) provide us a foundation for designing gradient-descent-type optimization algorithms to search for the optimal structure and source models that minimize the objective function **Y** as defined in (1.23). The convergence of such optimization algorithms depends upon the shape of the objective function (e.g. the number of local minima). For objective functions dominated by local minima, gradient- and/or Hessian-based optimization algorithms will have difficulty in converging to the global minimum unless the starting structure and source models are already very close to the global optimal. The shape of the objective function depends upon the data functionals used to quantify the discrepancies between the observed and the synthetic wave fields. Ideally, if the data functionals are strictly linear with respect to structure and source model parameters, we have a quadratic objective function and gradient- and/or Hessian-based optimization algorithms are guaranteed to converge to the global minimum from any starting structure and source models. In practice, the data functionals we use are rarely linear and the objective function can have more than one minimum. However, it is possible to find data functionals that are "weakly nonlinear" (Tarantola, 2005) through carefully conditioning the seismograms used in the inversion, in which gradient- and/or Hessian-based optimization algorithms are still effective in finding the global minimum. In this section, we analyze the properties of several data functionals that have been successfully used in solving seismological inverse problems and discuss their merits and shortcomings.

1.3.1 Differential Waveforms

Perhaps the most obvious data functional for waveform inversion is the difference between the observed and the synthetic seismograms at a set of sampling time t_n (e.g., Tarantola, 1988; Akcelik et al., 2002),

$$d^i_{srn} = \vec{u}^i_s(\mathbf{x}_r, t_n) - u^i_s(\mathbf{x}_r, t_n). \tag{1.59}$$

The seismogram perturbation kernel for this data functional is the Dirac delta function,

$$J^i_{srn}(t) = -\delta(t - t_n), \tag{1.60}$$

and the Fréchet kernels of this data functional with respect to mass density $\rho(\mathbf{x})$ and rate-of-relaxation tensor $\boldsymbol{\psi}(\mathbf{x}, t)$ are given directly by the Born approximation (Dahlen and Tromp, 1998). However, the validity of the Born approximation $\vec{u}^i_s(\mathbf{x}_r, t_n) \approx u^i_s(\mathbf{x}_r, t_n) + \delta u^i_s(\mathbf{x}_r, t_n)$ is limited by the weak-scattering condition (Wu, 2003). Using the data functional in (1.59) involves the linearization of the Fourier shift operator $\exp(i\omega t)$. For an observed waveform with centroid frequency ω_0, the Born approximation is valid only when

$$\omega_0 \Delta T^i_{srn} << 1, \tag{1.61}$$

where ΔT^i_{srn} is the travel-time shift relative to the synthetic waveform. In practice, the condition in (1.61) can be enforced by low-pass filtering the seismograms. The low-frequency waveforms can be used to constrain the long-wavelength features of the model before proceeding sequentially to higher frequencies and shorter wavelengths (Bunks et al., 1995). Such a "frequency-bootstrapping" approach can be effective in mitigating the nonlinearity of waveform inversions when the observed data have sufficient signal-to-noise ratio at lower frequencies.

An alternative data functional is the differences between the frequency-domain observed and synthetic seismograms at a set of sampling frequencies ω_n (e.g. Pratt et al., 1998; Pratt, 1999),

$$d^i_{srn} = \int dt \exp(-i\omega_n t) \left[\overline{u}^i_s(\mathbf{x}_r,t) - u^i_s(\mathbf{x}_r,t)\right]. \tag{1.62}$$

In this case, the seismogram perturbation kernel is given by the Fourier transform kernel

$$J^i_{srn}(t) = -\exp(-i\omega_n t). \tag{1.63}$$

The forward and the adjoint wave equations in (1.48)-(1.58) can all be solved in the frequency domain, which allows us to avoid the linearization of the Fourier shift operator effectively. But a drawback of this approach is that by using the Fourier transform we lose resolution in the time domain and misfit information from different temporal arrivals is mixed together, which may introduce nonlinear effects caused by the interference with different wave groups.

1.3.2 Cross-correlation Measurements

The data-synthetic travel-time shift of an isolated waveform in a time window $[t_n, t'_n]$ can be estimated by maximizing the cross-correlation between the observed and synthetic waveforms (e.g. Woodward and Masters, 1991),

$$d^i_{srn} = \Delta T^i_{srn}. \tag{1.64}$$

The perturbation kernel is proportional to the waveform derivative (Luo and Shuster, 1991; Zhao and Jordan, 1998),

$$J^i_{srn}(t) = -\frac{\partial_t u^i_s(\mathbf{x}_r,t)\,[H(t-t_n) - H(t-t'_n)]}{\int_{t_n}^{t'_n} dt\,|\partial_t u^i_s(\mathbf{x}_r,t)|^2}, \tag{1.65}$$

where $H(t)$ is the Heaviside function.

The amplitude anomaly of an isolated waveform in a time window $[t_n, t'_n]$ can also be estimated by waveform cross-correlation (Ritsema et al., 2002; Tromp et al., 2005). The relative amplitude anomaly can be defined as

$$d^i_{srn} = \Delta(\ln U^i_{srn}) \approx \Delta U^i_{srn}/U \tag{1.66}$$

and the perturbation kernel is proportional to the waveform (Dahlen and Baig, 2002),

$$J^i_{srn}(t) = \frac{u^i_s(\mathbf{x}_r,t)\,[H(t-t_n) - H(t-t'_n)]}{\int_{t_n}^{t'_n} dt\,[u^i_s(\mathbf{x}_r,t)]^2}. \tag{1.67}$$

For band-limited signals, the cross-correlation travel-time and amplitude anomalies usually provide good approximations to the phase and amplitude differences between the observed and synthetic waveforms around the dominant frequency (Gee and Jordan, 1992), but they may not characterize the differences in the shape of the waveforms (i.e., the frequency-dependence of phase and amplitude differences) very well.

1.3.3 Generalized Seismological Data Functionals (GSDF)

We can generalize the cross-correlation travel-time shift and relative amplitude anomaly to frequency-dependent quantities that completely describe the discrepancies between the observed and synthetic waveforms. In the frequency domain, the synthetic waveform can be mapped into the corresponding observed waveform by an exponential (Rytov) operator, $\exp[i\omega\Delta\tau_p(\omega) + i\Delta\tau_q(\omega)]$. The generalized seismological data functionals (Gee and Jordan, 1992) are observational approximations to the frequency-dependent phase-delay time $\Delta\tau_p(\omega)$ and amplitude-reduction time $\Delta\tau_q(\omega)$, obtained by time-windowing and narrow-band-filtering of the waveform cross-correlograms. The GSDF analysis has been successfully applied to tomographic inversions at different geographic scales (e.g. Gaherty et al., 1996; Katzman et al., 1998; Chen et al., 2007b) as well as to inversions for earthquake source parameters (e.g. McGuire et al. 2001; Chen et al., 2010b). Particularly, the linearization of GSDF misfit measurements depends on the Rytov approximation, which is valid for large accumulative phase-shifts as long as the phase perturbation per wavelength is small (Chernov, 1967; Snieder and Lomax, 1996). This is far less restrictive than the Born approximation, which requires small accumulative phase-shifts as shown in (1.61).

The GSDF data processing consists of several steps, as illustrated in Figure 1.1. We isolate the target wave group using an isolation filter $\tilde{u}_s^i(\mathbf{x}_r,t)$, which is obtained by windowing the complete synthetic seismogram,

$$\tilde{u}_s^i(\mathbf{x}_r,t) = W(t)u_s^i(\mathbf{x}_r,t). \tag{1.68}$$

In practice, we often apply a Tukey (i.e., cosine-tapered) window, which has a flat part in the middle of the window. The effect of applying such a time window is that the target waveform is as less distorted by the windowing operation as possible, while extraneous phases are excluded from the time window to reduce interference. We then cross-correlate the isolation filter with the complete synthetic seismogram and with the observed seismogram and we window the resulting synthetic and data cross-correlograms around the zero-lag. The windowed correlograms are then narrow band filtered at a set of frequencies ω_i. When certain conditions about windowing and narrow-band filtering are enforced (Chen et al., 2010a), the resulting narrow-band-filtered windowed correlograms can always be well matched by five-parameter Gaussian wavelets, which are cosine functions with frequencies at around ω_i and modulated by Gaussian envelopes. The differences in the phase and the amplitude between the synthetic and data Gaussian wavelets give us the estimates of the phase-delay time δt_p and amplitude-reduction time δt_q at each narrow-band-filtering frequency ω_i. A practical issue of the GSDF analysis is that the phase-delay measurements need to be corrected for possible cycle-skipping errors before they can be used in inversions. These cycle-skipping errors can usually be corrected by bootstrapping the phase from low frequencies to high frequencies (Ekström et al., 1997), which can be automated without too much difficulty.

The frequency-dependent GSDF measurements characterize the differences in the shape of the waveforms effectively. In Figure 1.2, we illustrate this point using an example. By correcting the phase and amplitude of the synthetic waveform using the GSDF measure-

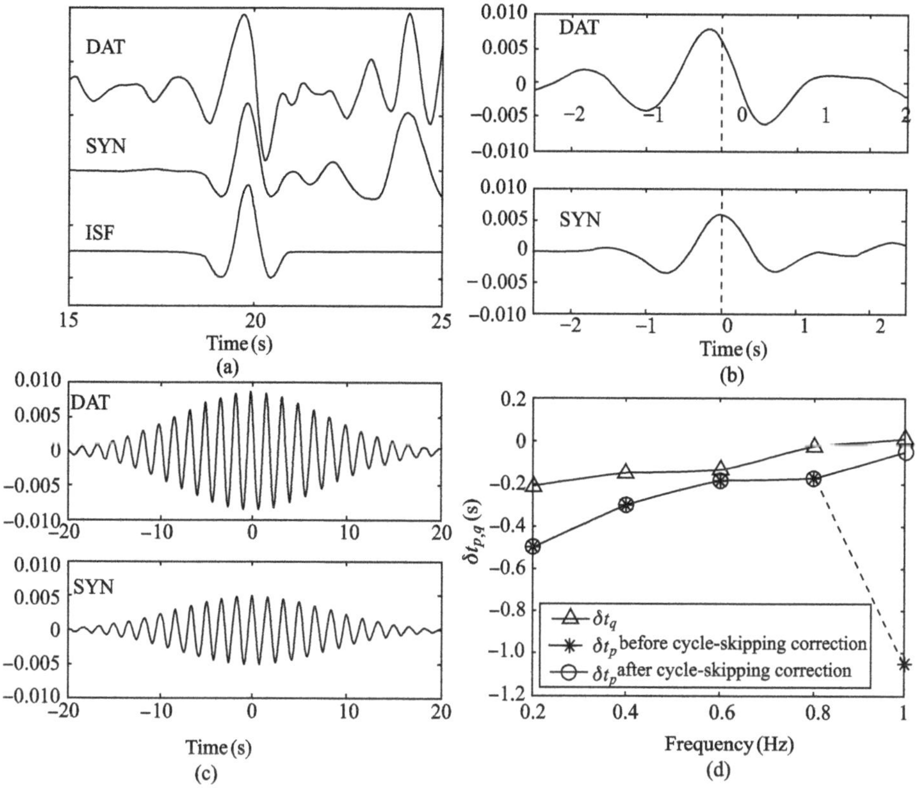

Fig. 1.1 An example of GSDF processing. (a) The observed seismogram (DAT), synthetic seismogram (SYN) and the isolation filter (ISF) for the S-wave waveform we are analyzing. (b) Cross-correlogram between the isolation filter and the observed seismogram (upper panel) and cross-correlogram between the isolation filter and the complete synthetic seismogram (lower panel). Windowed cross-correlograms are shown as dash lines. (c) Examples of narrow-band-filtered windowed cross-correlograms for the data (upper panel) and the synthetic (lower panel). Centre frequency of the narrow-band-filter is 0.6 Hz; half-width is 0.1 Hz. (d) GSDF measurements made at five sampling frequencies. Triangles, amplitude-reduction times; stars, phase-delay times before correcting for cycle-skipping errors; circles, phase-delay times after correcting for cycle-skipping errors.

ments made at five frequencies evenly distributed across the frequency band, we were able to recover the observed waveform almost perfectly.

Seismogram perturbation kernels for GSDF measurements were derived in Chen et al. (2010a). In this chapter, I give a much simplified derivation by assuming zero bandwidth for the narrowband filters, which results in a windowed Fourier analysis. The Fourier transform of the isolation filter can be expressed as

$$\hat{\tilde{u}}_s^i(\mathbf{x}_r,\omega) = \int dt \exp(-i\omega t) W(t) u_s^i(\mathbf{x}_r,t). \tag{1.69}$$

We can also express the left-hand-side of (1.69) using two frequency-dependent, time-like

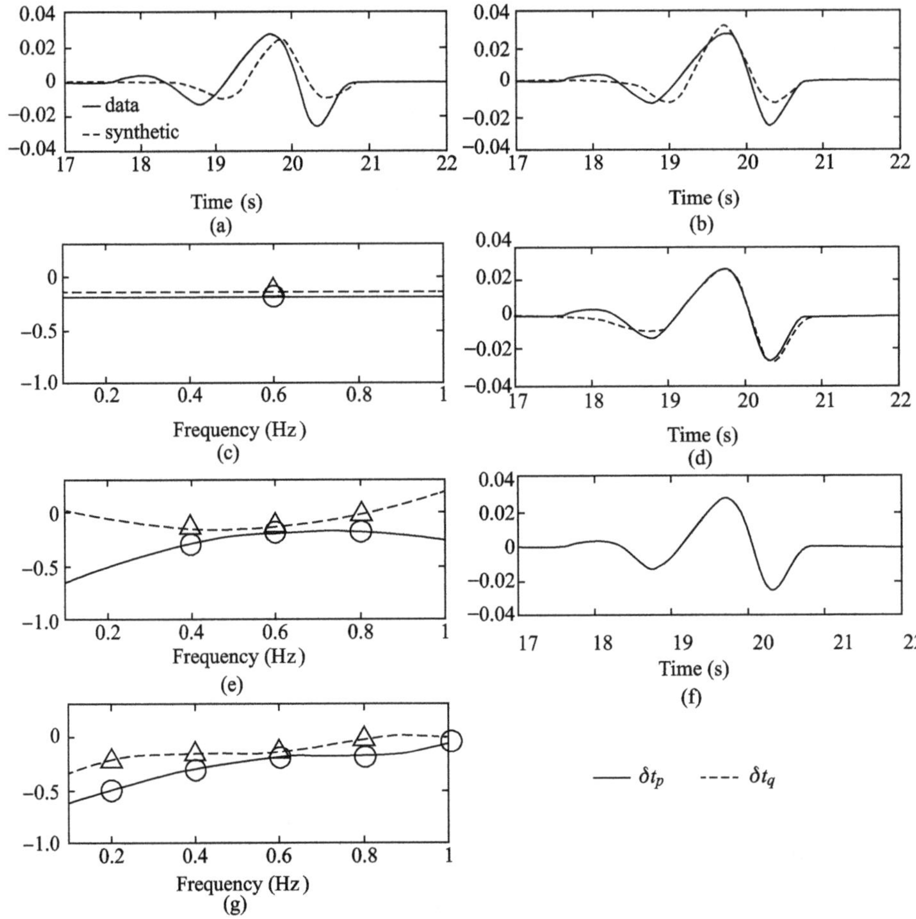

Fig. 1.2 An example of fitting the synthetic waveform (dash lines) to the observed waveform (solid lines) by correcting the phase and amplitude of the synthetic waveform using GSDF measurements of phase-delay (circles) and amplitude-reduction times (triangles). The observed and synthetic waveforms are the same as in Figure 1.1. (a) The original waveforms without perturbation. (b) and (c) correcting the synthetic waveform using GSDF measurements made at one sampling frequency 0.6Hz. (d) and (e) correcting synthetic waveform using GSDF measurements made at three sampling frequencies, 0.4, 0.6, and 0.8 Hz. (f) and (g) correcting the synthetic waveform using GSDF measurements made at five sampling frequencies 0.2, 0.4, 0.6, 0.8, and 1.0 Hz. Cubic splines are used to interpolate and extrapolate phase and amplitude perturbations to all other frequencies. This example demonstrates that minimizing the frequency-dependent GSDF measurements is equivalent to fitting the waveforms.

quantities, $\tau_{\mathrm{p}}(\mathbf{x}_r,\omega)$ and $\tau_{\mathrm{q}}(\mathbf{x}_r,\omega)$,

$$\exp\left[-i\omega\tau_p(\mathbf{x}_r,\omega)-\omega\tau_q(\mathbf{x}_r,\omega)\right]=\int dt\exp(-i\omega t)W(t)u_s^i(\mathbf{x}_r,t). \tag{1.70}$$

By linearizing the left-hand-side of (1.70), we obtain the variational derivatives of $\tau_{p,q}(\mathbf{x}_r,$

ω) with respect to the synthetic seismogram $u_s^i(\mathbf{x}_r,t)$ as follows:

$$\delta\tau_p(\mathbf{x}_r,\omega) = \int_{-\infty}^{\infty} dt\,\mathfrak{I}\left[\frac{W(t)e^{-i\omega t}}{-\hat{\hat{u}}_s^i(\mathbf{x}_r,\omega)\omega}\right]\delta u_s^i(\mathbf{x}_r,t), \tag{1.71}$$

$$\delta\tau_q(\mathbf{x}_r,\omega) = \int_{-\infty}^{\infty} dt\,\mathfrak{R}\left[\frac{W(t)e^{-i\omega t}}{-\hat{\hat{u}}_s^i(\mathbf{x}_r,\omega)\omega}\right]\delta u_s^i(\mathbf{x}_r,t). \tag{1.72}$$

The seismogram perturbation kernels for $\tau_{p,q}(\mathbf{x}_r,\omega)$ can then be expressed as

$$J_{psrn}^i(t) = \mathfrak{I}\left[\frac{W(t)e^{-i\omega t}}{-\hat{\hat{u}}_s^i(\mathbf{x}_r,\omega)\omega}\right], \tag{1.73}$$

$$J_{qsrn}^i(t) = \mathfrak{R}\left[\frac{W(t)e^{-i\omega t}}{-\hat{\hat{u}}_s^i(\mathbf{x}_r,\omega)\omega}\right]. \tag{1.74}$$

The Fréchet kernels of $\tau_{\mathrm{p,q}}(\mathbf{x}_r,\omega)$ with respect to seismic velocities can be obtained from the Euler-Lagrange equations (1.48)-(1.58), or from the adjoint method or the scattering-integral method as demonstrated in the following sections. Examples of the frequency-dependent Fréchet kernels for a direct-arriving P, the free-surface-reflected pP and pS for a source-receiver pair buried in the half-space model are shown in Figure 1.3. The sensitivity for the direct-arriving P-wave is not concentrated on the ray path but extends to a certain distance away from the ray path. For the kernels of the frequency-dependent phase-delay anomalies, we also see the "banana-doughnut" phenomena of vanishing sensitivity on the ray path (Marquering et al., 1999). The width of the first Fresnel zone and the spatial oscillation of the sensitivity depend on the frequency ω. The kernels for the surface-reflected phases pP and pS have similar characteristics to the direct-arriving P-wave. However, the widths of the first Fresnel zones for these surface-reflected phases are greater than the direct-arriving P-wave due to longer propagation distances than the direct P-wave.

In addition to GSDF analysis, there are a number of different techniques for extracting time- and frequency-dependent phase and amplitude misfit measurements. Holschneider et al. (2005) developed a technique for extracting frequency-dependent phase-delay, group-delay and amplitude anomalies by modeling the deformation of the wavelet transform of the analyzed seismogram. The continuous wavelet transform allows them to localize in the time-frequency domain and selectively fit portions of the seismogram. Using the Gabor transform, Fichtner et al. (2008) extended GSDF analysis to continuous time-frequency and quantified waveform differences using phase and envelope misfit functions that are continuous in time and frequency. The multi-taper technique (Thomson, 1982) uses orthogonal windows to provide statistically independent estimates of spectral phases and amplitudes of a signal. This approach was later extended by Lilly and Park (1995) to provide a formal uncertainty estimates for the extracted spectral parameters.

One major difficulty in effectively utilizing waveform data in seismic tomography is that unlike travel-time data, waveform data are much more nonlinear with respect to the seismic velocity model. In general, it has been recognized both in theoretical studies and in practical applications that the appropriate choice of data functionals plays an essential role in overcoming the nonlinearity of waveform inversions. Particularly, we need to

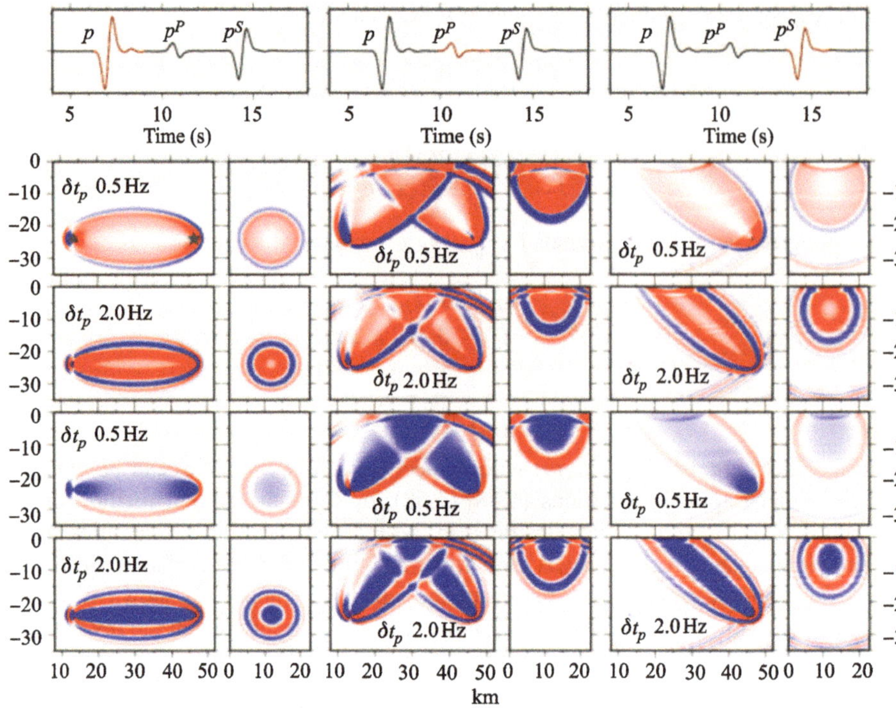

Fig. 1.3 Fréchet kernels for the direct-arriving P-wave (left column), surface reflected pP-wave (middle column) and pS-wave (right column) from an explosive source in a uniform half-space model. Plotted are the radial component of the synthetic seismogram, the isolation filters are highlighted in red, and the Fréchet kernels of GSDF $\delta t_{p,q}$ measurements at 0.5 Hz and 2.0 Hz with respect to P-wave speed α in the source-receiver vertical plane and in the transverse plane midway between the source and the receiver. The source (grey star) and the receiver (grey triangle) are buried at 24 km depth and have a distance of 32.2 km in between. In all the plots for the kernels, the color schemes are such that white represents zero; warm colors (yellow to orange to red) represent negative amplitudes indicating that a velocity increase leads to an advance in arrival time and an increase in amplitude and cool colors (light to dark blue) represent positive amplitudes indicating that a velocity increase leads to a delay in arrival time and a reduction in amplitude.

separate phase information, which is quasi-linearly related to seismic velocities, from the amplitude information, which can be highly nonlinear with respect to seismic velocities. Separation of different wave groups on the seismogram through localization in time domain and/or frequency domain allows us to reduce nonlinearity caused by interference in different wave groups and provides us a pathway to bootstrap our structure model through iteratively conditioning the data used in the inversion.

1.4 The Adjoint Method

A majority of previous work on full-wave seismic inversions (e.g. Tarantola, 1984; 1988; Pratt et al., 1998; 1999; Pratt and Shipp, 1999; Tromp et al., 2005) solves a simplified version of the variational problem defined by (1.23). A common simplification is to assume the seismic wave equation and its initial and boundary conditions have zero model residuals. In this case, the covariance functions $C_{q_s^i q_s^i}$, $C_{a_s^i a_s^i}$, $C_{b_s^i b_s^i}$ and $C_{c_s^i c_s^i}$ are set at zero, which corresponds to an infinite weight on the dynamic ground-motion model, i.e., the seismic wave equation and the associated initial and boundary conditions must be satisfied exactly. In the Euler-Lagrange equations (1.48)-(1.58), this assumption eliminates the direct coupling between the forward model, (1.55)-(1.58), and the adjoint wave-field ${}^{\dagger}\mathbf{u}_s(\mathbf{x},t)$, although the seismic source and structure parameters, $\mathbf{f}_s(\mathbf{x},t)$, $\mathbf{M}_s(\mathbf{x},t)$, $\rho(\mathbf{x})$ and $\boldsymbol{\psi}(\mathbf{x},t)$, still depend on the adjoint wave-field through (1.51)-(1.54). The resulting "strong-constraint" formulation is usually named "the adjoint method".

The adjoint method, which can be obtained from the Euler-Lagrange equations (1.48)-(1.58) by setting the right-hand-sides of (1.55)-(1.58) at zero, can also be derived using the Lagrange multiplier method (Liu, 2006), which is commonly employed in optimal control theory to solve optimization problems with strong constraints imposed by partial differential equations (e.g. Biegler et al., 2003). In the Lagrange multiplier method, an objective function defined in terms of a weighted summation of the misfit measurements is minimized and the seismic wave equation as well as its initial and boundary conditions are included into the objective function using Lagrange multipliers. Variations with respect to the Lagrange multipliers return the dynamic model (1.55)-(1.58) with the right-hand-sides set at zero. Variations with respect to the seismic source and structure parameters return a set of equations similar to (1.51)-(1.54) with the covariance functions set to the delta function. More detailed derivation of the adjoint method based on the Lagrange multiplier method can be found in Liu (2006).

The effects of neglecting errors in the dynamic models and/or the associated initial and boundary conditions have not yet been thoroughly investigated for full-wave seismic data assimilation problems. It has been found in data assimilation problems in meteorology and oceanography that one can obtain unphysical values for model parameters that compensate for errors neglected in dynamic models and/or their conditions (e.g. Zupanski and Zupanski, 2006 and references therein). It is clear that one will not be able to find the correct solution unless the approximations employed in the data assimilation procedure are valid. Deficiencies in dynamic models are usually difficult, if possible at all, to eliminate, but the weak-constraint variational approach formulated in (1.48)-(1.58) provides a systematic means to accommodate model residuals in full-wave seismic data assimilation processes.

1.4.1 An Example of Adjoint Travel-Time Tomography

To demonstrate the adjoint method and how it can be used for seismic tomography, an example of wave-equation travel-time tomography (Luo and Shuster, 1991) implemented using the adjoint method is given here. The Euler-Lagrange equations (1.48)-(1.58) with the right-hand-sides of (1.55)-(1.58) set at zero are solved iteratively to obtain a seismic velocity model from travel-time of the first-arriving *P*-waves.

The observed seismograms are from an active-source seismic survey at an undisclosed location in central Nevada, the United States. The region is dominated by typical Basin and Range geology with range fronts next to large valleys. The extension and crustal stretching produce large normal faults extending deep in the crust. Under favorable conditions, these deep fault structures can act as conduits for migration of geothermal fluids. The objective of the seismic survey is to image the fault structures that are often buried and don't have surface expressions. An initial velocity model is obtained using the travel-time of the first-arriving P-waves through wave-equation travel-time tomography based on the adjoint method. This initial velocity model is then used in depth migration to map detailed subsurface fault structures. The seismic line is about 5.36 km long. Eighty inline dynamite shots with 220-feet (67.1 meters) spacing were recorded by the full spread of 160 110-feet-spaced receivers.

The adjoint and forward wave equations (1.48)-(1.50) and (1.55)-(1.58) are solved using a three-dimensional fourth-order staggered-grid finite-difference method (Graves, 1996). The modeling volume, which is about 6-km long, 1-km wide and 3.4-km deep with its length parallel to the seismic line and its width perpendicular to the seismic line, is discretized into a uniform mesh with 5.6-meter grid-spacing, which results in about 122 million grid points. Each finite-difference wave-propagation simulation uses 2048 computing cores and takes around 15 minutes on an IBM Blue Gene/P supercomputer. A one-

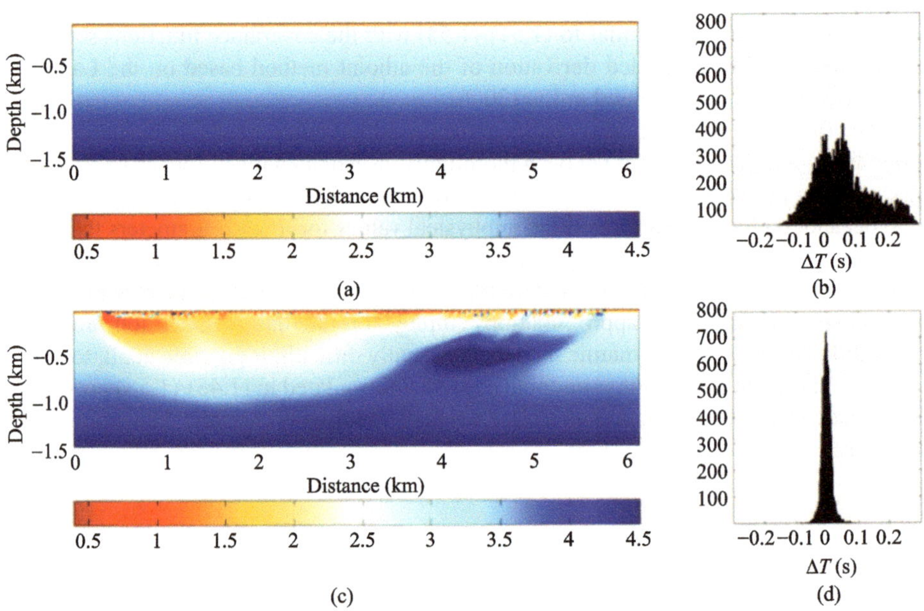

Fig. 1.4 An example of adjoint travel-time tomography for a seismic line in central Nevada, the United States. (a) A cross-section view of the P-velocity of the 1D starting model. The color bar indicates the range of the P-velocity. The seismic velocities increase smoothly with depth. (b) A histogram of the first-arrival travel-time misfit measurements for the starting model. The objective function for the starting model is $\chi^2 = 70.2(s^2)$. (c) A cross-section view of the P-velocity after 6 iterations of the adjoint method. (d) A histogram of the first-arrival travel-time misfit measurements for the improved model. The objective function for this model is $\chi^2 = 1.4(s^2)$.

dimensional seismic velocity model with P-wave speed increasing smoothly with depth is used as the starting model. A cross-section of the starting model along the seismic survey line is shown in Figure 1.4a. Synthetic seismograms are computed in the reference seismic velocity model using the finite-difference method and cross-correlation travel-time shifts between the observed and synthetic first-arriving P-waves are measured. The data functional used here is therefore given by (1.64) and the corresponding seismogram perturbation kernel is then given by (1.65). Bring (1.64) and (1.65) into the Euler-Lagrange equations (1.48)-(1.58) and solve them iteratively for 6 iterations, the P-wave speed model shown in Figure 1.4c is obtained. The maximum P-wave speed perturbation relative to the starting model is more than 65%. In each iteration, the synthetic seismograms and the cross-correlation travel-time measurements are re-computed for the reference seismic velocity model of the current iteration. After 6 iterations, the misfit functional χ^2, which is the summation of the squared travel-time shifts, is reduced by about 98%. Histograms of the travel-time shifts for the starting model and for the final model after 6 iterations are shown in Figure 1.4b and Figure 1.4d. This example demonstrates that through multiple iterations, the adjoint method can effectively account for the nonlinearity in travel-time tomography. The updated seismic velocity model, Figure 1.4c, provides more significantly improved fit to the observed waveforms than the 1D starting model, Figure 1.4a, and can serve as the starting model for iterative waveform inversions based on the adjoint method.

1.4.2 Review of Some Recent Adjoint Waveform Tomography

Recently, Tape et al. (2009; 2010) has adapted the adjoint method to image the crustal structure in Southern California using waveform recordings from local earthquakes. The spectral-element method (Komatitsch et al., 2004) was adopted to solve the 3D elastic wave equation in a crustal model for Southern California (Süss and Shaw, 2003). Frequency-dependent phase and amplitude misfit measurements extracted using the multitaper technique (Maggi et al., 2009) were used to quantify waveform misfits. After 16 iterations, strong heterogeneities with local perturbations up to 30% with respect to the 3D starting model were recovered. High-resolution features such as seismic velocity contrasts across faults due to compositional changes were recovered with high fidelity.

The adjoint method has also been adapted for continental-scale waveform tomography in Fichtner et al. (2009). The continuous time-frequency phase and envelope misfits (Fichtner et al., 2008) were used to quantify differences between observed and synthetic waveforms. The resulting seismic velocity model reveals structural features of Australasian upper-mantle in great detail. The same technique has been extended to infer radially anisotropic structure under Australasia in Fichtner et al. (2010). Detailed structural features with locally 2° lateral resolution were revealed. Strong radial anisotropy showing a clear ocean-continent dichotomy was recovered for depths above 150 km. For depths below 250 km, significant anisotropy with short wavelength ($<$ 500 km) was detected, indicating small-scale sublithospheric convection and/or a change in the dominant glide system of olivine.

1.5 The Scattering-Integral (SI) Method

The equations for updating the seismic structure models (1.53)-(1.54) involve a summation over the source index s. Considering the computational costs for numerically solving the three-dimensional forward and adjoint seismic wave equations in realistic geological media for each seismic source s, for tomography problems that involve a large number of distinct seismic sources, the Euler-Lagrange equations (1.48)-(1.58) for the weak-constraint formulation, as well as the adjoint method for the strong-constraint formulation, can become inefficient. The scattering-integral method, which can be considered as a variation of the generalized inverse with Hessian-based preconditioning, is particularly efficient when the number of receivers is larger than the number of seismic sources.

We introduce the Green's tensor for the adjoint system described by (1.48)-(1.50), $G^{ij}(\mathbf{x},t-t';\mathbf{x}')$, where following the convention in Aki and Richards (2002), the Green's tensor relates a unit impulsive force in direction j acting at location $\mathbf{x}'$ to the displacement response in direction i at location $\mathbf{x}$. The adjoint wave-field can then be expressed as

$$^{\dagger}u_s^i(\mathbf{x},t) = \int_V dV(\mathbf{x}') \int dt' \sum_j G^{ij}(\mathbf{x},t-t';\mathbf{x}')^{\dagger}f_s^j(\mathbf{x}',t'), \tag{1.75}$$

where the adjoint source field $^{\dagger}f_s^j(\mathbf{x}',t')$ is defined as

$$^{\dagger}f_s^j(\mathbf{x}',t') = -\sum_{rn} \delta(\mathbf{x}'-\mathbf{x}_r) J_{srn}^j(T-t') W_{d_{srn}^j d_{srn}^j} d_{srn}^j. \tag{1.76}$$

Evaluating the spatial integral on $\mathbf{x}'$, the adjoint wave-field can be expressed as

$$^{\dagger}u_s^i(\mathbf{x},t) = \sum_{r=1}^{N_r} \int dt' \sum_j G^{ij}(\mathbf{x},t-t';\mathbf{x}_r)^{\dagger}f_{sr}^j(t'), \tag{1.77}$$

where N_r is the total number of seismic receivers and the adjoint source field $^{\dagger}f_{sr}^j(t')$ is defined as

$$^{\dagger}f_{sr}^j(t') = -\sum_n J_{srn}^j(T-t') W_{d_{srn}^j d_{srn}^j} d_{srn}^j. \tag{1.78}$$

The receiver-side Green's tensor (RGT) $G^{ij}(\mathbf{x},t-t';\mathbf{x}_r)$ in (1.77) does not depend upon the seismic source. It can be constructed by calculating the space-time volumes of the wave-fields generated by three orthogonal unit impulsive point forces acting at the receiver location $\mathbf{x}_r$ (Zhao et al., 2006). We can now replace the N_s adjoint wave-propagation simulations for solving (1.48)-(1.50) for all seismic sources with $3N_r$ adjoint wave-propagation simulations for constructing the RGTs for all seismic receivers. For scalar wave equations (e.g., the acoustic wave equation), only N_r adjoint simulations are needed.

An important benefit of explicitly constructing and storing the RGTs is that they allow us to obtain an approximate Hessian for the objective function. The inverse of the approximate Hessian can be used as a preconditioner to speed up the convergence when solving the Euler-Lagrange equations iteratively. Bring (1.77) into the equations for updating structure parameters (1.53)-(1.54), we obtain

$$
\begin{aligned}
{}_{\gamma+1}\rho(\mathbf{x}) = {}_{\gamma}\rho(\mathbf{x}) - \int_V dV(\mathbf{x}')C_{\rho\rho}(\mathbf{x},\mathbf{x}') \sum_{s=1}^{N_s}\sum_{r=1}^{N_r}\sum_{nm}\Bigg[-\int dt J^m_{srn}(t) \int_0^T d\tau . \\
\sum_i G^{mi}(\mathbf{x}_r, t-\tau;\mathbf{x}')\partial_t^2 u_s^i(\mathbf{x}',\tau)\Bigg] W_{d^m_{srn}d^m_{srn}} d^m_{srn},
\end{aligned}
\tag{1.79}
$$

$$
\begin{aligned}
{}_{\gamma+1}\psi^{ijkl}(\mathbf{x},t) = {}_{\gamma}\psi^{ijkl}(\mathbf{x},t) - \int_V dV(\mathbf{x}') \int_0^T dt' C_{\psi^{ijkl}\psi^{ijkl}}(\mathbf{x},t;\mathbf{x}',t') \\
\sum_{s=1}^{N_s}\sum_{r=1}^{N_r}\sum_{nm}\Bigg[-\int dt'' J^m_{srn}(t'') \int_0^T d\tau \partial_j G^{mi}(\mathbf{x}_r, t''-t' \\
-\tau;\mathbf{x}')\partial_l u_s^k(\mathbf{x}',\tau)\Bigg] W_{d^m_{srn}d^m_{srn}} d^m_{srn},
\end{aligned}
\tag{1.80}
$$

where we have applied the reciprocity principle (Aki and Richards, 2002),

$$G^{ij}(\mathbf{x}',t;\mathbf{x}_r) = G^{ji}(\mathbf{x}_r,t;\mathbf{x}'). \tag{1.81}$$

Considering the Born approximation (Dahlen and Tromp, 1998), which provides the Fréchet kernels of the waveform with respect to structure parameters, and the seismogram perturbation kernel (1.30), which is the Fréchet kernel of the data functional with respect to the waveform, the terms in the square bracket in (1.79)-(1.80) are the Fréchet kernels of the data functional with respect to structure parameters $\rho(\mathbf{x})$ and $\boldsymbol{\psi}(\mathbf{x},t)$ obtained by applying the chain rule (Milne, 1980),

$$K^{\rho}_{d^m_{srn}}(\mathbf{x}) = -\int dt J^m_{srn}(t) \int_0^T d\tau \sum_i G^{mi}(\mathbf{x}_r,t-\tau;\mathbf{x})\partial_t^2 u_s^i(\mathbf{x},\tau), \tag{1.82}$$

$$K^{\psi^{ijkl}}_{d^m_{srn}}(\mathbf{x},t') = -\int dt J^m_{srn}(t) \int_0^T d\tau \partial_j G^{mi}(\mathbf{x}_r,t-t'-\tau;\mathbf{x})\partial_l u_s^k(\mathbf{x},\tau). \tag{1.83}$$

The temporal integral over τ in (1.82) and (1.83) is called the "scattering-integral" (Chen et al., 2007a). We can define a kernel matrix $\mathbf{A}$ that satisfies

$$\delta\mathbf{d} = \mathbf{A}\delta\mathbf{m}, \tag{1.84}$$

where $\mathbf{d}$ is a column-vector composed of all data functionals, $\mathbf{m}$ is a column-vector of spatially discretized structure parameters $\rho(\mathbf{x})$ and $\boldsymbol{\psi}(\mathbf{x},t)$ and $\mathbf{A}$ is a matrix with each row given by the spatially discretized Fréchet kernel of a data functional. The summations over s, r, n and m on the right-hand-sides of (1.79) and (1.80) can be expressed as $\mathbf{A}^T\mathbf{W}_{\mathbf{dd}}\mathbf{d}$ and an approximate Hessian of the objective function with respect to $\mathbf{m}$ is given by $\mathbf{A}^T\mathbf{W}_{\mathbf{dd}}\mathbf{A}+\mathbf{W}_{\mathbf{mm}}$. The Gauss-Newton normal equation can be expressed as

$$\left(\mathbf{A}^T\mathbf{W}_{\mathbf{dd}}\mathbf{A}+\mathbf{W}_{\mathbf{mm}}\right)\delta\mathbf{m} = \mathbf{A}^T\mathbf{W}_{\mathbf{dd}}\mathbf{d}, \tag{1.85}$$

and its least-square solution can be obtained by solving the linear system

$$\begin{bmatrix} \mathbf{W}_{\mathbf{dd}}^{1/2}\mathbf{A} \\ \mathbf{W}_{\mathbf{mm}}^{1/2} \end{bmatrix} \delta\mathbf{m} = \begin{bmatrix} \mathbf{W}_{\mathbf{dd}}^{1/2} \\ 0 \end{bmatrix} \tag{1.86}$$

via a relaxation method such as LSQR (Paige and Saunders, 1982). The structure parameters can then be updated by applying the $\boldsymbol{\delta}\mathbf{m}$ obtained by solving (1.86) to the reference model and it is possible to achieve quadratic convergence rate under suitable conditions.

Another benefit for explicitly computing and storing the RGTs is that they provide the Fréchet kernels of the data functional with respect to the seismic source parameters $\mathbf{f}_s$ and $\mathbf{M}_s$ (Zhao et al., 2006). Bring (1.77) into (1.51) and 1-(52), we obtain

$$\begin{aligned} {}_{\gamma+1}f_s^i(\mathbf{x},t) = {}_{\gamma}f_s^i(\mathbf{x},t) - \int_V dV(\mathbf{x}')\int_0^T dt' C_{f_s^i f_s^i}(\mathbf{x},t;\mathbf{x}',t') \\ \sum_{rn}\sum_j \left[\int dt'' J_{srn}^j(t'')G^{ij}(\mathbf{x}',t''-t';\mathbf{x}_r)\right] W_{d_{srn}^j d_{srn}^j} d_{srn}^j, \end{aligned} \tag{1.87}$$

$$\begin{aligned} {}_{\gamma+1}M_s^{ij}(\mathbf{x},t) = {}_{\gamma}M_s^{ij}(\mathbf{x},t) - \int_V dV(\mathbf{x}')\int_0^T dt' C_{M_s^{ij} M_s^{ij}}(\mathbf{x},t';\mathbf{x}',t') \\ \sum_{rn}\sum_m \left[-\int dt'' J_{srn}^m(t'')\partial_j G^{im}(\mathbf{x}',t''-t';\mathbf{x}_r)\right] W_{d_{srn}^m d_{srn}^m} d_{srn}^m. \end{aligned} \tag{1.88}$$

The terms in the square brackets in (1.87) and (1.88) provide the Fréchet kernels of the data functional with respect to the source parameters,

$$K_{d_{srn}^j}^{f_s^i}(\mathbf{x}',t') = \int dt J_{srn}^j(t)G^{ij}(\mathbf{x}',t-t';\mathbf{x}_r), \tag{1.89}$$

$$K_{d_{srn}^m}^{M_s^{ij}}(\mathbf{x}',t') = -\int dt J_{srn}^m(t)\partial_j G^{im}(\mathbf{x}',t-t';\mathbf{x}_r). \tag{1.90}$$

Since the RGTs do not depend upon the seismic source, the same set of RGTs can be used for different sources within the modeling volume and no additional wave-propagation simulations are needed. The Fréchet kernels needed for source parameter inversions can be generated by extracting a small, source-centered volume from the stored RGTs, which allows us to design automated, real-time or near-real-time source inversion algorithms (Zhao et al., 2006).

1.5.1 Full-Wave Tomography Based on SI

The first successful full-3D, full-wave seismic tomography using real waveform data from local earthquakes was presented in Chen et al. (2007b). This tomography study was based on the scattering-integral method. The forward wave-fields from the earthquakes and the RGTs were calculated using the fourth-order, staggered-grid finite-difference method

(Graves, 1996). Waveform misfits quantified using the frequency-dependent GSDF measurements (Chen et al., 2010a) were inverted to improve the 3D elastic starting model, the Southern California Earthquake Center (SCEC) Community Velocity Model Version 3.0 (CVM3.0) (Magistrale et al., 2000), in a 142-km-long, 84-km-wide and 26-km-deep volume around the Los Angeles Basin region. The revised 3D seismic velocity model, LAF3D, provides substantially better fit to the observed waveform data than the 3D starting model for frequencies up to 1.2 Hz. Map-views of the 3D starting model, CVM3.0, the revised model, LAF3D, and the model perturbation at different depths are shown in Figure 1.5.

The perturbations in *P*- and *S*-wave speeds with respect to the starting model are up to 10%. In the interior of the Los Angeles Basin, wave speeds are slightly higher in LAF3D than in CVM3.0, while at the edges of the basin wave speeds are slightly lower in LAF3D than in CVM3.0. This finding is consistent with the Harvard crust model for Southern California (Süss and Shaw, 2003), which is another 3D seismic velocity model derived primarily from industrial reflection/refraction profiles. The *P*-velocity in the Harvard model is slightly faster than CVM3.0 in the center of the basin and slightly slower

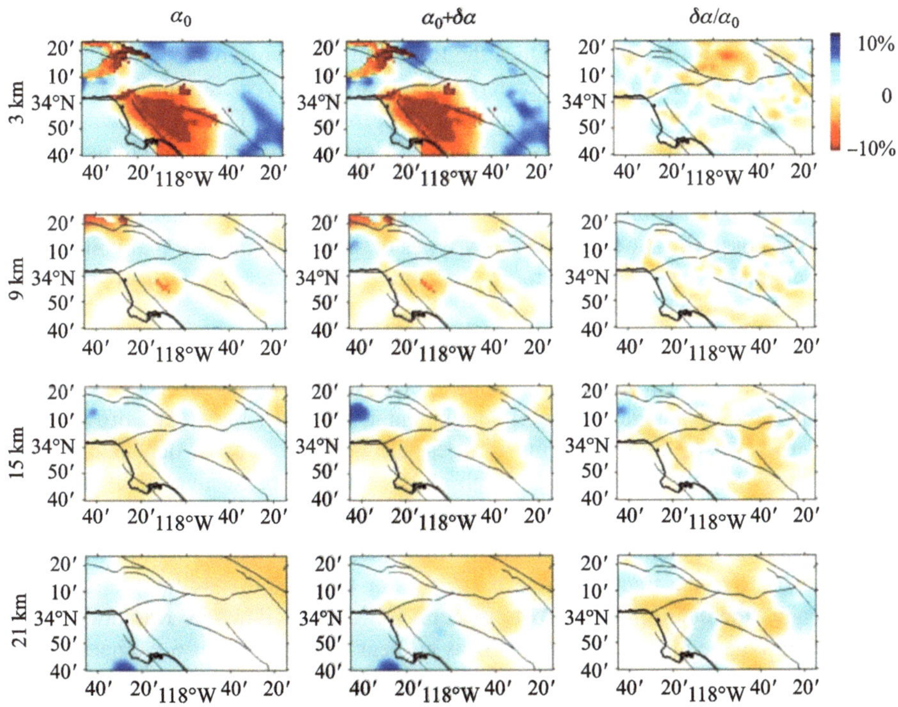

Fig. 1.5 Map views of the *P*-velocity model for the starting model CVM3.0 (left column), the improved model LAF3D (middle column), and the relative perturbation of LAF3D relative to CVM3.0 (right column). For the first and second columns, color indicates the variation with respect to the average velocity at each depth; for the third column, color indicates relative perturbation between LAF3D and CVM3.0.

than CVM3.0 on the borders of the basin. The inverted model perturbation seems to enhance the velocity gradient that already exists in the starting model. Particularly, the overall north-south gradient (from low to high) in both P and S velocities in the middle to lower crust is stronger in LAF3D than in CVM3.0. And the inverted P-velocity perturbation seems to enhance the west-east gradient in the lower crust. The inverted model perturbation has good spatial correlation with the distribution of major faults in the area. For example, there is a negative perturbation of about 1%-2% in P-velocity from about 3 km to about 9 km depth along the San Andreas Fault, suggesting that the P-velocity in CVM3.0 is too fast on this segment of the San Andreas Fault. Another example is that the Sierra Madre fault zone and the Santa Monica fault seem to coincide with the locations of some negative perturbations in both P and S velocities.

1.5.2 Earthquake Source Parameter Inversion Based on SI

Accurate and rapid estimation of earthquake source parameters is important both for seismic hazard analysis and for realistic interpretation of geological and tectonic structures. The scattering-integral method provides a very efficient means for assimilating seismic waveform observations into better estimations of earthquake source parameters using realistic 3D Earth structure models. Once the RGTs for all the receivers are computed and stored on disk, seismic source parameters can be estimated very rapidly by retrieving a small volume around the reference source location from the RGT database and no additional wave-propagation simulations are needed. Moreover, the capability to use realistic 3D structure models in computing the RGTs not only allows us to make more accurate estimates of classical source parameters (Zhao et al., 2006; Lee et al., 2011) but also opens up the possibility of estimating new types of source parameters that could lead to better understanding of regional tectonics and earthquake source physics and impose new constraints on seismic hazard analysis.

(a) An example of rapid CMT inversions based on SI

An important advantage of the SI implementation is that the RGTs provide exact Fréchet derivatives of the waveforms with respect to moment tensor densities for every grid point in the modeling volume, which allows for very rapid inversions of moment tensor solutions, even in complex 3D Earth structure models.

In general, the moment tensor $\mathbf{M}$ has 6 independent elements. For a purely deviatoric source, we require the trace of $\mathbf{M}$ to be zero. For a purely double-couple source, we further require the determinant of $\mathbf{M}$ to be zero. Following Kikuchi & Kanamori (1991), we represent a general moment tensor $\mathbf{M}$ as a linear combination of 6 elementary basis moment tensors $\mathbf{M}_m$,

$$\mathbf{M} = \sum_{m=1}^{6} a_m \mathbf{M}_m, \tag{1.91}$$

where a_m is the coefficient for the basis moment tensor $\mathbf{M}_m$ and the 6 basis moment tensors are given by

$$\mathbf{M}_1 = \begin{bmatrix} 0 & 1 & 0 \\ 1 & 0 & 0 \\ 0 & 0 & 0 \end{bmatrix}; \quad \mathbf{M}_2 = \begin{bmatrix} -1 & 0 & 0 \\ 0 & 1 & 0 \\ 0 & 0 & 0 \end{bmatrix}; \quad \mathbf{M}_3 = \begin{bmatrix} 0 & 0 & -1 \\ 0 & 0 & 0 \\ -1 & 0 & 0 \end{bmatrix};$$
$$\mathbf{M}_4 = \begin{bmatrix} 0 & 0 & 0 \\ 0 & 0 & -1 \\ 0 & -1 & 0 \end{bmatrix}; \quad \mathbf{M}_5 = \begin{bmatrix} 0 & 0 & 0 \\ 0 & -1 & 0 \\ 0 & 0 & 1 \end{bmatrix}; \quad \mathbf{M}_6 = \begin{bmatrix} 1 & 0 & 0 \\ 0 & 1 & 0 \\ 0 & 0 & 1 \end{bmatrix}. \tag{1.92}$$

We note that different from Kikuchi & Kanamori (1991) the coordinate system (x,y,z) for M_{ij} adopted in this study corresponds to (east, north, up) and the resulting basis moment tensors in equation (1.92) are different from those in Kikuchi & Kanamori (1991). The synthetic seismogram at receiver location $\mathbf{r}_\mathrm{R}$ due to a source at $\mathbf{r}_\mathrm{S}$ with a general moment tensor $\mathbf{M}$ can thus be expressed as a linear combination of the synthetics for the 6 basis moment tensors

$$u_k(\mathbf{r}_\mathrm{R},t;\mathbf{r}_\mathrm{S}) = \sum_{m=1}^{6} a_m g_{km}(\mathbf{r}_\mathrm{R},t;\mathbf{r}_\mathrm{S}), \tag{1.93}$$

where $g_{km}(\mathbf{r}_\mathrm{R},t;\mathbf{r}_\mathrm{S})$ is the kth component synthetic seismogram due to basis moment tensor $\mathbf{M}_m$ and is computed from the receiver Green tensor (RGT) by applying the reciprocity principle. Following Zhao et al. (2006), the displacement field from a point source located at $\mathbf{r}'$ with moment tensor M_{ij} can be expressed as (e.g. Aki & Richards 2000, equation 3.23)

$$u_k(\mathbf{r},t;\mathbf{r}') = M_{ij}\partial'_j G_{ki}(\mathbf{r},t;\mathbf{r}'), \tag{1.94}$$

where ∂'_j denotes the source-coordinate gradient with respect to $\mathbf{r}'$ and the Green's tensor $G_{ki}(\mathbf{r},t;\mathbf{r}')$ relates a unit impulsive force acting at location $\mathbf{r}'$ in direction $\hat{\mathbf{e}}_i$ to the displacement response at location $\mathbf{r}$ in direction $\hat{\mathbf{e}}_k$. Taking into account the symmetry of the moment tensor, we also have

$$u_k(\mathbf{r},t;\mathbf{r}') = \frac{1}{2}\left[\partial'_j G_{ki}(\mathbf{r},t;\mathbf{r}') + \partial'_i G_{kj}(\mathbf{r},t;\mathbf{r}')\right] M_{ij}. \tag{1.95}$$

Applying reciprocity of the Green's tensor

$$G_{ki}(\mathbf{r},t;\mathbf{r}') = G_{ik}(\mathbf{r}',t;\mathbf{r}), \tag{1.96}$$

equation (1.95) can be written as

$$u_k(\mathbf{r},t;\mathbf{r}') = \frac{1}{2}\left[\partial'_j G_{ik}(\mathbf{r}',t;\mathbf{r}) + \partial'_i G_{jk}(\mathbf{r}',t;\mathbf{r})\right] M_{ij}. \tag{1.97}$$

For a given receiver location $\mathbf{r} = \mathbf{r}_\mathrm{R}$, the receiver Green tensor (RGT) is a 3rd-order tensor defined as the spatial-temporal strain field,

$$H_{jik}(\mathbf{r}',t;\mathbf{r}_\mathrm{R}) = \frac{1}{2}\left[\partial'_j G_{ik}(\mathbf{r}',t;\mathbf{r}_\mathrm{R}) + \partial'_i G_{jk}(\mathbf{r}',t;\mathbf{r}_\mathrm{R})\right]. \tag{1.98}$$

Using this definition, the displacement recorded at receiver location $\mathbf{r}_R$ due to a source at $\mathbf{r}_S$ with moment tensor $\mathbf{M}$ can be expressed as

$$u_k(\mathbf{r}_R,t;\mathbf{r}_S) = M_{ij}H_{jik}(\mathbf{r}_S,t;\mathbf{r}_R) \quad \text{or} \quad \mathbf{u}(\mathbf{r}_R,t;\mathbf{r}_S) = \mathbf{M}:\mathbf{H}(\mathbf{r}_S,t;\mathbf{r}_R), \tag{1.99}$$

and the synthetic seismogram due to a source at $\mathbf{r}_S$ with the basis moment tensor $\mathbf{M}_m$ can be expressed as

$$\mathbf{g}_m(\mathbf{r}_R,t;\mathbf{r}_S) = \mathbf{M}_m : \mathbf{H}(\mathbf{r}_S,t;\mathbf{r}_R). \tag{1.100}$$

For a given receiver, the RGT can be computed through three wave-propagation simulations with a unit impulsive force acting at the receiver location $\mathbf{r}_R$ and pointing in the direction $\hat{\mathbf{e}}_k (k = 1,2,3)$ in each simulation and store the strain fields at all spatial grid points $\mathbf{r}'$ and all time sample t. The synthetic seismogram at the receiver due to any point source located within the modeling domain can be obtained by retrieving the strain Green's tensor at the source location from the RGT volume and then applying equation (1.99).

For each earthquake, we consider a random vector H composed of 6 source parameters: the longitude, latitude and depth of the centroid location $\mathbf{r}_S$, and the strike, dip and rake of the focal mechanism. We assume a uniform prior probability $P_0(H)$ over a sample space Ω_0, which is defined as a sub-grid in our modeling volume centered around the initial hypocenter location provided by the seismic network with grid spacing in three orthogonal directions given by a vector $\mathbf{h}_0$ and a focal mechanism space with the ranges given by $0^\circ \leqslant \text{strike} \leqslant 360^\circ$, $0^\circ \leqslant \text{dip} \leqslant 90^\circ$ and $-90^\circ \leqslant \text{rake} \leqslant 90^\circ$ and with angular intervals in strike, dip and rake specified by a vector $\boldsymbol{\theta}_0$. The strike, dip and rake values can be converted into the Cartesian components of the moment tensor $\mathbf{M}$ (Aki & Richards 2002), which can be subsequently converted into the 6 coefficients a_m through a simple algebraic manipulation. We apply Bayesian inference in three steps sequentially. In the first step, the likelihood function is defined in terms of waveform similarity between synthetic and observed seismograms. We quantify waveform similarity using a normalized correlation coefficient (NCC) defined as

$$NCC_n = \max_{\Delta t}\left[\int_{t_n^0}^{t_n^1} \bar{s}_n(t)s_n(t-\Delta t)dt \Bigg/ \sqrt{\int_{t_n^0}^{t_n^1} \bar{s}_n^2(t)dt \int_{t_n^0}^{t_n^1} s_n^2(t-\Delta t)dt}\right], \tag{1.101}$$

where n is the observation index, $\bar{s}_n(t)$ and $s_n(t)$ are the filtered observed seismogram and the corresponding synthetic seismogram, $[t_n^0, t_n^1]$ is the time window for selecting a certain phase on the seismograms for cross-correlation. We allow a certain time-shift Δt between the observed and synthetic waveforms. To prevent possible cycle-skipping errors, we restrict $|\Delta t|$ to being less than half of the shortest period. We assume a truncated exponential distribution for the conditional probability,

$$P(NCC_n | H_q) = \frac{\lambda_n \exp[-\lambda_n(1-NCC_n)]}{1-\exp(-2\lambda_n)}, \quad -1 < NCC_n \leqslant 1, \quad H_q \in \Omega_0, \tag{1.102}$$

where λ_n is the decay rate. Assuming the NCC observations are independent, the likelihood function can be expressed as

$$L_0\left(H\left|\bigcap_{n=1}^{N}NCC_n\right.\right)=\exp\left[-\sum_{n=1}^{N}\lambda_n\left(1-NCC_n\right)\right]\prod_{n=1}^{N}\left\{\lambda_n\left[1-\exp\left(-2\lambda_n\right)\right]^{-1}\right\},\tag{1.103}$$

where N is the total number of NCC observations. The posterior probability for the first step can then be expressed as

$$P_0\left(H\left|\bigcap_{n=1}^{N}NCC_n\right.\right)=\frac{P_0(H)\exp\left[-\sum\limits_{n=1}^{N}\lambda_n\left(1-NCC_n\right)\right]\prod\limits_{n=1}^{N}\left\{\lambda_n\left[1-\exp\left(-2\lambda_n\right)\right]^{-1}\right\}}{P_0\left(\bigcap\limits_{n=1}^{N}NCC_n\right)},\tag{1.104}$$

where

$$P_0\left(\bigcap_{n=1}^{N}NCC_n\right)=\sum_{q}P\left(\bigcap_{n=1}^{N}NCC_n\left|H_q\right.\right)P_0\left(H_q\right).\tag{1.105}$$

We note that the λ_n in front of $(1-NCC_n)$ in equation (1.104) can be used as a weighting factor for various purposes, such as to account for different signal-to-noise ratios in observed seismograms and to avoid the solution to be dominated by a cluster of closely spaced seismic stations. The probability for individual measurements

$$P_0\left(NCC_n\right)\propto\sum_{q}P\left(NCC_n\left|H_q\right.\right)P_0\left(H_q\right)\tag{1.106}$$

can be used for rejecting problematic observations. In practice, we only accept observations with

$$P_0\left(NCC_n\right)\geqslant Q_0.\tag{1.107}$$

A very low $P_0\left(NCC_n\right)$ indicates that the nth observed waveform cannot be fit well by any solutions in our sample space. This may be due to instrumentation problems or unusually high noise levels in the observed waveform data. In the second step, the posterior probability from the first step, equation (1.104), is used as the prior probability,

$$P_1\left(H\right)=P_0\left(H\left|\bigcap_{n=1}^{N}NCC_n\right.\right),\tag{1.108}$$

and the sample space for the second step, Ω_1, consists of source parameter vectors H that satisfy

$$P_1(H)\geqslant\bar{P}_1,\tag{1.109}$$

where $\bar{P}_1$ is a threshold used to reject source parameters with low probabilities. The sampling intervals for centroid location and focal mechanism are reduced to $\mathbf{h}_1$ and $\boldsymbol{\theta}_1$ respectively and synthetic seismograms for the new sample space Ω_1 are computed using equations (1.93) and (1.100). The likelihood function for the second step is defined in terms of the time-shift ΔT_n that maximizes the NCC observation as defined in equation (1.101). We assume an exponential distribution for the conditional probability

$$P(\Delta T_n | H_q) = \mu_n \exp(-\mu_n |\Delta T_n - \Delta T_M|), \quad H_q \in \Omega_1, \tag{1.110}$$

where μ_n is the decay rate and ΔT_M is the median of all ΔT_n observations. Assuming the ΔT_n observations are independent, the likelihood function for the second step can be expressed as

$$L_1\left(H \middle| \bigcap_{n=1}^{N} \Delta T_n\right) = \exp\left(-\sum_{n=1}^{N} \mu_n |\Delta T_n - \Delta T_M|\right) \prod_{n=1}^{N} \mu_n \tag{1.111}$$

and the posterior probability for the second step can be expressed as

$$P_1\left(H \middle| \bigcap_{n=1}^{N} \Delta T_n\right) = P_1(H) \exp\left(-\sum_{n=1}^{N} \mu_n |\Delta T_n - \Delta T_M|\right) \prod_{n=1}^{N} \mu_n \Big/ P_1\left(\bigcap_{n=1}^{N} \Delta T_n\right), \tag{1.112}$$

where

$$P_1\left(\bigcap_{n=1}^{N} \Delta T_n\right) = \sum_q P\left(\bigcap_{n=1}^{N} \Delta T_n \middle| H_q\right) P_1(H_q). \tag{1.113}$$

We note that like λ_n, the decay rate μ_n in equation (1.112) can also be used as a weighting factor. The probability for individual observation

$$P_1(\Delta T_n) \propto \sum_q P(\Delta T_n | H_q) P_1(H_q) \tag{1.114}$$

can be used for controlling observation quality, and we only accept observations with

$$P_1(\Delta T_n) \geqslant Q_1. \tag{1.115}$$

In the third step, the posterior probability from the second step, equation (1.112), is used as the prior probability,

$$P_2(H) = P_1\left(H \middle| \bigcap_{n=1}^{N} \Delta T_n\right). \tag{1.116}$$

The sample space for the third step, Ω_2, consists of source parameter vectors H that satisfy

$$P_2(H) \geqslant \bar{P}_2, \tag{1.117}$$

where $\bar{P}_2$ is our second threshold for rejecting source parameters with low probabilities. The sampling intervals for centroid location and focal mechanism are further reduced to $\mathbf{h}_2$ and $\boldsymbol{\theta}_2$ respectively and synthetic seismograms for the new sample space Ω_2 are computed. The likelihood function in the third step is defined in terms of the amplitude anomaly (Ritsema et al., 2002),

$$A_n = \int_{t_n^0}^{t_n^1} \bar{s}_n(t) s_n(t - \Delta T_n) dt \Big/ \int_{t_n^0}^{t_n^1} s_n(t) s_n(t - \Delta T_n) dt, \tag{1.118}$$

where ΔT_n is the time-shift that maximizes the NCC observation. We assume an exponential distribution for the conditional probability,

$$P(A_n|H_q) = \gamma_n \exp(-\gamma_n |\ln(A_n) - \ln(A_M)|), \quad H_q \in \Omega_2, \tag{1.119}$$

where A_M is the median of all A_n observations. Assuming the amplitude anomaly observations are independent, the likelihood function can be expressed as

$$L_2\left(H\left|\bigcap_{n=1}^{N} A_n\right.\right) = \exp\left(-\sum_{n=1}^{N} \gamma_n |\ln(A_n) - \ln(A_M)|\right)\prod_{n=1}^{N}\gamma_n, \tag{1.120}$$

and the posterior probability for the third step can be expressed as

$$P_2\left(H\left|\bigcap_{n=1}^{N} A_n\right.\right) = P_2(H)\exp\left(-\sum_{n=1}^{N} \gamma_n |\ln(A_n) - \ln(A_M)|\right)\prod_{n=1}^{N}\gamma_n \Big/ P_2\left(\bigcap_{n=1}^{N} A_n\right), \tag{1.121}$$

where

$$P_2\left(\bigcap_{n=1}^{N} A_n\right) = \sum_q P\left(\bigcap_{n=1}^{N} A_n \left|H_q\right.\right) P_2(H_q). \tag{1.122}$$

The probability for individual observations

$$P_2(A_n) \propto \sum_q P(A_n|H_q) P_2(H_q) \tag{1.123}$$

can be used to reject problematic observations, and we require

$$P_2(A_n) \geqslant Q_2. \tag{1.124}$$

After the third step is completed, the source parameter vector H_M that maximizes the posterior probability in equation (1.121) is selected as our optimal estimate. The centroid time is estimated as

$$T_S = \Delta T_M, \tag{1.125}$$

where ΔT_M is the median of all time-shift observations ΔT_n measured for the optimal source parameter vector H_M. The scalar seismic moment is estimated as

$$M_0 = A_M, \tag{1.126}$$

where A_M is the median of all A_n observations measured for the optimal source parameter vector H_M.

An important advantage of the Bayesian approach is that instead of a single best solution, the complete posterior probability density on the sample space is obtained, which allows for formal estimation of the uncertainties associated with the derived source parameters. In Figure 1.6, examples of the marginal probabilities for source parameters strike, dip, rake and depth are shown for the 3 September, 2002 Yorba Linda earthquake in California, US.

(b) Review of finite source parameter inversions based on SI

The lowest-order representation of a finite rupture is the finite moment tensor (FMT) (Chen et al., 2005), which contains the second-order polynomial moments of $\mathbf{M}_s(\mathbf{x},t)$ in

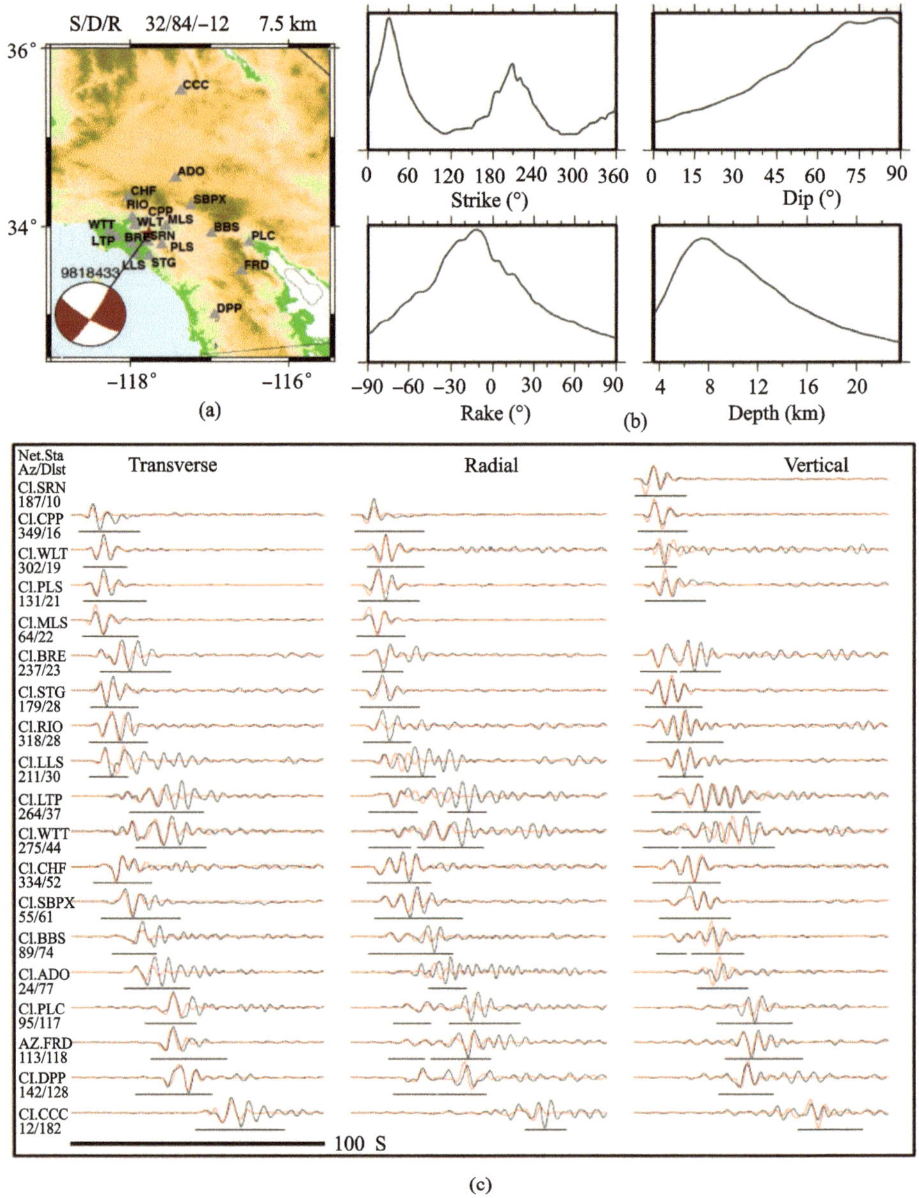

Fig. 1.6 An example of CMT inversion based on the SI method. (a) The map shows epicenter of the 3 September, 2002 M_w 4.3 Yorba Linda earthquake (the star) and the best-fit double-couple solution (the red beachball). Stations that have waveforms being selected for our inversion are shown in gray triangles. (b) The marginal probability densities for strike, dip, rake and depth obtained after the grid-search step based on Bayesian inference. (c) Examples of the waveforms selected for our CMT inversion. The black lines are observed seismograms and the red lines are synthetic seismograms computed using the optimal best-fit double-couple solution. The black bars below the seismograms indicate the waveforms selected for CMT inversion.

addition to the zeroth- and first-order polynomial moments included in the centroid moment tensor (CMT) representation. The FMT resolves fault-plane ambiguity of the CMT and provides several additional parameters of seismological interest, including the characteristic length, width, and duration of the faulting, as well as the directivity vector of the fault slip. FMT parameters have been successfully recovered for global large earthquakes ($M_w > 6$) using Green tensors computed in a 1D Earth structure model (McGuire et al., 2001). For regional small to medium-sized earthquakes ($2.5 < M_L < 5$), Green tensors computed in a 1D structure model are usually not accurate enough to warrant full recovery of FMT parameters due to stronger small-scale 3D heterogeneities in the crust. Some recent experiments in Southern California have shown that by adopting more realistic 3D structure models, it is feasible to recover FMT parameters for small to medium-sized earthquakes (Chen et al., 2005; Chen et al., 2010b).

Figure 1.7 shows a 3D rendering of the source mechanisms for some small earthquakes around the Los Angeles Basin area. The two nodal planes of the CMT solutions are now represented as two intersecting disks, whose sizes are proportional to the likelihood of

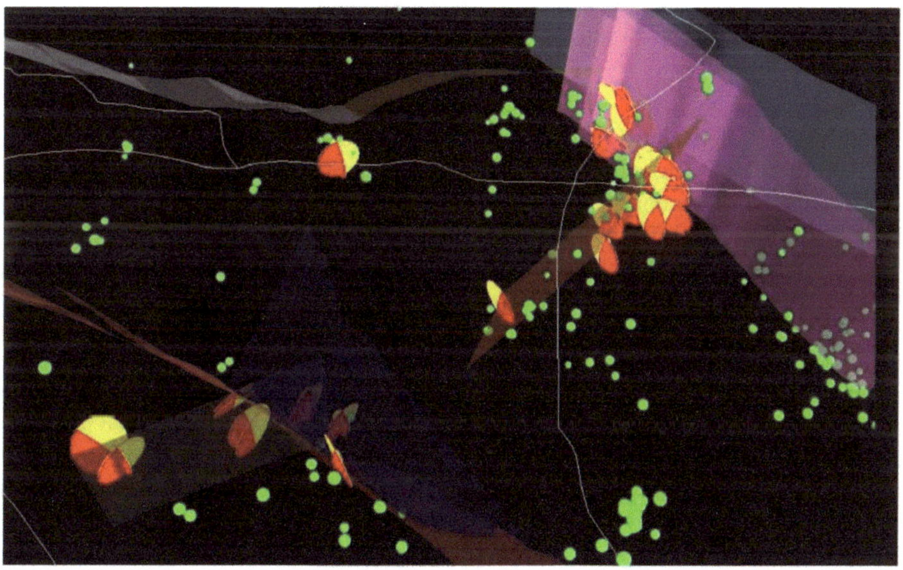

Fig. 1.7 Projection of a 3-D rendering of the focal mechanisms for earthquakes in the Los Angeles region. Focal mechanisms are represented as two intersecting disks, whose sizes are proportional to the likelihood of being the actual fault plane. Extensive quadrants are painted in yellow and compressive quadrants are painted in red. Major faults in this area are represented as colored transparent surfaces. The surfaces denoted as Fontana Seismicity Bestfit (FSB) and Yorba Linda Extruded (YLE) are linear least-square fit to the hypocentres of the earthquakes in the Fontana trend and the Yorba Linda cluster, respectively. Locations and orientations of the faults as well as the FSB and YLE surfaces are obtained from the SCEC Community Fault Model (CFM-A). Background seismicity is shown in green spheres. Major highways in this area are plotted in white solid lines. The 3-D rendering was done using the Java3D-based interactive 4-D visualization software SCEC-VDO developed by the SCEC Undergraduate Studies in Earthquake Information Technology (UseIT) interns under the supervision of Sue Perry (perry@usc.edu).

being the actual fault plane as determined from the FMT parameters (Chen et al., 2010b). The procedure does not rely on detecting directivity effects; therefore it is applicable to any types of earthquakes. For some earthquakes, the fault planes selected by this automated FMT inversion procedure were confirmed by the distributions of relocated aftershock hypocenters (Chen et al., 2010b). In regions where there are no precisely relocated aftershocks or for earthquakes with few aftershocks, this method provides the most convenient means for resolving fault plane ambiguity.

The success of those experiments suggests that improved versions of more accurate 3D structure models will eventually allow the automated recovery of FMTs for most small to medium-sized earthquakes in well-instrumented regions. By recovering FMT parameters for regional small to medium-sized earthquakes in a routine manner, we can provide clearer pictures on regional tectonics, reduce the uncertainties in inversions for crustal stress fields based on earthquake source mechanism data (Gephart, 1985; Michael, 1987; Gepart, 1990; Yin, 1996) and calculate Coulomb stress changes and provide a more robust description of probabilistic seismic hazard based on the stress transfer model (Harris, 1998; McCloskey et al., 2003; Steacy et al., 2005; Parsons, 2005).

1.6 Discussion

The advantage of the full-wave seismic data assimilation methodology over classical methods used in solving seismological inverse problems is two-fold.

First, the data functionals defined in (1.5) and explained in more detail in Section **1.3** allow us to treat different types of misfit measures between observation and model-prediction (e.g., travel-time, amplitude and waveform) in a unified and self-consistent framework based on the wave equation. Misfit measures commonly used in classical methods, such as travel-times of body waves, dispersion curves of surface waves and frequencies of the earth's free oscillation, usually have quite different back-propagation mechanisms based on approximations appropriate for different types of phases on the seismogram. In the full-wave methodology, they can all be back-propagated to provide constraints on the seismic source and structure models through the use of the wave equation and the seismogram perturbation kernel. Explicit expressions of the seismogram perturbation kernels for several classical types of misfit measures are presented in Section **1.3** and in Chen et al. (2007a). Moreover, this general wave-equation-based framework opens up the possibility of fully utilizing the complete seismic waveform data, which contain much more information about the propagation medium and the excitation source than classical data functionals such as travel-time or amplitudes of a limited set of wave arrivals on the seismogram. Such full waveform information, which can be decoded using the full-wave methodology, can potentially provide us seismic source and structure models with much higher resolution than classical methods.

Second, in the full-wave methodology, either the adjoint method or the scattering-integral implementation, the mechanism for back-propagating misfit information to modifications of source and structure models is exact, therefore eliminating possible errors caused by approximate back-propagation mechanisms commonly found in classical methods and improving the accuracy of the obtained source and structure models. A widely used classical back-propagation method is based on the ray theory for the wave equation (e.g., Beylkin, 1985). In such ray-theoretic framework, the earthquake wave-fields and the adjoint wave-fields in the adjoint method or the receiver-side Green tensors (RGTs) in the scattering-integral implementation are usually approximated using solutions of the eikonal equation, which provides an approximation to the wave equation and is valid at high frequencies (i.e., the Fresnel zone size is much smaller than the scale of the heterogeneities). In the full-wave methodology, the Fréchet kernels of the objective function and the Fréchet kernels of the data functional are all computed using purely numerical solutions to the seismic wave equation, thereby eliminating possible errors caused by kernel inaccuracy.

The implementation of full-wave seismic data assimilation methodology is far from trivial. The computational efficiency of different types of implementations depends upon the overall problem geometry, particularly on the ratio of sources to receivers, as well as trade-offs in computational resources, such as the relative costs of compute cycles to data storage. The waveform data can potentially provide much more detailed information about the seismic source and the Earth structure, but to effectively utilize waveform information we need to overcome the nonlinearity associated with waveform inversions.

1.6.1 Computational Challenges

The computational challenges of full-wave seismic data assimilation largely lie in (a) the solution of the three-dimensional seismic wave equations (forward and adjoint) for computing the synthetic seismograms and the gradient and/or Hessian of the objective function, and (b) the optimization algorithms used for finding the optimal structure and/or source models that minimize the objective function.

Both the scattering-integral implementation and the adjoint method require a large number of wave-propagation simulations for problems of practical interest. To make full-wave seismic data assimilation feasible on a practical time scale, the efficiency of the underlying seismic wave equation solver becomes paramount, which poses significant computational challenges in terms of both the design of the numerical algorithms and their implementations on homogeneous/heterogeneous multi-core computer clusters.

In recent years, seismologists have developed parallel computer codes using a variety of numerical algorithms to solve the seismic wave equation (e.g., Graves, 1996; Akcelik et al., 2002; Komatitsch and Tromp, 2002; Komatitsch et al., 2004). At present, most of the parallel seismic wave-propagation simulation codes cannot exploit multiple cores beyond running additional MPI processes per node. A general impression is that once the message-passing library is made "SMP-aware" (i.e., message-passing between processors on the same node can bypass the network, therefore reducing network performance bottlenecks), the single-level, flat message-passing-only programming model may continue to be the best overall approach to a scalable design. However, despite the success of the single-level, message-passing-only approach, there is growing evidence that future seismic wave-propagation simulation software will have to address increasing node-level complexities by introducing another level of parallelism. Message-passing may continue to be the dominant inter-node programming model in future high-performance scientific applications, but it may not be sufficient for driving all the cores on a single node for much longer, at least at the single-level, flat message-passing model. Seismologists and computer scientists are now exploring new node-level programming models that can effectively combine with the inter-node message-passing model and incorporate this new programming model into the next-generation seismic wave equation solvers.

The adjoint method and the scattering-integral method are two different variations of the generalized inverse and they are physically equivalent. However, their computational requirements are quite different and which one is more computationally efficient depends on the overall setup of the inversion problem, as well as trade-offs in computational resources (Chen et al., 2007a).

In the adjoint method, both the forward wave-field $\mathbf{u}_s$ and the adjoint wavefield ${}^{\dagger}\mathbf{u}_s$ are source-specific, therefore $2N_s$ simulations (N_s forward and N_s adjoint) are needed for one calculation of the gradients of the objective function. The temporal correlation between the forward and the adjoint wave-fields in (1.53)-(1.54) is independent of the spatial location $\mathbf{x}$, and the total number of operations is proportional to $N_V N_s$, where N_V is the total number of spatial grid points. If the conjugate-gradient algorithm (Press, 1999) is adopted for minimizing the objective function, each iteration requires one evaluation of the gradient to determine the conjugate-direction in the model space, followed by a line search to determine the optimal step length to take in the conjugate direction. There are a variety of algorithms for line search (Box et al., 1969) and its implementation certainly affects the number of simulations needed. A cubic interpolation scheme for the line search (Tromp et al., 2005) requires an additional N_s forward and N_s adjoint simulations and the total number of time-integration operations is proportional to $2N_V N_s$. During the time-integration operations, we need simultaneous access to both the forward and the adjoint wave-fields and one of these two wave-fields needs to be time-reversed. If we choose to store the forward wave-field or the adjoint wave-field on disk and read it back during the time integration, the total number of simulations is $4N_s$ and the disk storage requirement is the space-time volume of one wave-field, which is proportional to $N_V N_T$, where N_T is the total number of time steps for the simulation. To avoid storing the complete space-time wave-fields and the associated I/O cost, for non-dissipative wave equations, another approach is to reconstruct the time-reversed forward wave-field or adjoint wave-field on the fly. In this case, we need to store only the last time-frame of the forward or the adjoint wave-field as a function of space, which has a volume proportional to N_V, and additional $2N_s$ simulations are needed to recover the time-reversed forward or adjoint wave-field by solving the forward or adjoint wave-equation backwards in time, taking the final state of the wave-field as its initial condition (Gauthier et al., 1986). The total number of simulations is then $6N_s$.

In the scattering-integral implementation, the calculation of the Fréchet kernels for individual data functionals requires simultaneous access to both the forward wave-field from the source and the Green tensor from the receiver. We can compute the RGTs once for each receiver, store them on disk (at cost $\sim N_V N_T$), and use it for every source in the dataset, which can be efficient when the number of sources is larger than the number of receivers. A complete calculation of the data specific kernels requires $3N_r$ simulations (considering 3 orthogonal unit forces at each receiver) and one forward simulation for each source – a total $3N_r + N_s$ simulations, plus storage cost proportional to $3N_rN_VN_T$. The number of time integration operations is determined by the convolution between the forward wave-field from the source and the RGT and is proportional to $2N_t$, where N_t is the number of time steps within the support of the seismogram perturbation kernel, the number of spatial grid points N_V and the total number of seismograms N_u used in the inversion. For three-component receivers, $N_u \leqslant 3N_rN_s$. Owing to noise and instrumental reasons for eliminating seismograms, N_u is usually much less than this bound. In the scattering-integral method, the quadratic objective function is minimized using the Gauss-Newton algorithm, which requires additional computing time (though not too much compared with the wave-propagation simulations) to solve the linear system in (1.86).

1.6.2 Nonlinearity

The nonlinearity of the structural inverse problem, which includes multiple scattering (higher-order terms in the Born series), is accounted for through iteration, in which the reference structure model is updated at the beginning of each iteration and the Green's tensors (or the adjoint wave-fields in the adjoint method) are re-computed using the updated structure model. This approach accounting for nonlinearity is different from inverse scattering based on reconstructing higher-order terms in the Born or Rytov series (e.g., Tsihrintzis and Devaney, 2000ab), in which the higher-order terms are computed using the same Green's tensor as the first-order term. The higher-order terms in the Born/Rytov series can be difficult to compute except for simple structure models for which analytical solutions of the Green's tensors are available.

The convergence of the optimization algorithm depends upon the shape (number of local minima) of the objective function used to quantify the misfits between observations and model predictions. Objective functions defined in terms of the energy of differential waveforms (Tarantola, 1984; 1988) are dominated by local minima because the waveform itself is a very nonlinear function of the velocity model. In this case, conjugate-gradient or other types of gradient- and/or Hessian-based algorithms will have difficulty in converging unless the starting model is already very close to the global minimum. The gradient and/or Hessian, even though they are computed exactly, are not very useful if the starting model is far from the global minimum.

There are a number of different strategies for treating this problem in the full-wave domain, for instance, using a wave-equation travel-time tomography (Luo and Shuster, 1991) to align the phases before doing a waveform inversion. One strategy that has been developed substantially is based on a scale-decomposition approach or "frequency bootstrapping" (i.e., starting from low-frequency waveform data to improve large-scale structure and move to higher frequencies while the model improves (Bunks et al., 1995; Akcelik et al., 2002; Sirgue and Pratt, 2004)), which can allow a descent algorithm to stay in the neighborhood of the global minimum or the "basin of attraction". This type of approach may fail in practice if the waveform data do not have sufficient signal-to-noise ratio at low frequencies that would allow us to bootstrap from a starting model far away from the global minimum. In this case, we have a scale gap in our approach.

Another strategy is to re-parameterize waveform information into another form that is more linear with respect to structure model. Experiences from decades of travel-time tomography show that objective functions defined in terms of differential travel-times are much smoother than those defined in terms of differential waveforms, thus being more suitable for gradient-descent-type optimization algorithms. To better capture waveform information, Gee and Jordan (1992), Holschneider et al. (2005), among others, proposed to use frequency-dependent phase-delay and amplitude anomalies (GSDF) to re-parameterize waveform misfit. The exact Fréchet derivatives of these misfit measurements, as well as the gradient and/or Hessian of the objective function, can be computed by composing the seismogram perturbation kernel and the Born kernel through the chain rule (Chen et al., 2007a).

The linearization of the frequency-dependent phase and amplitude misfits (GSDF measurements) is closely related to the Rytov approximation, which can handle large accumulative phase-shifts much better than the Born approximation. Unlike the Born approximation, the Rytov approximation is not limited to the "small phase-shift" constraint (weak scattering); instead, it is limited by the "small-angle" (forward-scattering) constraint. The Rytov approximation has been widely used in long-distance propagations with primarily forward-scattering or small-angle scattering such as the line-of-sight propagation of optical or radio waves (e.g., Chernov, 1960; Tatarskiĭ , 1971; Ishimaru, 1999) and the diffraction tomography (e.g., Devaney, 1981; Wu and Toksöz, 1987). In Chen et al. (2007b), frequency-dependent phase-delay time measured on direct-arriving $P-$ and $S-$waves (forward-scattered or transmitted waves) was inverted and in one Gauss-Newton iteration, more than 80% of the large phase-shifts, which can be larger than 1/4 cycle, between observed and synthetic waveforms were corrected (Figures 18 and 19 in Chen et al., 2007b), resulting in perturbations up to 10% relative to the starting model. In Tape et al. (2009; 2010), the authors inverted frequency-dependent phase-delays measured mostly on surface waves and obtained nearly 30% perturbations relative to their starting model after 16 conjugate-gradient iterations. These experiments suggest that objective functions defined in terms of the frequency-dependent phase-delay measurements for forward-scattered waves are smoother (i.e., have larger basins of attraction) with respect to seismic velocities than objective functions defined in terms of differential waveforms. If the objective function is relatively smooth, gradient and/or Hessian-based optimization algorithms may not necessarily require the starting model to be very close to the global minimum.

1.7 Summary

In this article, the generalized inverse for the full-wave seismic data assimilation problem has been derived, which provides a unified methodology for iteratively improving our estimates of seismic source and/or Earth structure models. The starting point was the Bayes' theorem where the observations, the dynamic model and the poorly known seismic source and structure parameters were introduced with assumptions of Gaussian priors, which led to an objective function in the quadratic form. By applying the Hamilton's principle, we derived the Euler-Lagrange equations, which allow us to obtain an optimal solution using gradient-descent-type optimization algorithms. The adjoint method and the scattering-integral method, which have been successfully applied in full-3D, full-wave seismic tomography and earthquake source parameter inversions, can be considered as different implementation of the same physical principle based on the generalized inverse. The design of appropriate data functionals is an important issue. The frequency-dependent phase and amplitude misfit measurements made on localized waveforms can not only quantify waveform misfits effectively but also reduce the nonlinearity associated with the waveform inversion. Some recent applications of full-wave seismic tomography at different geographic scales and earthquake source parameter inversions were reviewed and some challenging issues related to the computational implementation and the nonlinearity of waveform inversion were discussed.

References

Akcelik, V., Biros, G., and Ghattas, O. (2002). Parallel multiscale Gauss-Newton-Krylov methods for inverse wave propagation. Proceedings of the 2002ACM/IEEE conference on Supercomputing, pages 1-15.

Aki, K. and Richards, P. G. (2002). Quantitative Seismology. Second edition. University Science Books.

Arnold, V. (1989). Mathematical Methods of Classical Mechanics. Springer-Verlag.

Bengtsson, L., Ghil, M., and Källén, E. (1981). Dynamic Meteorology: Data Assimilation Methods. Springer.

Bennett, A. (1992). Inverse Methods in Physical Oceanography. Cambridge University Press.

Beylkin, G. (1985). Imaging of discontinuities in the inverse scattering problem by inversion of a causal generalized radon transform. Journal of Mathematical Physics, 26(1): 99-108.

Biegler, L. (2003). Large-scale PDE-constrained Optimization. Springer.

Box, M., Davies, D., Swann, W., and Imperial Chemical Industries, l. (1969). Nonlinear Optimization Techniques. Mathematical and Statistical Techniques for Industry. Published for Imperial Chemical Industries Ltd by Oliver & Boyd.

Bunks, C., Saleck, F. M., Zaleski, S., and Chavent, G. (1995). Multiscale seismic waveform inversion. Geophysics, 60(5): 1,457-1,473.

Chen, P., Jordan, T. H., and Lee, E.-J. (2010a). Perturbation kernels for generalized seismological data functionals (GSDF). Geophysical Journal International, 183(2): 869-883.

Chen, P., Jordan, T. H., and Zhao, L. (2005). Finite moment tensor of the 3 September 2002 Yorba Linda earthquake. Bulletin of the Seismological Society of America, 95(3): 1,170-1,180.

Chen, P., Jordan, T. H., and Zhao, L. (2007a). Full three-dimensional tomography: a comparison between the scattering-integral and adjoint-wavefield methods. Geophysical Journal International, 170(1): 175-181.

Chen, P., Jordan, T. H., and Zhao, L. (2010b). Resolving fault plane ambiguity for small earthquakes. Geophysical Journal International, 181(1): 493-501.

Chen, P., Zhao, L., and Jordan, T. H. (2007b). Full 3D tomography for the crustal structure of the Los Angeles region. Bulletin of the Seismological Society of America, 97(4): 1,094-1,120.

Chernov, L. (1967). Wave propagation in a random medium. Dover Publications.

Dahlen, F. A. and Baig, A. M. (2002). Fréchet kernels for body-wave amplitudes. Geophysical Journal International, 150: 440-466.

Dahlen, F. A. and Tromp, J. (1998). Theoretical Global Seismology. Princeton University Press.

Daley, R. (1993). Atmospheric Data Analysis. Cambridge University Press.

Devaney, A. J. (1981). Inverse-scattering theory within the Rytov approximation. Optics Letters, 6(8): 374-376.

Ekström, G., Tromp, J., and Larson, E. (1997). Measurements and global models of surface wave propagation. Journal of Geophysical Research, 102(B4): 8,137-8,157.

Evensen, G. (2009). Data Assimilation: The Ensemble Kalman Filter. Springer.

Fichtner, A., Kennett, B. L. N., Igel, H., and Bunge, H.-P. (2007). Full waveform tomography for radially anisotropic structure: new insights into present and past states of the Australasian upper mantle. Earth and Planetary Science Letters, 290(3-4): 270-280.

Fichtner, A., Kennett, B. L. N., Igel, H., and Bunge, H.-P. (2008). Theoretical background for continental- and global-scale full-waveform inversion in the time-frequency domain. Geophysical Journal International, 175(2): 665-685.

Fichtner, A., Kennett, B., Igel, H., and Bunge, H. (2009). Full seismic waveform tomography for upper-mantle structure in the Australasian region using adjoint methods. Geophysical Journal International, 179(3): 1,703-1,725.

Gaherty, J. B., Jordan, T. H., and Gee, L. S. (1996). Seismic structure of the upper mantle in a central pacific corridor. Journal of Geophysical Research, 101(B10): 22,291-22,309.

Gauthier, O., Virieux, J., and Tarantola, A. (1986). Two-dimensional nonlinear inversion of seismic waveforms: numerical results. Geophysics, 51(7): 1,387-1,403.

Gee, L. S. and Jordan, T. H. (1992). Generalized seismological data functionals. Geophysical Journal International, 111(2): 363-390.

Gephart, J. W. (1985). Principal stress directions and the ambiguity in fault plane identification from focal mechanisms. Bulletin of the Seismological Society of America, 75(2): 621-625.

Gephart, J. W. (1990). Stress and the direction of slip on fault planes. Tectonics, 9(4): 845-858.

Graves, R. W. (1996). Simulating seismic wave propagation in 3D elastic media using staggered-grid finite differences. Bulletin of the Seismological Society of America, 86(4): 1,091-1,106.

Harris, R. A. (1998). Introduction to special section: stress triggers, stress shadows, and implications for seismic hazard. Journal of Geophysical Research, 103(B10): 24,347-24,358.

Holschneider, M., Diallo, M. S., Kulesh, M., Ohrnberger, M., Luck, E., and Scherbaum, F. (2005). Characterization of dispersive surface waves using continuous wavelet transforms. Geophysical Journal International, 163(2): 463-478.

Ishimaru, A. (1999). Wave Propagation and Scattering in Random Media. IEEE Press.
Kalnay, E. (2003). Atmospheric Modeling, Data Assimilation, and Predictability. Cambridge University Press.
Katzman, R., Zhao, L., and Jordan, T. (1998). High-resolution, two-dimensional vertical tomography of the central pacific mantle using ScS reverberations and frequency-dependent travel times. Journal of Geophysical Research, 103(B8): 17,933-17,971.
Komatitsch, D. and Tromp, J. (2002). Spectral-element simulations of global seismic wave propagation – I Validation. Geophysical Journal International, 149(2): 390-412.
Komatitsch, D., Liu, Q., Tromp, J., Süss, P., Stidham, C., and Shaw, J. H. (2004). Simulations of ground motion in the Los Angeles basin Based upon the spectral-element method. Bulletin of the Seismological Society of America, 94(1): 187-206.
Lee, E-J, Chen, P., Jordan, T. H. and Wang, L. (2011). Rapid centroid moment tensor inversion in a three-dimensional earth structure model for earthquakes in Southern California. Geophysical Journal International, in press.
Lilly, J. M. and Park, J. (1995). Multiwavelet spectral and polarization analyses of seismic records. Geophysical Journal International, 122(3): 1,001-1,021.
Liu, Q. (2006). Spectral-element simulations of 3-D seismic wave propagation and applications to source and structural inversions. PhD thesis, California Institute of Technology.
Luo, Y. and Shuster, G. (1991). Wave-equation traveltime inversion. Geophysics, 56(5): 645-653.
Maggi, A., Tape, C., Chen, M., Chao, D., and Tromp, J. (2009). An automated time-window selection algorithm for seismic tomography. Geophysical Journal International, 178: 257-281.
Magistrale, H., Day, S., Clayton, R. W., and Graves, R. W. (2000). The SCEC Southern California reference three-dimensional seismic velocity model version 2. Bulletin of the Seismological Society of America, 90(6B): S65-S76.
Malanotte-Rizzoli, P. (1996). Modern Approaches to Data Assimilation in Ocean Modeling. Elsevier.
Marquering, H., Dahlen, F. A., and Nolet, G. (1999). Three-dimensional sensitivity kernels for finite-frequency traveltimes: the banana-doughnut paradox. Geophysical Journal International, 137(3): 805-815.
McCloskey, J., Nalbant, S. S., Steacy, S., Nostro, C., Scotti, O., and Baumont,D. (2003). Structural constraints on the spatial distribution of aftershocks. Geophysical Research Letters, 30: 1,610-1,613.
McGuire, J. J., Zhao, L., and Jordan, T. H. (2001). Teleseismic inversion for the second-degree moments of earthquake space–time distributions. Geophysical Journal International, 145: 661-678.
Michael, A. J. (1987). Use of focal mechanisms to determine stress: a control study. Journal of Geophysical Research, 92(B1): 357-368.
Milne, R. (1980). Applied Functional Analysis: An Introductory Treatment. Pitman.
Paige, C. C. and Saunders, M. A. (1982). Algorithm 583 LSQR: Sparse linear equations and least squares problems. ACM Transactions on Mathematical Software (TOMS), 8(2): 195-209.
Parsons, T. (2005). Significance of stress transfer in time-dependent earthquake probabil-

ity calculations. Journal of Geophysical Research, 110(B05S02): 1-20.
Pratt, R. G. (1999). Seismic waveform inversion in the frequency domain, part 1: theory and verification in a physical scale model. Geophysics, 64(3): 888-901.
Pratt, R. G. and Shipp, R. M. (1999). Seismic waveform inversion in the frequency domain, part 2: fault delineation in sediments using cross-hole data. Geophysics, 64(3): 902-914.
Pratt, R. G., Shin, C., and Hicks, G. J. (1998). Gauss-Newton and full Newton methods in frequency-space seismic waveform inversion. Geophysical Journal International, 133(2): 341-362.
Press,W. (1999). Numerical recipes in C: the art of scientific computing. Cambridge University Press.
Rhie, J. and Romanowicz, B. (2004). Excitation of earth's continuous free oscillations by atmosphere–ocean–seafloor coupling. Nature, 431(7,008): 552-556.
Ritsema, J., Rivera, L., Komatitsch, D., Tromp, J., and van Heijst, H. (2002). Effects of crust and mantle heterogeneity on PP/P and SS/S amplitude ratios. Geophysical Research Letters, 29(10): 72-1-4.
Süss, M. P. and Shaw, J. H. (2003). P wave seismic velocity structure derived from sonic logs and industry reflection data in the Los Angeles basin, California. Journal of Geophysical Research, 108(B3): 2,170.
Schulz, M. (2006). Control Theory in Physics and Other Fields of Science: Concepts, Tools, and Applications. Springer.
Simon, D. (2006). Optimal State Estimation: Kalman, H [infinity] and Nonlinear Approaches. Wiley-Interscience.
Sirgue, L. and Pratt, R. G. (2004). Efficient waveform inversion and imaging: a strategy for selecting temporal frequencies. Geophysics, 69(1): 231-248.
Snieder, R. and Lomax, A. (1996). Wavefield smoothing and the effect of rough velocity perturbations on arrival times and amplitudes. Geophysical Journal International, 125(3): 796-812.
Steacy, S., Nalbant, S. S., McCloskey, J., Nostro, C., Scotti, O., and Baumont, D. (2005). Onto what planes should coulomb stress perturbations be resolved? Journal of Geophysical Research, 110(B05S15): 1-14.
Tape, C., Liu, Q., Maggi, A., and Tromp, J. (2009). Adjoint tomography of the Southern California crust. Science, 325(5943): 988-992.
Tape, C., Liu, Q., Maggi, A., and Tromp, J. (2010). Seismic tomography of the Southern California crust based on spectral-element and adjoint methods. Geophysical Journal International, 180(1): 433-462.
Tarantola, A. (1984). Inversion of seismic reflection data in the acoustic approximation. Geophysics, 49(8): 1,259-1,266.
Tarantola, A. (1988). Theoretical background for the inversion of seismic waveforms including elasticity and attenuation. Pure and Applied Geophysics, 128(1.2): 365-399.
Tarantola, A. (2005). Inverse Problem Theory and Methods for Model Parameter Estimation. Society for Industrial and Applied Mathematics, illustrated edition.
Tatarskiĭ, V., Oceanic, U. S. N., Administration, A., and (U.S.) N. S. F. (1971). The effects of the turbulent atmosphere on wave propagation. Israel Program for Scientific Translations.

Thomson, D. (1982). Spectrum estimation and harmonic analysis. Proceedings of the IEEE, 70(9): 1,055-1,096.

Tromp, J., Tape, C., and Liu, Q. (2005). Seismic tomography, adjoint methods, time reversal and banana-doughnut kernels. Geophysical Journal International, 160(1): 195-216.

Tsihrintzis, G. A. and Devaney, A. J. (2000a). Higher order (nonlinear) diffraction tomography: Inversion of the Rytov series. IEEE Transactions on Information Theory, 46(5): 1,748-1,761.

Tsihrintzis, G. A. and Devaney, A. J. (2000b). Higher-order (nonlinear) diffraction tomography: Reconstruction algorithms and computer simulation. IEEE Transactions on Image Processing, 9(9): 1,560-1,572.

Wikle, C. and Berliner, L. (2007). A Bayesian tutorial for data assimilation. PhysicaD: Nonlinear Phenomena, 230(1.2): 1-16.

Woodward, R. and Masters, G. (1991). Global upper mantle structure from long period differential travel times. Journal of Geophysical Research, 96(B4): 6,351-6,377.

Wu, R. (2003). Wave propagation, scattering and imaging using dual-domain one way and one-return propagators. Pure and Applied Geophysics, 160: 509-539.

Wu, R. and Toksöz, M. (1987). Diffraction tomography and multisource holography applied to seismic imaging. Geophysics, 52(1): 11-25.

Wunsch, C. (1996). The Ocean Circulation Inverse Problem. Cambridge University Press.

Yin, Z.-M. (1996). An improved method for the determination of the tectonic stress field from focal mechanism data. Geophysical Journal International, 125(3): 841-849.

Zhao, L. and Jordan, T. H. (1998). Sensitivity of frequency-dependent traveltimes to laterally heterogeneous, anisotropic earth structure. Geophysical Journal International, 133: 683-704.

Zhao, L., Chen, P., and Jordan, T. H. (2006). Strain green's tensors, reciprocity, and their applications to seismic source and structure studies. Bulletin of the Seismological Society of America, 96(5): 1,753-1,763.

Zhao, L., Jordan, T. H., Olsen, K. B., and Chen, P. (2005). Fréchet kernels for imaging regional earth structure based on three-dimensional reference models. Bulletin of the Seismological Society of America, 95(6): 2,066-2,080.

Zupanski, D. and Zupanski, M. (2006). Model error estimation employing an ensemble data assimilation approach. Monthly Weather Review, 134: 1,337-1,354.

Author Information

Po Chen

Department of Geology and Geophysics, University of Wyoming,

Wyoming 82071, USA

E-mail: pochengeophysics@gmail.com

Chapter 2
One-Return Propagators and the Applications in Modeling and Imaging

Ru-Shan Wu, Xiao-Bi Xie, and Shengwen Jin

This chapter reviews the one-return method for calculating seismic wave propagation in complex acoustic and elastic models. We give a new, intuitive derivation of the one-return approximation using sequential thin-slab transmission/reflection operators. This derivation can reach the same formulation as the De Wolf approximation, and in the same time provide an efficient implementation of the method. The method is based on the multiple-forescattering-single-backscattering approximation. It neglects the internal reverberations (internal multiples) but can take into account all forward scattering phenomena, such as focusing/defocusing, diffraction, refraction, interference, and model primary reflections from heterogeneities. One-return method can be implemented using an iterative marching algorithm shuttling between the space and wavenumber domains. For models where reverberation and resonance scattering can be neglected, this method provides an accurate and highly efficient algorithm, especially for large velocity models at high-frequencies. It has been used for reservoir AVO simulations in heterogeneous visco-elastic media with moderate elastic parameter variations (perturbations less than 40%). In seismic imaging, one-return propagators can migrate turning waves and duplex waves (or prism waves) and therefore, image steeply dipping structures and vertical overhangs. For waveform inversion and migration velocity analysis, the one-return propagator is very efficient for calculating sensitivity kernels that involve reflections from interfaces. We also reviewed recent progress in overcoming the limitations and shortcomings of the one-way, one-return methods. The one-way propagator with super-wide angle interpolates the wavefronts from two orthogonal one-way propagators to reconstruct the accurate wavefront up to nearly the full angle range, and was used to image overhanging salt flanks. One-return boundary element method was developed to deal with strong-contrast elastic media. It can generate synthetics with only primary reflections from the salt boundaries, and therefore, be used to study the artifacts caused by salt multiples in subsalt images. However, the current version uses Green's functions in homogeneous media. One-return modeling in heterogeneous, strong-contrast media is still a challenging problem.

Keywords: Elastic wave modeling, One way propagator, One-return approximation, Seismic imaging

2.1 Introduction

One-way approximation for wave propagation has been introduced and widely used as propagators in forward and inverse problems of scalar, acoustic and elastic waves (e.g., Claerbout, 1970, 1976; Landers and Claerbout, 1972; Flatté and Tappert, 1975; Corones, 1975; Tappert, 1977; McCoy, 1977; Hudson, 1980; Ma, 1982; Wales and McCoy, 1983; Fishman and McCoy, 1984, 1985; Wales, 1986; McCoy and Frazer, 1986; Collins, 1989, 1993; Collins and Westwood, 1991; Stoffa et al., 1990; Fisk and McCarter, 1991; Wu and Huang, 1992; Zhang, 1993; Ristow and Ruhl, 1994; Wu, 1994, 1996, 2003; Wu and Xie, 1994; Wu and Jin, 1997; Xie and Wu, 1998, 2001, 2005;Grimbergen et al., 1998; Van Stralen et al., 1998; Wild and Hudson, 1998; Jin et al., 1998, 1999, 2002; Jin and Wu, 1999; Huang et al., 1999a,b; Thomson, 1999, 2005; De Hoop et al., 2000; Lee at al., 2000; Wild et al., 2000; Wu et al., 2000a,b; Le Rousseau and de Hoop, 2001; Wu and Wu, 2001; Han and Wu, 2005; Zhang et al., 2005, 2006). The great advantages of one-way propagation methods are the fast speed of computation, often being several orders of magnitudes faster than the full-wave finite difference and finite element methods, and the huge saving in internal memory.

The successful extension and applications of one-way elastic wave propagation methods has stimulated the research interest in developing similar theory and techniques for reflected or backscattered wave calculation. There are several approaches in extending the one-way propagation method to include backscattering and multiple scattering calculations, such as the generalized Bremmer series (GBS) approach (Corones, 1975; De Hoop, 1996; Wapenaar, 1996, 1998; van Stralen et al., 1998; Thomson, 1999, 2005; Le Rousseau and de Hoop, 2001) and the generalized screen propagator (GSP) approach (Wu, 1994, 1996, 2003; Wu and Xie, 1993, 1994; Wu et al., 1995; Wild and Hudson, 1998; de Hoop et al., 2000; Wild et al., 2000; Xie et al., 2000; Xie and Wu, 2001, 2005; Wu et al., 2007). The key difference between these approaches is how to define a reference Green's function in the multiple scattering series. The GBS approach adopts an asymptotic solution of the acoustic or elastic wave equation in the heterogeneous medium as the Green's function, i.e. the one-way propagator in the preferred direction. The multiple scattering series is based on the interaction of Green's field (incident field) with the medium heterogeneities. The GSP approach, on the other hand, does not use asymptotic solutions. Instead, the approach uses the multiple-forward-scattering (MFS) corrected one-way propagator as the Green's function. When the backscattered field, calculated at each thin-slab, is propagated in the backward direction, it uses the same MFS corrected one-way propagator. In Wu et al. (2007), detailed analysis and comparison of different approximations can be found. In this chapter, we will focus on the GSP approach based on the De Wolf approximation (one-return approximation).

In the generalized screen approach, Wu and Huang (1995) introduced a wide-angle modeling method for backscattered acoustic waves using the multiple-forward-scattering approximation and a phase-screen propagator. Xie and Wu (1995; 2001) extended the complex screen method to include the calculation of backscattered elastic waves under the small-angle approximation. Wu (1996) and Wu et al., (2007) derived a more general theory for acoustic and elastic waves using the De Wolf approximation, and the theory provided two versions of algorithms: the thin-slab method and the complex-screen method. Wu and Wu (1999, 2003a) introduced a fast implementation of the thin-slab method and a second order improvement for the complex-screen method. Wu and Xie (2009) discussed some practical applications of this method. In the rest part of this chapter, we review the one-return method and discuss its applications in seismic modeling, imaging and inversion with numerical examples.

2.2 Primary-Only Modeling and One-Return Approximation

"*Primaries*" are referred to as the primary transmitted or reflected waves, which involve only single interaction at each interface (transmission or reflection) or at each small-scale volumetric heterogeneity (forward or backscattering) inside a complex medium. The concept of primary transmission and reflection is closely related to the acquisition (observation) geometry. Mathematically, primary transmission and reflection (T/R) can be defined with respect to certain preferred direction. In surface seismic reflection survey, z-direction is conveniently defined as the preferred direction (globally). In smoothly inhomogeneous media, the preferred direction can be also defined as being along the ray-direction. In this chapter, we will discuss only the globally preferred direction (z-direction) in wave-equation-based methods.

The mathematical basis of primary-only modeling is the multiple-forescattering-single-backscattering (MFSB) approximation, which is also called one-way-one-return (OWOR) approximation. The physics behind this approximation is the neglect of multiple scattering between the forward and backward directions (with respect to the preferred direction), i.e. the reverberations. The OWOR is totally wave-equation-based, not the high-frequency asymptotics. This approach needs the separation of the forward-propagated and backward-propagated waves in the wave equation. The separation can be done by either decomposing the scattered wavefield or factorizing the scattering operator into two parts: the forward part and the backward part. The wavefield decomposition leads to a marching wave equation method which may include the coupling between the forward and backward waves (e.g., Weston, 1989), while the operator decomposition (factorization) forms the theoretical basis of one-way or one-return modeling. In the latter approach, the one-way-one-return approximation can be formulated as the first order term in a reordered multiple scattering series (the De Wolf series). In this way, the multiple backscattering and reverberations can be calculated by multiple sweeps of the one-return algorithm. In this chapter we limit the treatment to the approach of operator decomposition.

For operator decomposition, the formulation can be based on either PDE (partial differential equation) or the integral equation. For the PDE approach, the operator factorization results in a square-root operator and many approximations for one-way propagators were derived based on the PsDO (pseudo-differential operator) theory (Fishman and McCoy, 1984, 1985; Zhang, 1993; Thomson, 1999, 2005; De Hoop et al., 2000; Zhang et al., 2005, 2006, 2007). The wave equation solution after the factorization can be casted into generalized Bremmer series (De Hoop, 1996; De Hoop et al., 2000). For the integral equation approach, the approximation can be derived from the De Wolf series and its first approximation (De Wolf approximation or one-return approximation) (De Wolf, 1971, 1985; Wu, 1994, 1996, 2003; Xie and Wu, 2001; Wu et al., 2007). In this chapter, we will follow the integral equation approach and the De Wolf approximation.

Integral equation, and differential equation are different forms of wave equation, and describe mathematically the same physical phenomena: wave motion and propagation. Integral wave equation, such as the Lippmann-Schwinger equation, can be derived from the differential wave equation using the Green's function. The solution of the Lippmann-Schwinger equation can be casted into a Born series, of which the first order term is called the Born approximation. The Born approximation is among the simplest solutions and widely used in wave scattering problems. The approximation is a weak scattering approximation, and only valid when the scattered field is much smaller than the incident field. This implies that the heterogeneities are weak and the propagation distance is short (or the scattering volume is small). The valid region of the Born approximation for the forward scattering is very different from that for the backscattering. Scattered fields in the Born approximation are simple summation of contributions from all parts of the scattering volume, and scattering at each part is independent of other parts. Therefore, the Born approximation does not obey energy conservation, In the forward direction, scattered fields from different sections along the propagation path arrive at the receiving point in phase with the incident field, so that they will be coherently superposed. This leads to the linear increase of the total field, which is a serious divergence problem for forward scattering. However, the total observed field in the backward direction is the sum of all the backscattered fields from all the scatterers since there is no incident wave in the backward direction. In addition, the volume of the scatterers of coherent stacking for backscattered waves is limited due to the travel time differences. For this reason, backscattering does not have the catastrophic divergence and the validity region for the Born approximation is much larger than the case of forward scattering. The De Wolf approximation, which is a multiple-forescattering-single-backscattering (MFSB) approximation, has been introduced to overcome the limitation of the Born approximations in long-range forward propagation.

The mathematical basis for the Born series and De Wolf series as the solution of the Lippmann-Schwinger equation (integral wave equation) has been detailed in Wu et al. (2007). Here we adopt a more intuitive approach to show the physics and calculation procedure of the method. Despite of starting from the physical intuition, we can reach the same mathematical formulation if we make the limiting process using the corresponding notations.

For an arbitrary heterogeneous media of a large volume, as shown in Figure 2.1a, we can slice the volume into numerous thin-slabs transversal to the propagation direction (preferred direction). Shown in Figure 2.1b and Figure 2.1c are examples of individual thin-slabs for a scalar medium (velocity is the only parameter), and an elastic medium. Assume each thin-slab is thin enough so that the Born approximation can be used for the scattering calculation. In Figure 2.1a, u_0 is the incident field, U is the scattered field and the total field will be denoted by $u = u_0 + U$. The one-return approximation uses the scattering operator spitting, so the scattered waves by the thin-slab are divided into the forescattered wave U_f and the backscattered wave U_b. At the thin-slab exit, we add the scattered wave to the incident wave, forming the total forward field $u_f = u_0 + U_f$. This

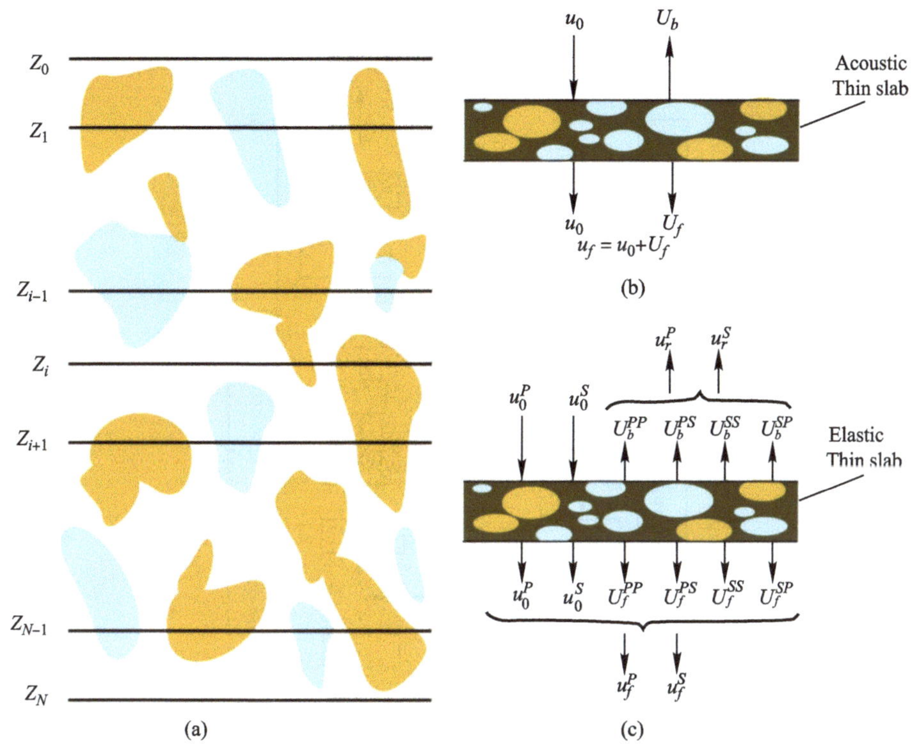

Fig. 2.1 (a) Cartoon showing the thin-slab decomposition of a velocity model, where a heterogeneous medium is sliced into an N-stack of thin-slabs; (b) An acoustic (scalar wave) thin-slab, where the incident plus the forescattered waves form the transmitted wave and the back-scattered wave composes the reflected wave; (c) An elastic thin-slab and the different scattered waves generated due to the interaction between incident P and S waves and the thin-slab. The superscripts PP, PS, SS and SP denote the P-to-P, P-to-S, S-to-S and S-to-P scatterings.

new total forward field will be used as the updated incident field for the next thin-slab. For an elastic thin slab shown in Figure 2.1c, the superscripts denote different wave types (P or S) and conversion modes (P-to-P, P-to-S, S-to-S and S-to-P scatterings). Similar to the scalar wave case, after interacting with an elastic thin slab, all forward scattered waves together with the incident waves form the forward P and S waves at the exit side of the slab (the incident waves for the next thin-slab), while at the entrance side, backscattered fields form the reflected P and S waves. We first discuss the scalar wave case to demonstrate the principle of one-return approximation. In the next section, we will summarize the formulation for the elastic wave case.

Figure 2.2 shows a cartoon illustrating the marching algorithm of the one-return approximation. For the lth thin-slab, we define a slab transmission operator T_l and a slab reflection operator (backscattering operator) R_l. Together they form the split operators in the forward and backward directions. The transmission operator can be expressed explicitly by the forescattering operator F_l,

$$T_l = 1 + F_l. \tag{2.1}$$

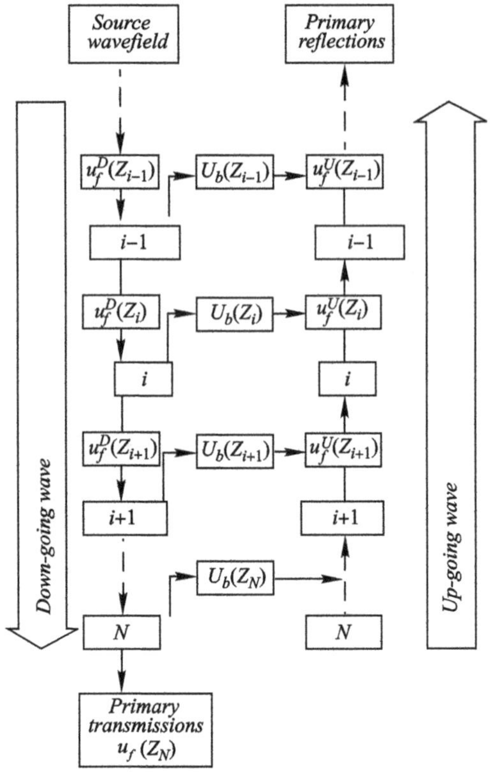

Fig. 2.2 Cartoon illustrating the double-sweep procedure of the one-return modeling. In the downgoing sweep, after interacting with each thin-slab, the transmitted P- and S-wavefields are used as the input for the next thin-slab. This process updates the incident wavefields slab-by-slab. During the downgoing sweep, all backscattered wavefields at each thin-slab are stored in the memory. In the upgoing sweep, those stored reflections are retrieved and propagated to the surface using the sequential transmission operators similar to the downgoing sweep.

With the split thin-slab operators, we can write out the forescattering updated incident field u_f (the transmitted wavefield) and the backscattered field U_b at each slab entrance as (see Figure 2.2)

$$
\begin{gathered}
u_f(z_0) = u_0 \quad U_b(z_0) = R_1 u_f(z_0) \\
u_f(z_1) = T_1 u_f(z_0) \quad U_b(z_1) = R_2 u_f(z_1) \\
\cdots \\
u_f(z_i) = T_i u_f(z_{i-1}) = \left\{ \prod_{l=1}^{i} T_l \right\} u_0, \quad U_b(z_i) = R_{i+1} u_f(z_i) = R_{i+1} \left\{ \prod_{l=1}^{i} T_l \right\} u_0 \\
\cdots \\
u_f(z_N) = T_N u_f(z_{N-1}), \quad U_b(z_{N-1}) = R_N u_f(z_{N-1})
\end{gathered}
\tag{2.2}
$$

where the z coordinate is the depth (refer to Figure 2.1a). If we define a one-way forward propagator as a *sequential thin-slab transmission operator*,

$$\mathbf{P}(z_j,z_i)=\prod_{l=i}^{j}T_l, \tag{2.3}$$

we can write the transmitted wave at the bottom of the medium as

$$u_t(z_N)=\mathbf{P}(z_N,z_0)u_0(z_0) \tag{2.4}$$

and the primary reflected (backscattered) field at the top of the medium as the summation of the single backscattered waves of all the thin-slabs, propagated to the top by one-way propagators defined in (2.3).

$$\begin{aligned} u_r &= \sum_{i=0}^{N-1}\mathbf{P}(z_0,z_i)U_b(z_i)=\sum_{i=0}^{N-1}\mathbf{P}(z_0,z_i)R_{i+1}u_f(z_i) \\ &= \sum_{i=0}^{N-1}\mathbf{P}(z_0,z_i)R_{i+1}\mathbf{P}(z_i,z_0)u_0(z_0). \end{aligned} \tag{2.5}$$

In the case of local Born approximation, thin-slab transmission operator is

$$T_l=G_l\left\{1+\chi_f\right\} \tag{2.6}$$

where G_l is the local Green's operator (in the background media) in the l-th thin-slab, "1" is the identity operator, and χ_f is scattering potential to the forward half-space. Applying to an incident field u_0, the transmitted field is

$$\begin{aligned} u_f(z_l) &= T_l u_0(z_{l-1})=G_l\left\{1+\chi_f\right\}u_0(z_{l-1}) \\ &= u_0(z_l)+k^2\int_{thin\text{-}slab}d^3\mathbf{r}'G_l\left(\mathbf{x},z_l;\mathbf{r}'\right)\chi_f\left(\mathbf{r}'\right)u_0\left(\mathbf{r}'\right). \end{aligned} \tag{2.7}$$

The thin slab reflection operator is

$$R_l=G_l\chi_b \tag{2.8}$$

where χ_b is the scattering potential in the backward half-space. The one-way propagator defined in (2.3) is the *multiple forescattering Green's function*

$$G_f(z_i,z_0)=\prod_{l=1}^{i}T_l=\prod_{l=1}^{i}G_l\left\{1+\chi_f(z_l)\right\}. \tag{2.9}$$

After expanding the product and reordering the series, we reach the De Wolf approximation (one-return approximation) (see equation 28 of Wu et al., 2007),

$$G_f(z_i,z_0)=\sum_{m=0}^{i}\left[G_0\chi_f\right]^m G_0 \quad 0\leqslant m\leqslant i \tag{2.10}$$

where G_0 is the free-space (background) Green's function between any two thin-slabs

$$G_0(z_j,z_i) = \prod_{l=i}^{j} G_l(z_l).$$

In the same way, we can derive the updated incident field in terms of multiple forward scattering series

$$u_f(z_i) = \sum_{m=0}^{i} [G_0\chi_f]^m u_0 \quad 0 \leqslant m \leqslant i. \tag{2.11}$$

When taking the limit of infinitely thin-slab, i.e. $\Delta z \to 0$, we reach the integral form of one-return approximation (De Wolf approximation),

$$u(\mathbf{x},z) = u_f(\mathbf{x},z) + k^2 \int_V d^3\mathbf{r}' G_f(\mathbf{x},z;\mathbf{r}')\chi_b(\mathbf{r}')u_f(\mathbf{r}'). \tag{2.12}$$

From the above derivation, we see that iterative application of thin-slab propagators can derive the De Wolf approximation (MFSB approximation). Furthermore, it provides a very efficient implementation of the MFSB (multiple-forescattering-single-backscattering). In the next section, we will concentrate on the elastic one-return modeling.

2.3 Elastic One-Return Modeling

As seen from the previous section, once we derive the thin-slab transmission and reflection operators, one-return modeling (De Wolf approximation) can be implemented by a double-sweep of marching algorithm with iterative application of thin-slab operator. In the following, we summarize the formulations of elastic thin-slab operator. The method belongs to a more general approach of generalized screen propagators (GSP).

In a linear general heterogeneous medium, the elastic dynamic equation can be expressed as (Aki and Richards, 1980)

$$-\omega^2\rho(\mathbf{x})\mathbf{u}(\mathbf{x}) = \nabla\cdot\sigma(\mathbf{x}), \tag{2.13}$$

where $\mathbf{x}$ is the location, ω the frequency, $\mathbf{u}$ the displacement vector, σ the stress tensor (dyadic), $\nabla\cdot$ the divergence operator and ρ the density of the medium, and there is no body force in the medium. The stress-displacement relation is

$$\sigma(\mathbf{x}) = \mathbf{c}(\mathbf{x}) : \varepsilon(\mathbf{x}) = \frac{1}{2}\mathbf{c} : (\nabla\mathbf{u} + \mathbf{u}\nabla), \tag{2.14}$$

where $\mathbf{c}$ is the elastic constant tensor of the medium, ε is the strain field, $\mathbf{u}\nabla$ stands for the transpose of $\nabla\mathbf{u}$ (the gradient operator) and " : " stands for double scalar product of tensors defined through $(\mathbf{ab}){:}(\mathbf{cd}) = (\mathbf{b}\cdot\mathbf{c})(\mathbf{a}\cdot\mathbf{d})$. Equation (2.13) can be written as a wave equation of the displacement field:

$$-\omega^2\rho(\mathbf{x})\mathbf{u}(\mathbf{x}) = \nabla\cdot\left[\frac{1}{2}\mathbf{c} : (\nabla\mathbf{u} + \mathbf{u}\nabla)\right]. \tag{2.15}$$

Using perturbation theory, the elastic parameters and the total wavefield can be decomposed as

$$\rho(\mathbf{x}) = \rho_0 + \delta\rho(\mathbf{x}),$$
$$\mathbf{c}(\mathbf{x}) = \mathbf{c}_0 + \delta\mathbf{c}(\mathbf{x}),$$
$$\mathbf{u}(\mathbf{x}) = \mathbf{u}^0(\mathbf{x}) + \mathbf{U}(\mathbf{x}), \tag{2.16}$$

where ρ_0 and $\mathbf{c}_0$ are the parameters of the background medium, $\delta\rho$ and $\delta\mathbf{c}$ are the corresponding perturbations, $\mathbf{u}^0$ is the incident field and $\mathbf{U}$ is the scattered field, then (2.15) can be rewritten as a wave equation about $\mathbf{U}$

$$-\omega^2\rho_0\mathbf{U} - \nabla\cdot\left[\frac{1}{2}\mathbf{c}_0 : (\nabla\mathbf{U} + \mathbf{U}\nabla)\right] = \mathbf{F} \tag{2.17}$$

with

$$\mathbf{F} = \omega^2\delta\rho\,\mathbf{u} + \nabla\cdot[\delta\mathbf{c} : \varepsilon] \tag{2.18}$$

as an equivalent body force due to scattering.

We choose z-axis as the primary propagation direction and use $\mathbf{u} = \mathbf{u}^f$, the updated forward propagated field, as the incident wave for the thin-slab in the current marching step. Substituting the equivalent body force (2.18) and equation (2.16) into equation (2.17), we have the scattered wavefield generated by the interaction between the incident wave and the heterogeneity,

$$\mathbf{U}(\mathbf{x}_T, z^*) = \int_V d^3\mathbf{x}' \left\{\delta\rho\omega^2\mathbf{u}^f(\mathbf{x}'_T, z) + \nabla\cdot\left[\delta\mathbf{c} : \varepsilon^f(\mathbf{x}'_T, z)\right]\right\}\cdot\mathbf{G}^0\left(\mathbf{x}_T, z^*; \mathbf{x}'_T, z\right), \tag{2.19}$$

where z^* is the observation depth, $\mathbf{G}^0$ is the Green's function in the background model, $\mathbf{x} = (\mathbf{x}_T, z)$, with $\mathbf{x}_T$ and z as horizontal and vertical coordinates.

2.3.1 Local Born Approximation

Following the derivations of Wu et al. (2007), we can express the scattered displacement field for a thin-slab in the horizontal wavenumber domain using the local Born approximation. Note that the local incident field $\mathbf{u}^f$ and ε^f have the footprints of the heterogeneities in the previous thin-slabs, so the formulation is not a global Born approximation. However, we assume the thin-slab is thin enough so the local incident field will not be influenced by the local heterogeneities in the current thin-slab. In this way, we can apply the Born approximation locally within the thin-slab. After Fourier transform along the transversal plane with respect to $\mathbf{x}_T$ in equation (2.19), we obtain the local Born expression in the horizontal wavenumber domain:

$$\mathbf{U}(\mathbf{K}_T, z^*) = \int_{z'}^{z_1} dz \iint d^2\mathbf{x}_T \left\{\delta\rho\omega^2\mathbf{u}^f(\mathbf{x}_T, z) + \nabla\cdot\left[\delta\mathbf{c} : \varepsilon^f(\mathbf{x}_T, z\right]\right\}\cdot\mathbf{G}^0\left(\mathbf{K}_T, z^*; \mathbf{x}_T, z\right), \tag{2.20}$$

where the thin slab is located between z' and z_1, $\mathbf{K}_T$ is the horizontal wavenumber vector, and

$$\mathbf{G}^0(z^*,\mathbf{k}_T;z,\mathbf{x}_T) = \frac{ik_\alpha^2}{2\rho_0\omega^2}\hat{k}_\alpha\hat{k}_\alpha\frac{1}{\gamma_\alpha}e^{i\mathbf{k}_\alpha\cdot\mathbf{r}} + \frac{ik_\beta^2}{2\rho_0\omega^2}\left(\mathbf{I}-\hat{k}_\beta\hat{k}_\beta\right)\frac{1}{\gamma_\beta}e^{i\mathbf{k}_\beta\cdot\mathbf{r}}, \tag{2.21}$$

with $\mathbf{I}$ as the unit dyadic, and

$$\gamma_\alpha = \sqrt{k_\alpha^2 - K_T^2},$$

$$\gamma_\beta = \sqrt{k_\beta^2 - K_T^2}, \tag{2.22}$$

where $k_\alpha = \omega/\alpha_0$ and $k_\beta = \omega/\beta_0$ are P and S wavenumbers with α_0 and β_0 as the P and S wave background velocities in the thin-slab, respectively,

$$\hat{k}_\alpha = \frac{1}{k_\alpha}(\mathbf{K}_T,\gamma_\alpha), \tag{2.23}$$

$$\hat{k}_\beta = \frac{1}{k_\beta}\left(\mathbf{K}_T,\gamma_\beta\right), \tag{2.24}$$

and are unit vectors along P and S wave propagation directions.

For isotropic media,

$$\delta\mathbf{c}(\mathbf{x}) : \varepsilon(\mathbf{x}) = \delta\lambda(\mathbf{x})\,|\varepsilon|\,\mathbf{I} + 2\delta\mu(\mathbf{x})\varepsilon(\mathbf{x}). \tag{2.25}$$

Substituting (2.21) into (2.20), we can derive the dual-domain expressions for scattered displacement fields in isotropic elastic media.

For P to P scattering:

$$\begin{aligned}\mathbf{U}^{PP}(K_T,z^*) = {}& \frac{ik_\alpha^2}{2\gamma_\alpha}\int_{z'}^{z_1} dz e^{ik_z^\alpha(z^*-z)}\cdot\left\{\hat{k}_\alpha\hat{k}_\alpha\cdot\iint d^2\mathbf{x}_T e^{-i\mathbf{K}_T\cdot\mathbf{x}_T}\frac{\delta\rho(\mathbf{x}_T,z)}{\rho}\mathbf{u}_\alpha^f(\mathbf{x}_T,z)\right.\\ & -\hat{k}_\alpha\iint d^2\mathbf{x}_T e^{-i\mathbf{K}_T\cdot\mathbf{x}_T}\frac{\delta\lambda(\mathbf{x}_T,z)}{\lambda+2\mu}\frac{1}{ik_\alpha}\nabla\cdot\mathbf{u}_\alpha^f(\mathbf{x}_T,z)\\ & \left.-\hat{k}_\alpha\left(\hat{k}_\alpha\hat{k}_\alpha\right):\iint d^2\mathbf{x}_T e^{-i\mathbf{K}_T\cdot\mathbf{x}_T}\frac{\delta\mu(\mathbf{x}_T,z)}{\lambda+2\mu}\frac{1}{ik_\alpha}\varepsilon_\alpha^f(\mathbf{x}_T,z)\right\},\end{aligned} \tag{2.26}$$

with $k_z^\alpha = +\gamma_\alpha$ for forescattering and $k_z^\alpha = -\gamma_\alpha$ for backscattering, and $\hat{k}_\alpha = (\mathbf{K}_T, k_z^\alpha)/k_\alpha$. Note that we replaced ρ_0, λ_0, μ_0 in denominators by $\rho = \rho_0 + \delta\rho$, $\lambda = \lambda_0 + \delta\lambda$ and $\mu = \mu_0 + \delta\mu$. This replacement is the result of asymptotic matching between the Born approximation for large-angle scattering and the h-f asymptotic travel-time (phase) for forward propagation. It is proved (Wu and Wu, 2003a) that with this replacement (asymptotic matching), the phase-shift in the exact forward direction is accurate and the phase error for small angles is reduced compared with the Born approximation. In the meanwhile, the phase error for large angle scattering is much smaller than that of the phase screen approximation.

In (2.26) $\mathbf{u}_\alpha^f(\mathbf{x}_T,z)$, $\nabla\cdot\mathbf{u}_\alpha^f(\mathbf{x}_T,z)$ and $\varepsilon_\alpha^f(\mathbf{x}_T,z)$ can be calculated by a dual domain method:

$$\mathbf{u}_\alpha^f(\mathbf{x}_T,z) = \frac{1}{4\pi}\iint d^2\mathbf{K}'_T e^{i\mathbf{K}'_T\cdot\mathbf{x}_T} u_\alpha^0(\mathbf{K}'_T)e^{i\gamma'_\alpha(z-z')} \tag{2.27}$$

$$\frac{1}{ik_\alpha}\nabla\cdot\mathbf{u}^f_\alpha(\mathbf{x}_T,z)=\frac{1}{4\pi}\iint d^2\mathbf{K}'_T e^{i\mathbf{K}'_T\cdot\mathbf{x}_T}\hat{k}'_\alpha\cdot\mathbf{u}^0_\alpha(\mathbf{K}'_T)e^{i\gamma'_\alpha(z-z')}, \tag{2.28}$$

$$\begin{aligned}\frac{1}{ik_\alpha}\varepsilon^f_\alpha(\mathbf{x}_T,z)&=\frac{1}{4\pi}\iint d^2\mathbf{K}'_T e^{i\mathbf{K}'_T\cdot\mathbf{x}_T}\frac{1}{2}[\hat{k}'_\alpha\mathbf{u}^0_\alpha(\mathbf{K}'_T)+\hat{k}'_\alpha\mathbf{U}^0_\alpha(\mathbf{K}'_T)]e^{i\gamma'_\alpha(z-z')}\\&=\frac{1}{4\pi}\iint d^2\mathbf{K}'_T e^{i\mathbf{K}'_T\cdot\mathbf{x}_T}\hat{k}'_\alpha\hat{k}'_\alpha u^0_\alpha(\mathbf{K}'_T)e^{i\gamma'_\alpha(z-z')},\end{aligned} \tag{2.29}$$

where $u^0_\alpha(\mathbf{K}'_T)=\left|\mathbf{u}^0_\alpha(\mathbf{K}'_T)\right|$ and $\hat{k}'_\alpha=(\mathbf{K}'_T,\gamma'_\alpha)/k_\alpha$.

For P to S scattering:

$$\begin{aligned}\mathbf{U}^{PS}(\mathbf{K}_T,z^*)=\frac{ik^2_\beta}{2\gamma_\beta}\int_{z'}^{z_1}dze^{ik^\beta_z(z^*-z)}\Big\{&(\mathbf{I}-\hat{k}_\beta\hat{k}_\beta)\cdot\iint d^2\mathbf{x}_T e^{-i\mathbf{K}_T\cdot\mathbf{x}_T}\frac{\delta\rho(\mathbf{x}_T,z)}{\rho}\mathbf{u}^f_\alpha(\mathbf{x}_T,z)\\&-(\mathbf{I}-\hat{k}_\beta\hat{k}_\beta)\cdot\left[\hat{k}_\beta\cdot\iint d^2\mathbf{x}_T e^{-i\mathbf{K}_T\cdot\mathbf{x}_T}2\frac{\delta\mu(\mathbf{x}_T,z)}{\mu}\frac{1}{ik_\beta}\varepsilon^f_\alpha(\mathbf{x}_T,z)\right]\Big\},\end{aligned} \tag{2.30}$$

where $\hat{k}_\beta=(\mathbf{K}_T,k^\beta_z)/k_\beta$.

For S to P scattering:

$$\begin{aligned}\mathbf{U}^{SP}(\mathbf{K}_T,z^*)=\frac{ik^2_\alpha}{2\gamma_\alpha}\int_{z'}^{z_1}dze^{ik^\alpha_z(z^*-z)}\Big\{&\hat{k}_\alpha\hat{k}_\alpha\cdot\iint d^2\mathbf{x}_T e^{-i\mathbf{K}_T\cdot\mathbf{x}_T}\frac{\delta\rho(\mathbf{x}_T,z)}{\rho}\mathbf{u}^f_\beta(\mathbf{x}_T,z)\\&-\left(\frac{k_\alpha}{k_\beta}\right)\hat{k}_\alpha\left(\hat{k}_\alpha\hat{k}_\alpha\right):\iint d^2\mathbf{x}_T e^{-i\mathbf{K}_T\cdot\mathbf{x}_T}2\frac{\delta\mu(\mathbf{x}_T,z)}{\mu}\frac{1}{ik_\beta}\varepsilon^f_\beta(\mathbf{x}_T,z)\Big\}.\end{aligned} \tag{2.31}$$

For S to S scattering:

$$\begin{aligned}\mathbf{U}^{SS}(\mathbf{K}_T,z^*)=\frac{ik^2_\beta}{2\gamma_\beta}\int_{z'}^{z_1}dze^{ik^\beta_z(z^*-z)}\Big\{&(\mathbf{I}-\hat{k}_\beta\hat{k}_\beta)\cdot\iint d^2\mathbf{x}_T e^{-i\mathbf{K}_T\cdot\mathbf{x}_T}\frac{\delta\rho(\mathbf{x}_T,z)}{\rho}\mathbf{u}^f_\beta(\mathbf{x}_T,z)\\&-(\mathbf{I}-\hat{k}_\beta\hat{k}_\beta)\cdot\left[\hat{k}_\beta\cdot\iint d^2\mathbf{x}_T e^{-i\mathbf{K}_T\cdot\mathbf{x}_T}2\frac{\delta\mu(\mathbf{x}_T,z)}{\mu}\frac{1}{ik_\beta}\varepsilon^f_\beta(\mathbf{x}_T,z)\right]\Big\}.\end{aligned} \tag{2.32}$$

In equations (2.31) and (2.32), $\mathbf{u}^f_\beta(\mathbf{x}_T,z)$ and $\varepsilon^f_\beta\ (\mathbf{x}_T,z)$ can be calculated by

$$\mathbf{u}^f_\beta(\mathbf{x}_T,z)=\frac{1}{4\pi^2}\iint d^2\mathbf{K}'_T e^{i\mathbf{K}'_T\cdot\mathbf{x}_T}\mathbf{u}^0_\beta(\mathbf{K}'_T)e^{i\gamma'_\beta(z-z')} \tag{2.33}$$

$$\frac{1}{ik_\beta}\varepsilon^f_\beta(\mathbf{x}_T,z)=\frac{1}{4\pi^2}\iint d^2\mathbf{K}'_T e^{i\mathbf{K}'_T\cdot\mathbf{x}_T}\frac{1}{2}\left[\hat{k}'_\beta\mathbf{u}^0_\beta(\mathbf{K}'_T)+\mathbf{u}^0_\beta(\mathbf{K}'_T)\hat{k}'_\beta\right]e^{i\gamma'_\beta(z-z')}, \tag{2.34}$$

where $\hat{k}'_\beta=(\mathbf{K}'_T,\gamma'_\beta)/k_\beta$.

2.3.2 The Thin Slab Approximation

From equations (2.26) to (2.34), we see that the leading-order interaction between incident fields and heterogeneities is expressed in three-dimensional volume integrals. Also the scattered and incident wavenumbers are coupled with each other. So the computation of these equations is still intensive. In this section, parts of the integration over z in the equations are analytically estimated. Assume that the slab for each marching step is thin enough that the parameters (velocity and density) can be approximately taken as invariant along z, the integration with respect to z in equation (2.26) can be calculated as

$$\int_{z'}^{z_1} dz e^{ik_z^\alpha(z^*-z)+i\gamma'_\alpha(z-z')} = \begin{cases} \Delta z e^{i(\gamma_\alpha+\gamma'_\alpha)\Delta z/2}\operatorname{sinc}\left[\frac{\gamma_\alpha-\gamma'_\alpha}{2}\Delta z\right] & \text{for forescattering } (z^*=z_1), \\ \Delta z e^{i(\gamma_\alpha+\gamma'_\alpha)\Delta z/2}\operatorname{sinc}\left[\frac{\gamma_\alpha+\gamma'_\alpha}{2}\Delta z\right] & \text{for backscattering } (z^*=z'). \end{cases} \tag{2.35}$$

We see that the integration over z has been done analytically; however, γ_α and γ'_α are still coupled, which prevents the fast computation of the thin-slab method. To decouple γ_α and γ'_α, we neglect the angular variation of amplitude factors but keep the phase information untouched by taking the approximation $\gamma_\alpha = \gamma'_\alpha = k_\alpha$ for the amplitude factors in equation (2.35). This assumption is valid for the case where the small-angle scattering is dominant, and therefore the direction of the scattered waves are not far from the incident direction. Under this approximation, equation (2.35) becomes

$$\int_{z'}^{z_1} dz e^{ik_z^\alpha(z^*-z)+i\gamma'_\alpha(z-z')} \approx \begin{cases} \Delta z e^{i(\gamma_\alpha+\gamma'_\alpha)\Delta z/2} & \text{for forescattering } (z^*=z_1), \\ \Delta z e^{i(\gamma_\alpha+\gamma'_\alpha)\Delta z/2}\operatorname{sinc}(k_\alpha\Delta z) & \text{for backscattering } (z^*=z'). \end{cases} \tag{2.36}$$

For the scattered fields P-S, S-P and S-S, similar approximations can be obtained as follows. For P-S or S-P scattering,

$$\begin{aligned} &\int_{z'}^{z_1} dz e^{ik_z^\beta(z^*-z)+i\gamma'_\alpha(z-z')} \\ &\approx \begin{cases} \Delta z e^{i(\gamma'_\alpha+\gamma_\beta)\Delta z/2}\operatorname{sinc}[(k_\alpha-k_\beta)\Delta z/2] & \text{for forescattering } (z^*=z_1), \\ \Delta z e^{i(\gamma'_\alpha+\gamma_\beta)\Delta z/2}\operatorname{sinc}[(k_\alpha+k_\beta)\Delta z/2] & \text{for backscattering } (z^*=z'). \end{cases} \end{aligned} \tag{2.37}$$

For S-S scattering,

$$\int_{z'}^{z_1} dz e^{ik_z^\beta(z^*-z)+i\gamma'_\beta(z-z')} \approx \begin{cases} \Delta z e^{i(\gamma_\beta+\gamma'_\beta)\Delta z/2} & \text{for forescattering } (z^*=z_1), \\ \Delta z e^{i(\gamma_\beta+\gamma'_\beta)\Delta z/2}\operatorname{sinc}(k_\beta\Delta z) & \text{for backscattering } (z^*=z'). \end{cases} \tag{2.38}$$

After integration over z, the integration over transverse plane $\mathbf{x}_T$ in equations (2.26) – (2.34) can be carried out by the FFT. In order to further expedite the computation, we can group the scattered field equations (2.26) to (2.34) into $\mathbf{U}^P(\mathbf{K}_T,z^*) = \mathbf{U}^{PP}(\mathbf{K}_T,z^*) +$

$\mathbf{U}^{SP}(\mathbf{K}_T, z^*)$, and $\mathbf{U}^S(\mathbf{K}_T, z^*) = \mathbf{U}^{PS}(\mathbf{K}_T, z^*) + \mathbf{U}^{SS}(\mathbf{K}_T, z^*)$, i.e.

$$\begin{aligned}\mathbf{U}^P(\mathbf{K}_T, z^*) = {} & \frac{ik_\alpha^2}{2\gamma_\alpha} e^{i\gamma_\alpha \Delta z/2} \Delta z \hat{k}_\alpha \left\{ \hat{k}_\alpha \cdot \iint d^2\mathbf{x}_T e^{-i\mathbf{K}_T\cdot\mathbf{x}_T} \frac{\delta\rho(\mathbf{x}_T)}{\rho} \left[\eta^{PP} \mathbf{u}_\alpha^f(\mathbf{x}_T) + \eta^{SP} \mathbf{u}_\beta^f(\mathbf{x}_T) \right] \right. \\ & - \iint d^2\mathbf{x}_T e^{-i\mathbf{K}_T\cdot\mathbf{x}_T} \frac{\delta\lambda(\mathbf{x}_T)}{\lambda + 2\mu} \frac{1}{ik_\alpha} \nabla \cdot \left[\eta^{PP} \mathbf{u}_\alpha^f(\mathbf{x}_T) \right] \\ & - \left(\hat{k}_\alpha \hat{k}_\alpha \right) : \iint d^2\mathbf{x}_T e^{-i\mathbf{K}_T\cdot\mathbf{x}_T} \frac{2\delta\mu(\mathbf{x}_T)}{\lambda + 2\mu} \frac{1}{ik_\alpha} \left[\eta^{PP} \varepsilon_\alpha^f(\mathbf{x}_T) \right. \\ & \left. \left. + \eta^{SP} \varepsilon_\beta^f(\mathbf{x}_T) \right] \right\},\end{aligned} \tag{2.39}$$

$$\begin{aligned}\mathbf{U}^S(\mathbf{K}_T, z^*) = {} & \frac{ik_\beta^2}{2\gamma_\beta} e^{i\gamma_\beta \Delta z/2} \Delta z (\mathbf{I} - \hat{k}_\beta \hat{k}_\beta) \cdot \left\{ \iint d^2\mathbf{x}_T e^{-i\mathbf{K}_T\cdot\mathbf{x}_T} \frac{\delta\rho(\mathbf{x}_T)}{\rho} \left[\eta^{PS} \mathbf{u}_\alpha^f(\mathbf{x}_T) \right. \right. \\ & \left. + \eta^{SS} \mathbf{u}_\beta^f(\mathbf{x}_T) \right] - \hat{k}_\beta \cdot \iint d^2\mathbf{x}_T e^{-i\mathbf{K}_T\cdot\mathbf{x}_T} \frac{2\delta\mu(\mathbf{x}_T)}{\mu} \frac{1}{ik_\beta} \left[\eta^{PS} \varepsilon_\alpha^f(\mathbf{x}_T) \right. \\ & \left. \left. + \eta^{SS} \varepsilon_\beta^f(\mathbf{x}_T) \right] \right\},\end{aligned} \tag{2.40}$$

where z^* ($z^* = z'$ or $z^* = z_1$) indicates the position of the receiver plane. The modulation factors η^{PP}, η^{SP}, η^{PS} and η^{SS} are

$$\eta^{PP} = \begin{cases} 1 & \text{for forescattering} \\ \operatorname{sinc}(k_\alpha \Delta z) & \text{for backscattering} \end{cases}, \tag{2.41}$$

$$\eta^{PS} = \eta^{SP} = \begin{cases} \operatorname{sinc}[(k_\alpha - k_\beta)\Delta z/2] & \text{for forescattering} \\ \operatorname{sinc}[(k_\alpha + k_\beta)\Delta z/2] & \text{for backscattering} \end{cases}, \tag{2.42}$$

$$\eta^{SS} = \begin{cases} 1 & \text{for forescattering} \\ \operatorname{sinc}(k_\beta \Delta z) & \text{for backscattering} \end{cases}. \tag{2.43}$$

Note that the factors $e^{i\gamma_\alpha(z-z')}$ and $e^{i\gamma_\beta(z-z')}$ have been replaced by $e^{i\gamma_\alpha \Delta z/2}$ and $e^{i\gamma_\beta \Delta z/2}$ for calculating the background fields. The phase matching (asymptotic matching) has been applied in equations (2.39) and (2.40).

2.3.3 Small-Angle Approximation and the Screen Propagator

In the thin-slab method (2.39) and (2.40), we need to calculate the incident displacement and strain fields in each marching step. We can further simplify the calculation by using the complex-screen approximation which is based on the small angle approximation. The approximation needs to be made in the wavenumber domain. The wavenumber domain formulation can be obtained by substituting equations (2.29), (2.34) into equations (2.26) and (2.30)-(2.32).

$$\mathbf{U}^{PP}(\mathbf{K}_T,\mathbf{K}_T') = \frac{i}{2\gamma_\alpha}k_\alpha^2 u_0^P \hat{k}_\alpha \left[(\hat{k}_\alpha\cdot\hat{k}_\alpha')\frac{\delta\rho(\tilde{\mathbf{k}})}{\rho} - \frac{\delta\lambda(\tilde{\mathbf{k}})}{\lambda+2\mu} - (\hat{k}_\alpha\cdot\hat{k}_\alpha')^2\frac{2\delta\mu(\tilde{\mathbf{k}})}{\lambda+2\mu}\right], \tag{2.44}$$

$$\mathbf{U}^{PS}(\mathbf{K}_T,\mathbf{K}_T') = \frac{i}{2\gamma_\beta}k_\beta^2\left(\mathbf{I}-\hat{k}_\beta\hat{k}_\beta\right)\cdot\mathbf{u}_0^P\left[\frac{\delta\rho(\tilde{\mathbf{k}})}{\rho} - 2\frac{\beta_0}{\alpha_0}\left(\hat{k}_\alpha\cdot\hat{k}_\beta'\right)\frac{\delta\mu(\tilde{\mathbf{k}})}{\mu}\right], \tag{2.45}$$

$$\mathbf{U}^{SP}(\mathbf{K}_T,\mathbf{K}_T') = \frac{i}{2\gamma_\alpha}k_\alpha^2\left(\mathbf{u}_0^S\cdot\hat{k}_\alpha\right)\hat{k}_\alpha\left[\frac{\delta\rho(\tilde{\mathbf{k}})}{\rho} - 2\frac{\beta_0}{\alpha_0}\left(\hat{k}_\beta\cdot\hat{k}_\alpha'\right)\frac{\delta\mu(\tilde{\mathbf{k}})}{\mu}\right], \tag{2.46}$$

$$\mathbf{U}^{SS}(\mathbf{K}_T,\mathbf{K}_T') = \frac{i}{2\gamma_\beta}k_\beta^2\left(\mathbf{I}-\hat{k}_\beta\hat{k}_\beta\right)\cdot\left\{\mathbf{u}_0^S\left[\frac{\delta\rho(\tilde{\mathbf{k}})}{\rho} - (\hat{k}_\beta\cdot\hat{k}_\beta')\frac{\delta\mu(\tilde{\mathbf{k}})}{\mu}\right]\right.$$
$$\left. -\hat{k}_\beta'\left(\mathbf{u}_0^S\cdot\hat{k}_\beta\right)\frac{\delta\mu(\tilde{\mathbf{k}})}{\mu}\right\}, \tag{2.47}$$

where $\mathbf{u}_0^P$ is the spectral field of the incident P wave, and $\delta\rho(\tilde{\mathbf{k}})$, $\delta\lambda(\tilde{\mathbf{k}})$ and $\delta\mu(\tilde{\mathbf{k}})$ are the three-dimensional Fourier transforms of medium perturbations, and $\tilde{\mathbf{k}} = \hat{k} - \hat{k}'$ is the exchange wavenumber with $\hat{k}'$ and $\hat{k}$ as incident and scattering wavenumber vectors, respectively.

Under small angle approximation, both incoming and outgoing wavenumbers have small transversal components K_T compared to the longitudinal component γ (vertical wavenumber). For an isotropic medium, the scattered fields can be expressed as (Xie and Wu, 2001)

$$\mathbf{U}_f^{PP}(\mathbf{K}_T,z_0) = -ik_\alpha\Delta z\hat{k}_\alpha\eta_f^{PP}\iint d\mathbf{x}_T' e^{-i\mathbf{K}_T\cdot\mathbf{x}_T'}u_0^P(\mathbf{x}_T',z_0)\frac{\delta\alpha(\mathbf{x}_T')}{\alpha_0}, \tag{2.48}$$

$$\mathbf{U}_f^{PS}(\mathbf{K}_T,z_0) = -ik_\beta\Delta z\eta_f^{PS}\hat{k}_\beta\times\left\{\hat{k}_\beta\times\iint d\mathbf{x}_T' e^{-i\mathbf{K}_T\cdot\mathbf{x}_T'}\mathbf{u}_0^P(\mathbf{x}_T',z_0)\left[\left(\frac{\beta_0}{\alpha_0}-\frac{1}{2}\right)\frac{\delta\rho(\mathbf{x}_T')}{\rho_0}\right.\right.$$
$$\left.\left. +2\left(\frac{\beta_0}{\alpha_0}\right)\frac{\delta\beta(\mathbf{x}_T')}{\beta_0}\right]\right\}, \tag{2.49}$$

$$\mathbf{U}_f^{SP}(\mathbf{K}_T,z_0) = -ik_\alpha\Delta z\eta_f^{SP}\hat{k}_\alpha\left\{\hat{k}_\alpha\cdot\iint d\mathbf{x}_T' e^{-i\mathbf{K}_T\cdot\mathbf{x}_T'}\mathbf{u}_0^S(\mathbf{x}_T',z_0)\left[\left(\frac{\beta_0}{\alpha_0}-\frac{1}{2}\right)\frac{\delta\rho(\mathbf{x}_T')}{\rho_0}\right.\right.$$
$$\left.\left. +2\left(\frac{\beta_0}{\alpha_0}\right)\frac{\delta\beta(\mathbf{x}_T')}{\beta_0}\right]\right\}, \tag{2.50}$$

$$\mathbf{U}_f^{SS}(\mathbf{K}_T,z_0) = -ik_\beta\Delta z\eta_f^{SS}\hat{k}_\beta\times\left\{\hat{k}_\beta\times\iint d\mathbf{x}_T' e^{-i\mathbf{K}_T\cdot\mathbf{x}_T'}\mathbf{u}_0^S(\mathbf{x}_T',z_0)\frac{\delta\beta(\mathbf{x}_T')}{\beta_0}\right\}, \tag{2.51}$$

$$\mathbf{U}_b^{PP}(\mathbf{K}_T,z_0) = -ik_\alpha \Delta z \hat{k}_\alpha \eta_b^{PP} \iint d\mathbf{x}_T' e^{-i\mathbf{K}_T\cdot\mathbf{x}_T'} u_0^P(\mathbf{x}_T',z_0) \frac{\delta Z_\alpha(\mathbf{x}_T')}{Z_{\alpha 0}}, \tag{2.52}$$

$$\mathbf{U}_b^{PS}(\mathbf{K}_T,z_0) = ik_\beta \Delta z \eta_b^{PS} \hat{k}_\beta \times \left\{ \hat{k}_\beta \times \iint d\mathbf{x}_T' e^{-i\mathbf{K}_T\cdot\mathbf{x}_T'} \mathbf{u}_0^P(\mathbf{x}_T',z_0) \left[\left(\frac{\beta_0}{\alpha_0} + \frac{1}{2} \right) \frac{\delta\rho(\mathbf{x}_T')}{\rho_0} + 2\left(\frac{\beta_0}{\alpha_0} \right) \frac{\delta\beta(\mathbf{x}_T')}{\beta_0} \right] \right\}, \tag{2.53}$$

$$\mathbf{U}_b^{SP}(\mathbf{K}_T,z_0) = ik_\alpha \Delta z \eta_b^{SP} \hat{k}_\alpha \left\{ \hat{k}_\alpha \cdot \iint d\mathbf{x}_T' e^{-i\mathbf{K}_T\cdot\mathbf{x}_T'} \mathbf{u}_0^S(\mathbf{x}_T',z_0) \left[\left(\frac{\beta_0}{\alpha_0} + \frac{1}{2} \right) \frac{\delta\rho(\mathbf{x}_T')}{\rho_0} + 2\left(\frac{\beta_0}{\alpha_0} \right) \frac{\delta\beta(\mathbf{x}_T')}{\beta_0} \right] \right\}, \tag{2.54}$$

$$\mathbf{U}_b^{SS}(\mathbf{K}_T,z_0) = ik_\beta \Delta z \eta_b^{SS} \hat{k}_\beta \times \left\{ \hat{k}_\beta \times \iint d\mathbf{x}_T' e^{-i\mathbf{K}_T\cdot\mathbf{x}_T'} \mathbf{u}_0^S(\mathbf{x}_T',z_0) \frac{\delta Z_\beta(\mathbf{x}_T')}{Z_{\beta 0}} \right\}, \tag{2.55}$$

where $Z_\alpha = \rho\alpha$ and $Z_\beta = \rho\beta$ are P and S wave impedances and $Z_{\alpha 0} = \rho_0\alpha_0$ and $Z_{\beta 0} = \rho_0\beta_0$ are their background values. η^{PP}, η^{PS}, η^{SP} and η^{SS} can be calculated with equations (2.41)-(2.43). The subscripts f and b denote the forward and backward scattered waves, respectively. From equations (2.48)-(2.55), it can be seen that transmitted waves $\mathbf{U}_f^{PP}$ and $\mathbf{U}_f^{SS}$ are controlled by P- and S-wave velocity perturbations and the reflected waves $\mathbf{U}_b^{PP}$ and $\mathbf{U}_b^{SS}$ are controlled by impedance perturbations.

In the forward direction, there are both incidence waves and scattered waves. The forward propagated wavefield $\mathbf{u}_f$ can be obtained as follows:

$$\mathbf{u}_f(\mathbf{x}_T,z_1) = \frac{1}{4\pi^2} \int d\mathbf{K}_T' [\mathbf{u}_f^P(\mathbf{K}_T',z_1) + \mathbf{u}_f^S(\mathbf{K}_T',z_1)] e^{i\mathbf{K}_T'\cdot\mathbf{x}_T}, \tag{2.56}$$

where

$$\mathbf{u}_f^P(\mathbf{K}_T',z_1) = e^{i\gamma_\alpha|z_1-z_0|} \left[\mathbf{u}_0^P(\mathbf{K}_T',z_0) + \mathbf{U}_f^{PP}(\mathbf{K}_T',z_0) + \mathbf{U}_f^{SP}(\mathbf{K}_T',z_0) \right], \tag{2.57}$$

$$\mathbf{u}_f^S(\mathbf{K}_T',z_1) = e^{i\gamma_\beta|z_1-z_0|} \left[\mathbf{u}_0^S(\mathbf{K}_T',z_0) + \mathbf{U}_f^{SS}(\mathbf{K}_T',z_0) + \mathbf{U}_f^{PS}(\mathbf{K}_T',z_0) \right]. \tag{2.58}$$

The backward propagated wavefield $\mathbf{u}_b$ is composed of pure scattered waves and can be obtained as follows:

$$\mathbf{u}_b(\mathbf{x}_T,z_1) = \frac{1}{4\pi^2} \int d\mathbf{K}_T' [\mathbf{u}_b^P(\mathbf{K}_T',z_1) + \mathbf{u}_b^S(\mathbf{K}_T',z_1)] e^{i\mathbf{K}_T'\cdot\mathbf{x}_T}, \tag{2.59}$$

where

$$\mathbf{u}_b^P(\mathbf{K}_T',z_1) = \left[\mathbf{U}_b^{PP}(\mathbf{K}_T',z_0) + \mathbf{U}_b^{SP}(\mathbf{K}_T',z_0) \right], \tag{2.60}$$

$$\mathbf{u}_b^S(\mathbf{K}_T',z_1) = \left[\mathbf{U}_b^{SS}(\mathbf{K}_T',z_0) + \mathbf{U}_b^{PS}(\mathbf{K}_T',z_0) \right]. \tag{2.61}$$

In equations (2.56) and (2.59) $\mathbf{u}_f^P$ and $\mathbf{u}_f^S$ are forward propagated P and S waves at the exit side of the slab, and $\mathbf{u}_b^P$ and $\mathbf{u}_b^S$ are backscattered P and S waves at the entrance side of the slab. They are calculated by summing up the incidence waves $\mathbf{u}_0^P$ and $\mathbf{u}_0^S$, and scattered waves $\mathbf{U}_f^{PP}$, $\mathbf{U}_f^{SP}$, $\mathbf{U}_f^{SS}$, $\mathbf{U}_f^{PS}$, $\mathbf{U}_b^{PP}$, $\mathbf{u}_b^{SP}$, $\mathbf{U}_b^{SS}$ and $\mathbf{U}_b^{PS}$ using equations (2.57)-(2.58) and (2.60)-(2.61). These scattered fields can be calculated using equations (2.26)-(2.32) if thin slab approximation is adopted. In the latter case, the scattering patterns η can be calculated using either zero or first order approximations (for details see Wu et al., 2007).

2.3.4 Numerical Implementation

One-return treatment of the propagator

Equations (2.39) and (2.40) (for the thin-slab approximation), or (2.56) and (2.59) (for the screen approximation) provide the solution to calculate the response of a thin slice of the model to the incident wave. To calculate the response of the entire model to the incidence wave, we use the multiple forward scattering and single back scattering approximations described in Section 2.2. A marching method is adopted for updating the wavefield as illustrated in Figure 2.2. A *double-sweep* method is used to calculate the transmitted and reflected (backscattered) wavefield. We first calculate the down-going wavefield from the source to the bottom of model (down sweep). For each depth step, the interaction of the forward (here it is downward) field with the thin-slab is updated and propagated to the next depth using equations (2.56)-(2.58), and is used as the incidence for the next depth. In a marching way, all multiple forward scatterings are included in the down-going wave. At the same time, at each depth, the back scattered wavefields are calculated using equations (2.59)-(2.61) and recorded in the memory or disk. After the down-going propagation sweeps the entire model, we use the same propagator for the second time to calculate the up-going wave from the bottom to the top of the model (up sweep). At each depth, the up-going waves are updated by the thin-slab transmission operator. In addition, the previously recorded back scattered waves are retrieved at each depth and added to the up-going field. After the up-sweep of the entire model from the bottom to the top, we obtain all the back scattered waves under the one-return approximation. By using the marching algorithm in the double sweeps, all multiple forward scattered and single back scattered waves are included in the results. If we calculate and keep the records of the back scattered fields in the up-sweep too, then we can model the third and higher-order multiples in the subsequent sweeps.

Dual domain implementation

In calculating scattering waves, i.e., equations (2.26)-(2.32) or equations (2.48)-(2.55), interactions between the incident wave and the perturbations of elastic parameters are calculated in the space domain. However, the forward propagation of the wavefield in the background velocity is calculated in the wavenumber domain. The inverse Fourier transform in equations (2.56) and (2.59) transform the wavefield from wavenumber domain to the space domain, while the Fourier transform in equations (2.26)-(2.32) or equations (2.48)-(2.55) transfer the wavefield from space domain to the wavenumber domain. In both domains, the calculations are localized, forming a very efficient algorithm.

2.3.5 Elastic, Acoustic and Scalar Cases

The equations presented in the previous sections are for elastic wave case, where both P- and S-waves are multi-component vector waves. Under the thin-slab approximation, the interactions between the medium and waves involve tensor (strain field) operations. Due to the symmetric properties of these tensors, there are only 6 independent components for each tensor. If complex screen approximation is adopted, the interactions involve only vector operations. The equations for acoustic and scalar waves are not included in the current chapter. However, the interested readers are referred to the papers by *Wu* (1996) and *Wu et al.* (2007).

2.4 Applications of One-Return Propagators in Modeling, Imaging and Invcrsion

2.4.1 Applications to Modeling

First we show the applications of one-return approximation to the calculation of synthetic seismograms in elastic wave modeling. Figure 2.3a shows a 3D French model, where the parameters of the background medium are α_0 =3.6 km/s, β_0 =2.08 km/s and ρ_0 = 2.2 g/cm^3, and the grey colored structure has a perturbation of -10% for both P- and S-wave velocities. A Ricker wavelet with a dominant frequency of 10 Hz is used. For the 8th-order 3D elastic finite-difference method (Yoon, 1996), the spacing interval is 20 m. The actual grid size used is $250 \times 250 \times 250$ including 25 grids of absorbing boundary for each face of the model. Time interval used is 0.001 sec and 2500 time steps are calculated. The calculations took approximately 28 hours. For thin-slab method, the spacing interval used is 20 m in transversal plane, which is the same as that used for the finite-difference method. But a fine grid size of 5 m is used in propagation direction. We did the same calculation on the same machine using the thin-slab propagator. It took 2.7 hours. We see that the two methods agree with each other fairly well, but the thin-slab is about 10 times faster than the finite-difference method.

Shown in Figure 2.4 are primary waves calculated in the 2D acoustic SEG/EAGE salt model using the one-return method. These snapshots are composed of down- and up-going waves. We see that the primary reflections from major sediment/salt boundaries and from the interfaces in the sedimentary layers are properly modeled.

Figure 2.5 shows the comparisons of reflection coefficients calculated by the thin-slab method and the exact solutions (Reflection/Transmission theoretical calculation). The upper panel corresponds to P wave and the lower panel to S wave incidences respectively. The perturbation of the bottom layer is 20% with respect to the top layer for both P and S wave velocities. We can see the good agreement between these two methods.

Figure 2.6 gives one application example of the one-return modeling to the reservoir AVO (amplitude variation with offsets) calculations. It models the combined effects of random scattering and intrinsic attenuation in oil/gas reservoir AVO. In practice, the geo-

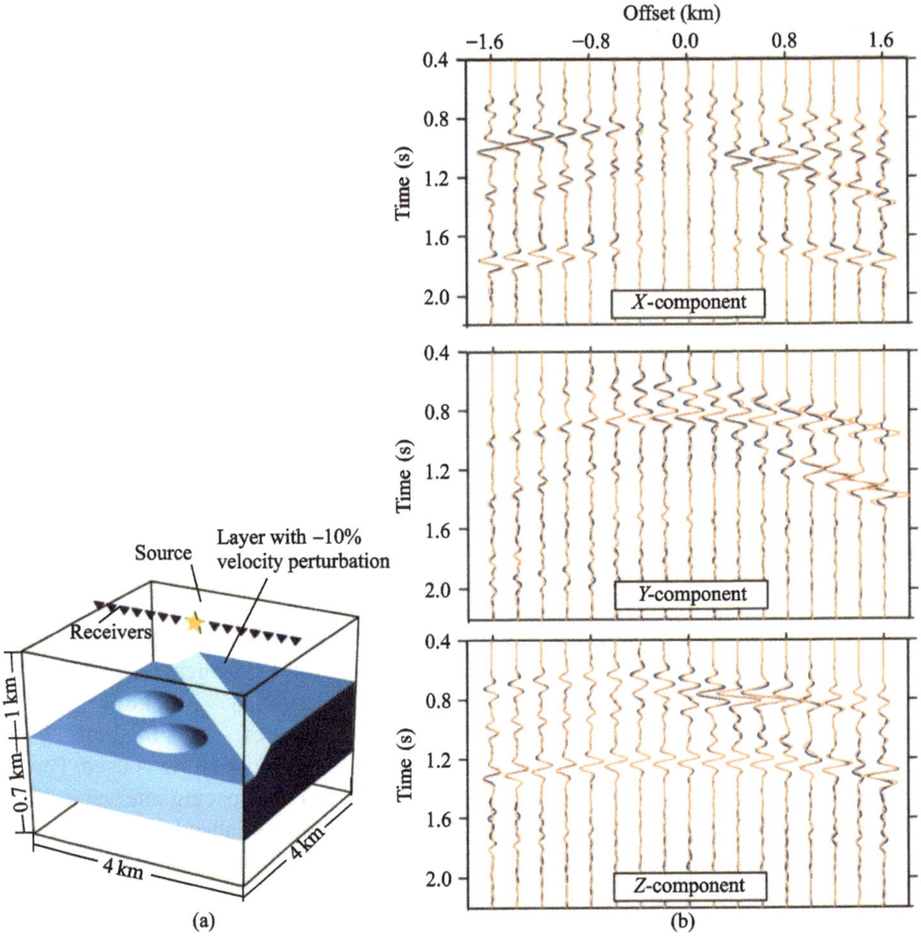

Fig. 2.3 Synthetic examples calculated in a 3D velocity model. Illustrated in (a) is the velocity model modified from the French model (French 1974), and in (b) are comparisons between the synthetic seismograms calculated using the one-return method (blue lines) and the finite-difference method (red lines).

logic models may contain arbitrary spatial variations in P- and S-wave quality factors, as well as density and velocities. The Q values vary from 10 to 150 for different reservoir rocks. The correlation lengths of the random field for perturbing Q and elastic parameters are the same. The rms values are 4% for elastic parameters and 25% for Q. The source and receiver array is located in shale and 1200 m away from the interface. The dotted lines in Figure 2.6 correspond to the homogeneous cases with constant Q's. The one-return modeling method can be applied to the complicated geological setting and the calculation is quite efficient. We see from the calculated AVO that intrinsic attenuation mainly affects the absolute reflected amplitudes and heterogeneities in Q's and elastic parameters affect local amplitude fluctuation with offset. AVO responses of the target interfaces have been significantly deformed due to both the heterogeneities and intrinsic attenuation.

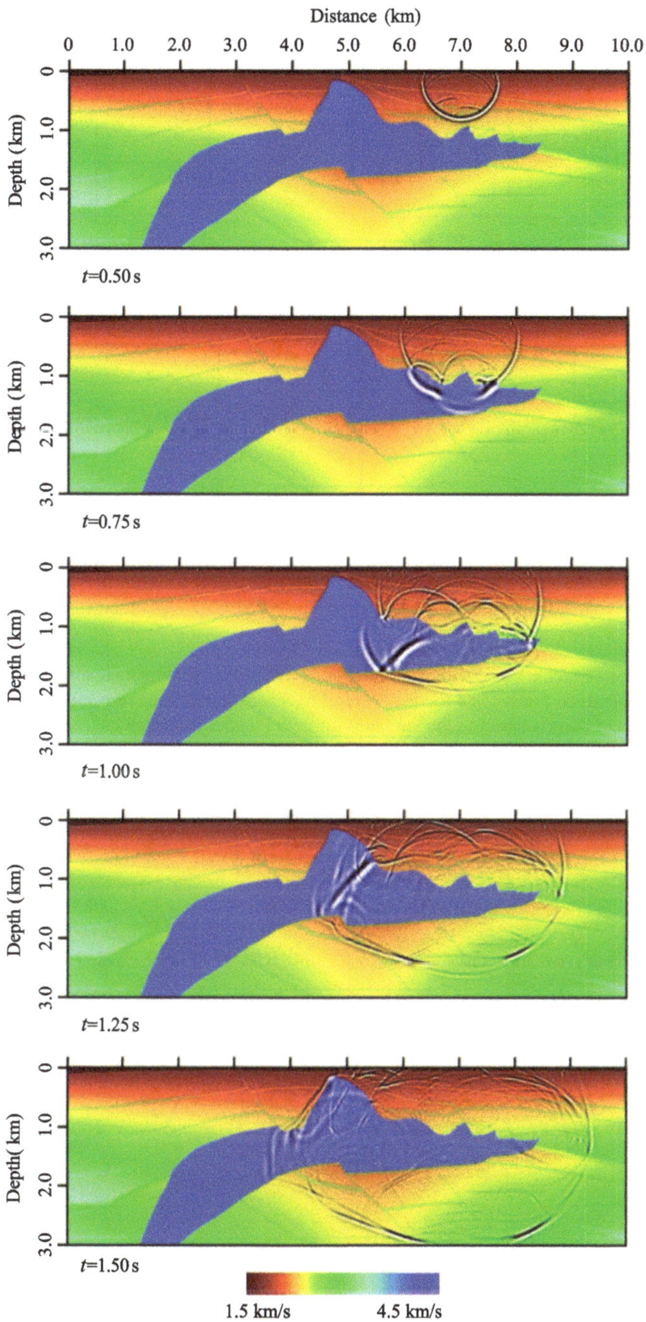

Fig. 2.4 Snapshots from the 2D acoustic SEG/EAGE salt model. The source is an 18 Hz Ricker wavelet. The snapshots are obtained by combining the up- and down-going waves.

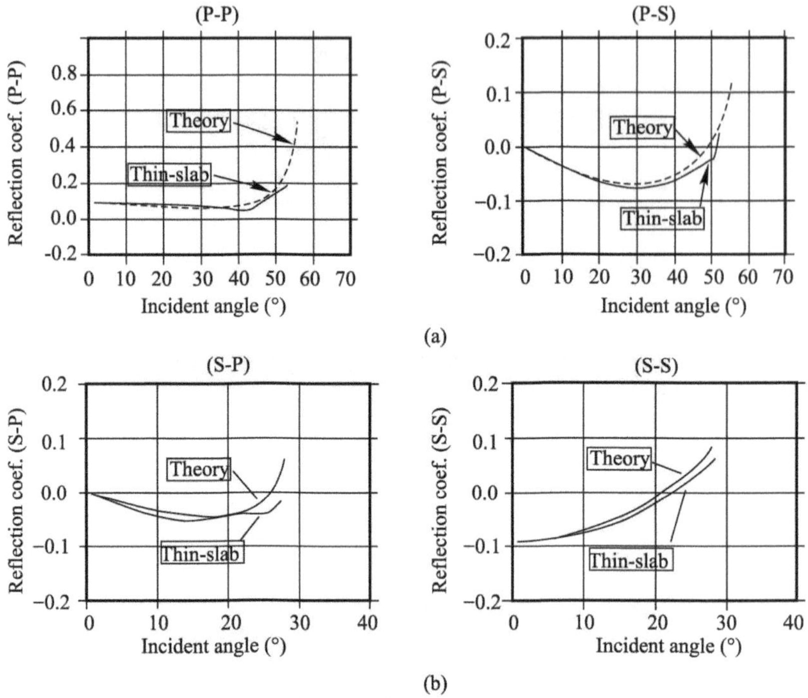

Fig. 2.5 Comparisons of reflection coefficients calculated by the thin-slab method and the exact solutions. The upper panel corresponds to P wave and the lower panel to S wave incidences respectively. The perturbation of the bottom layer is 20% with respect to the top layer for both P and S wave velocities.

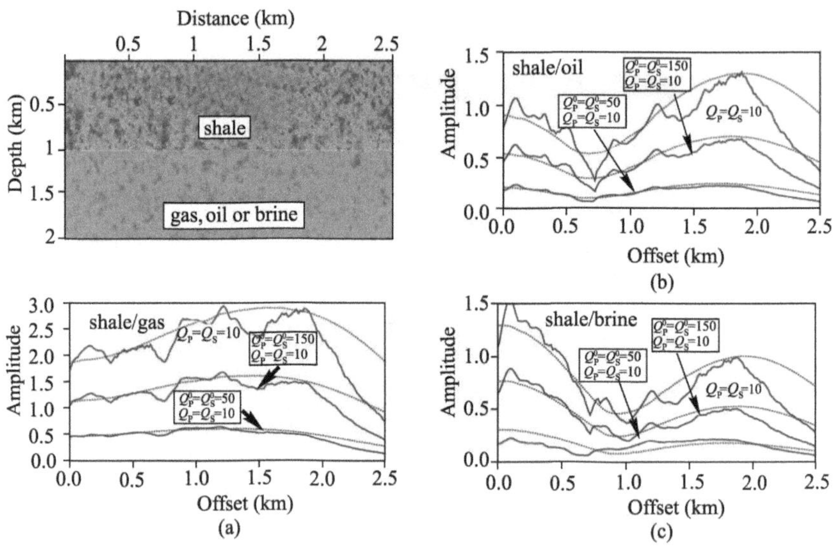

Fig. 2.6 An example showing the one-return modeling applied to the reservoir AVO (amplitude variation with offsets) calculation. The upper left panel shows a reservoir model. The formation is anelastic and heterogeneous. The rest three panels show AVO's for various kinds of interfaces: (a) shale/gas, (b) shale/oil, and (c) shale/brine. Three different background Q'_S (Q = infinity, 150, 50) are given to shale. The sand has background $Q_P = Q_S = 10$. The correlation lengths of the random field for perturbing Q and elastic parameters are the same. The rms values are 4% for elastic parameters and 25% for Q.

2.4.2 One-Return Propagators Used in Migration Imaging

Conventional one-way wave equation migration handles images of primary reflections, but neglects turning waves, multiples and other multiple reflections such as duplex waves. Duplex waves exist in the geologic structures with vertical features such as faults and flanks of salt bodies. In this case, primary reflections of steep reflectors may not be recorded in a limited acquisition aperture. Therefore, one-way wave equation migration cannot produce the image of such events. On the other hand, doubly reflected duplex waves may be recorded and should be taken into consideration in the migration. Based on the concept of multiple fore-scattering and single back-scattering (MFSB) a one-return wave equation migration that extrapolates waves with a single returning point can be applied to the duplex waves. The principle of one-return propagator with double sweeps has been shown in Section 2.2. Followed by a properly designed imaging condition, one-return migration produces the depth image of primary reflections, the same as conventional one-way wave equation migration, as well as the image from the contribution of turning waves and duplex waves that one-way wave equation fails to handle (Jin et al., 2006; Xu and Jin, 2006).

A conventional one-way propagator only calculates waves propagated downward (down-going or back propagated up-going waves), while the one-return propagator can handle waves reflected from a interface then propagate upward. Thus, it can generate different source and receiver wave modes including P_S^D, P_S^U, P_G^D and P_G^U (see Figure 2.7), where subscripts S and G denote source and receiver (geophone) side waves, and superscripts D and U denote down- and up-going waves. Accordingly, the corresponding partial images generated by these wave modes are (Jin et al., 2006)

$$I_{DD} = \sum_{\omega} P_S^D P_G^{D^*}, \tag{2.62}$$

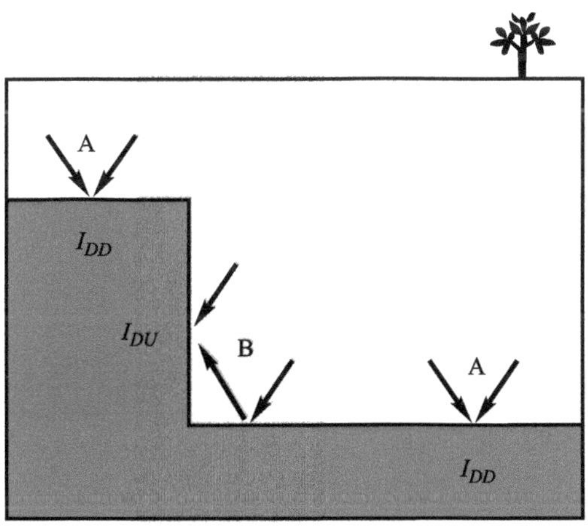

Fig. 2.7 Illustration of imaging conditions for different combinations of source and receiver waves. I_{DD} is the conventional imaging condition for down-/down-going waves of source and receiver sides. I_{DU} is an imaging condition for down-/up-going and up-/down-going waves of source and receiver sides.

$$I_{UU} = \sum_{\omega} P_S^U P_G^{U^*} \tag{2.63}$$

and

$$I_{DU} = \sum_{\omega} P_S^D P_G^{U^*} + \sum_{\omega} P_S^U P_G^{D^*}, \tag{2.64}$$

where "*" denotes applying complex conjugate. In these equations, I_{DD} is the image for down-/down-going waves for both source and receivers and it is the same as the traditional one-way wave equation migration (refer to Figure 2.7). I_{DU} is the image for down-/up- and up-/down-going waves for both source and receivers, and it handles the duplex waves. I_{UU} is the image for up-/up-going waves. This image condition actually handles multiples but will not be covered in this chapter. The final image can be obtained by summing up these partial image volumes. Here, we use synthetic examples to demonstrate the unique capability of this approach to image the steep and overhang structures.

The one-return propagator can be used to migrate duplex waves and prism waves. Shown in Figure 2.8a is a simple two-layer model with a vertical fault. The wave path labeled with "A" is the direct arrival; labeled with "B" is the primary reflection and labeled with "C" is a doubly bounced reflection against the vertical fault wall. The double bouncing reflection is also called duplex wave which is a special kind of prism wave. Illustrated in Figure 2.8b is the wavefield snapshot at 2.4 second simulated using finite-difference forward modeling. Different phases can be clearly seen from this snapshot. Shown in

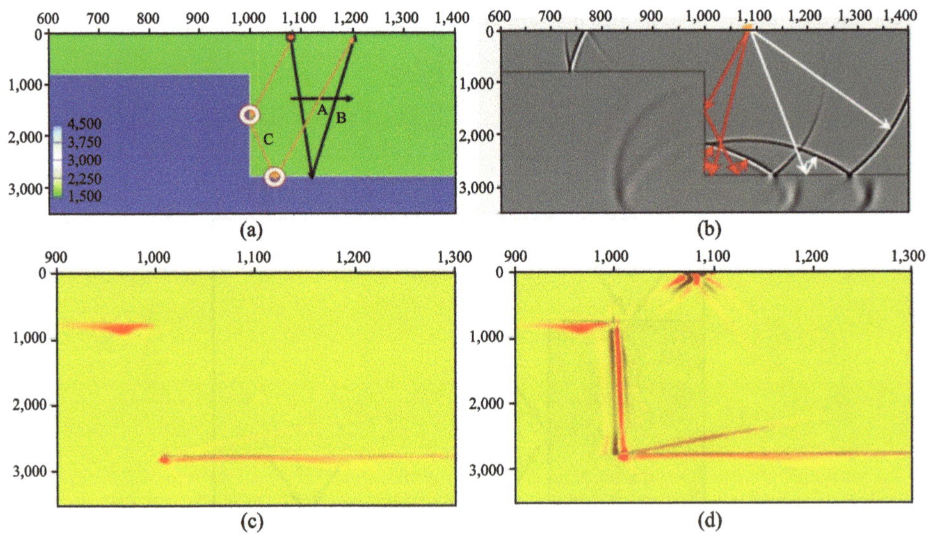

Fig. 2.8 Image vertical fault wall using one-return propagator. (a) A two-layer velocity model with a vertical fault wall. The direct arrival, the primary reflection and the doubly bounced reflection against the vertical fault wall are all illustrated in the figure. (b) The wavefield snapshot at 2.4 second. (c) Depth image using the conventional one-way wave-equation-based propagator. (d) Depth image using the one-return propagator. Note that the vertical fault wall has been properly reconstructed.

Figure 2.8c is the depth image which is calculated using the one-way wave equation migration and from a single shot record. The vertical fault structure is totally missing from this image. The depth image in Figure 2.8d is calculated using the one-return wave equation migration and from the same shot record. The vertical fault wall is fully reconstructed by duplex waves (Jin et al., 2006).

The one-return propagator can be also used to migrate turning waves that propagate beyond 90 degrees, thus to image overhang structure. Shown in Figure 2.9a is a typical salt model with a steeply dipping overhang. Note that primary reflections propagate through salt body while strong overturned waves exist in the sediments. Figure 2.9b illustrates the depth image generated using only one-way propagator. Due to the lack of contributions from turning waves, the image amplitude of salt overhang is very weak. Shown in Figure 2.9c is the partial image generated by the contributions from turning waves using one-return wave equation migration. Illustrated in Figure 2.9d is the final depth image by summing up images from both one-way propagator and one-return propagator. In this way, the image amplitude of overhang salt flank is significantly enhanced (Xu and Jin, 2006).

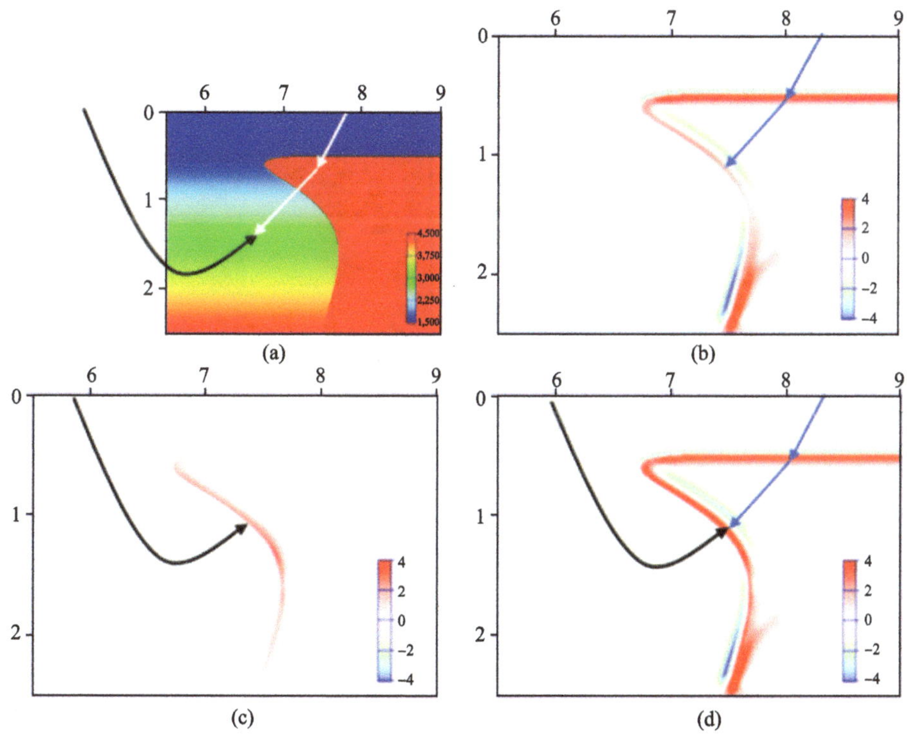

Fig. 2.9 Imaging the overhang salt flank using both one-way and one-return propagators. (a) A typical salt model with a steeply dipping salt overhang, noticing that primary reflections propagate through salt body while strong overturned waves exist in the sediments; (b) Depth image using one-way wave equation migration; (c) Partial image contributed by turning waves using one-return wave equation migration; (d) Final depth image by summing up both (b) and (c).

The next example is for the BP 2004 Molar benchmark dataset. Illustrated in Figure 2.10a is the central part of the velocity model which consists of a deeply rooted salt body with sedimentary inclusions which present a lot of challenges in depth imaging. Figure 2.10b shows depth image using one-way wave-equation-based propagator. The boundary of the deeply rooted salt body is not imaged. Shown in Figure 2.10c is the partial image calculated using the one-return wave equation migration. By including the contributions from both the prism waves and turning waves, the image of the steeply dipping salt boundary is clearly seen. Finally, shown in Figure 2.10d is the full image of one-return wave equation migration. The boundaries from both the top of the salt crown and the deeply rooted salt body are well reconstructed (Jin et al., 2006; Xu and Jin, 2006, 2007).

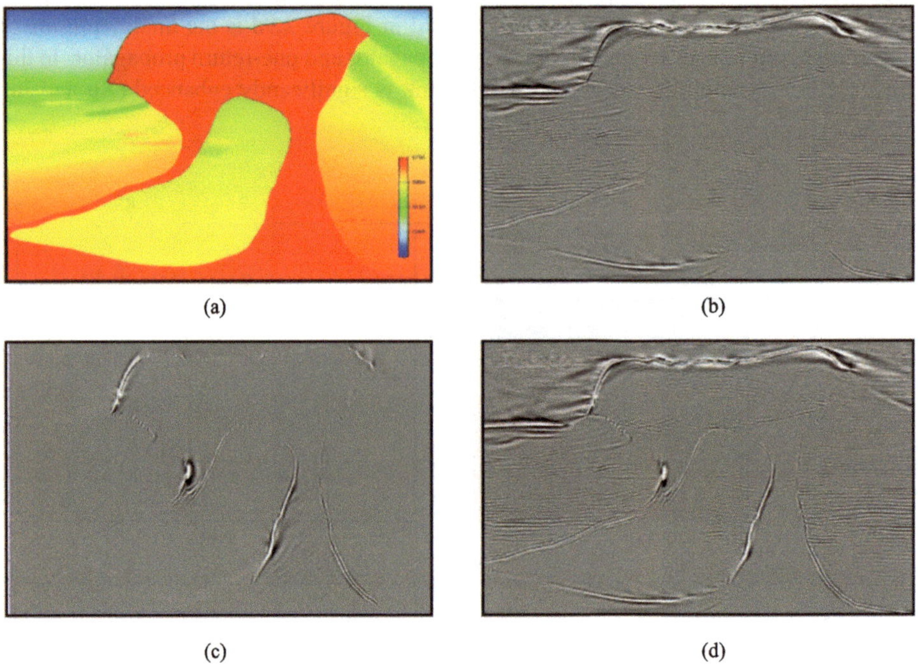

Fig. 2.10 One-return wave equation migration for the BP 2004 Molar benchmark dataset. (a) The central part of the velocity model; (b) Depth image using only the one-way wave-equation-based propagator; (c) Partial image using the one-return wave equation migration; (d) the full image by summing up both (b) and (c). (Synthetic data courtesy of BP).

2.4.3 Calculate Finite-Frequency Sensitivity Kernels Used in Velocity Inversion

In seismic migration, a correct velocity model plays an important role in obtaining high quality image. The process to update the velocity model based on the migration image is the migration velocity analysis (MVA), which is a special type of velocity tomography. Comparing to the conventional tomography, where the information regarding the velocity model error is extracted from the data domain (seismograms), the MVA extracts the information from inconsistency of the depth image. The most commonly used inconsistency is the residual moveout (RMO) in different types of common image gathers. To update the velocity model, the most important part in MVA is converting the observed residual moveout into velocity corrections and back-projecting them into the model space for velocity updating. In the past, this was dominated by the ray-tracing-based tomography method which assumes an infinitely high frequency. Recently, the sensitivity of finite-frequency signals to velocity model has been investigated by researchers working in different fields (Woodward, 1992; Vasco et al., 1995; Dahlen et al., 2000; Zhao, et al., 2000; Skarsoulis and Cornuelle, 2004; Spetzler and Snieder, 2004; Sava and Biondi, 2004; Jocker, et al., 2006; and Buursink and Routh, 2007; Fliedner et al., 2007). Finite-frequency sensitivity kernels have been calculated and used for solving many tomography problems with great success. The major obstacle that prevents this method from being used in migration velocity analysis is that these finite-frequency sensitivity kernels are mostly derived for transmitted waves (e.g., travel time delays or amplitude fluctuations in seismograms). On the contrary, the seismic migration extracts the information regarding the velocity error from the reflectivity image, which is related to the calculation of reflect waves. Based on the Born and Rytov approximations, Xie and Yang (2007, 2008) formulated the sensitivity kernel for the shot-record prestack depth migration. This sensitivity kernel relates the observed RMO in depth image to the velocity correction in the model. This is a wave-equation-based method which avoids many disadvantages of the ray-based tomography. The one-return method is ideal in calculating sensitivity kernels involving reflection. The single frequency travel time sensitivity kernel is composed of two parts, the down-going leg and the up-going leg.

$$K_D^F(\mathbf{r},\mathbf{r}_S,\mathbf{r}_I,\omega)=\operatorname{imag}\left[2k_0^2\frac{G_D(\mathbf{r};\mathbf{r}_S,\omega)\,G(\mathbf{r};\mathbf{r}_I,\omega)}{G_D(\mathbf{r}_I;\mathbf{r}_S,\omega)}\right], \tag{2.65}$$

$$K_U^F(\mathbf{r},\mathbf{r}_S,\mathbf{r}_I,\omega)=\operatorname{imag}\left[2k_0^2\frac{G_U^*(\mathbf{r};\mathbf{r}_S,\omega)\,G(\mathbf{r};\mathbf{r}_I,\omega)}{G_U^*(\mathbf{r}_I;\mathbf{r}_S,\omega)}\right], \tag{2.66}$$

where ω is frequency, $k_0=\omega/v(\mathbf{r})$ is the background wavenumber, $v(\mathbf{r})$ is the background velocity, G is the Green's function, subscripts U and D are for up- and down-going waves, $\mathbf{r}$ is the space location, $\mathbf{r}_S$ and $\mathbf{r}_I$ are the source and image locations, and imag $(\cdot)$ denotes taking imaginary part. In equations (2.65) and (2.66), the sensitivity kernels are composed by the correlation of two Green's functions, one radiated from the source and the other from the image point, normalized by the source wavefield at the image location.

The major difference between the up- and down-going legs is that in prestack depth migration, the up-going wave is obtained from the time-reversed reflect wave. The time reversal in the time domain is equivalent to the complex conjugate in the frequency domain. Thus complex conjugate is applied in the up-going leg. The broadband sensitivity kernel can be obtained by integrating the single frequency sensitivity kernel over frequencies

$$K_D^B(\mathbf{r},\mathbf{r}_S,\mathbf{r}_I) = \int \frac{A(\omega)}{\omega} K_D^F(\mathbf{r},\mathbf{r}_S,\mathbf{r}_I,\omega)d\omega, \tag{2.67}$$

$$K_U^B(\mathbf{r},\mathbf{r}_S,\mathbf{r}_I) = \int \frac{A(\omega)}{\omega} K_U^F(\mathbf{r},\mathbf{r}_S,\mathbf{r}_I,\omega)d\omega \tag{2.68}$$

where $A(\omega)$ is a factor related to the source spectrum (Xie and Yang, 2008). Finally, the sensitivity kernel for the residual moveout $R(\mathbf{r}_I,\mathbf{r}_S)$ can be obtained as follows:

$$R(\mathbf{r}_I,\mathbf{r}_S) = -\frac{v_0(\mathbf{r}_I)}{2\cos[\theta(\mathbf{r}_I,\mathbf{r}_S)]}\delta t(\mathbf{r}_I,\mathbf{r}_S) \tag{2.69}$$

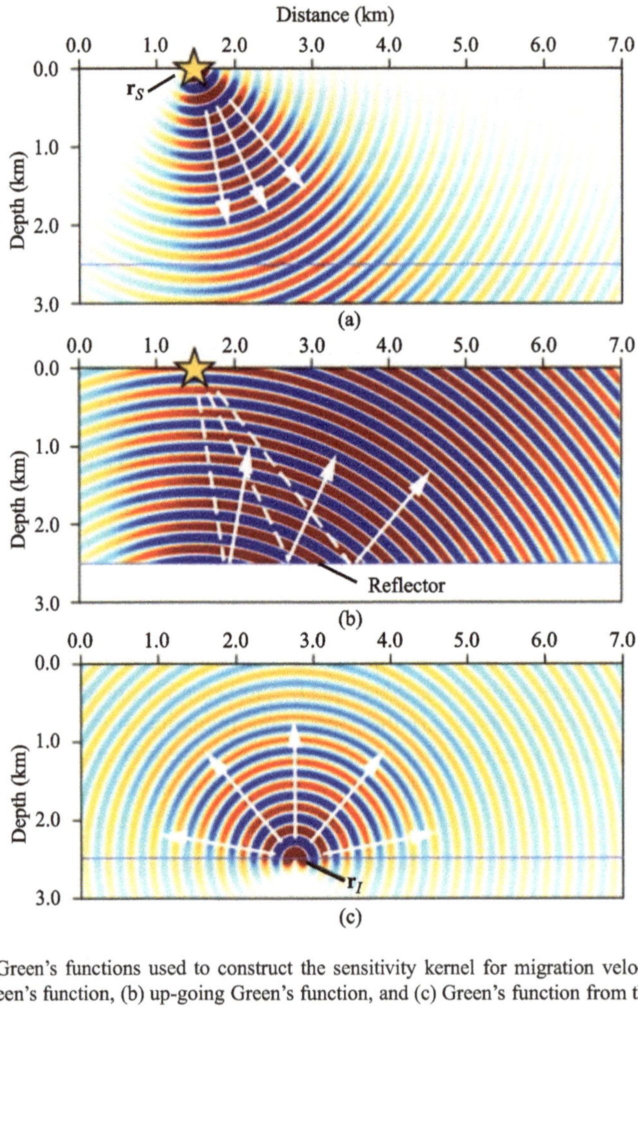

Fig. 2.11 The Green's functions used to construct the sensitivity kernel for migration velocity analysis. (a) Down-going Green's function, (b) up-going Green's function, and (c) Green's function from the image point.

with the factor $v_0(\mathbf{r}_I)/2\cos[\theta(\mathbf{r}_I,\mathbf{r}_S)]$ being a local time-depth convertor and $\delta t = \delta t_D + \delta t_U$, where

$$\delta t_D(\mathbf{r}_I,\mathbf{r}_S) = \int_V m(\mathbf{r}')K_D^B(\mathbf{r}',\mathbf{r}_S,\mathbf{r}_I)dv', \tag{2.70}$$

$$\delta t_U(\mathbf{r}_I,\mathbf{r}_S) = \int_V m(\mathbf{r}')K_U^B(\mathbf{r}',\mathbf{r}_S,\mathbf{r}_I)dv'. \tag{2.71}$$

Equations (2.69)-(2.71) form the velocity inversion system. The cartoon in Figure 2.11 illustrates the three Green's functions used to calculate the sensitivity kernel for velocity updating, where Figure 2.11a and Figure 2.11b is for down- and up-going waves, and Figure 2.11c is for waves radiated from the image point. The up-going Green's function in Figure 2.11b is calculated using the one-return method.

Shown in Figure 2.12a is a 10 Hz broadband sensitivity kernel for migration geometry calculated using the one-return method. The source side sensitivity kernel is similar to that for transmitted waves. However, the receiver side kernel is quite different. Near the source location and immediately above the image point, there are two sensitive regions where the velocity model affects the RMO most. As comparisons, shown in Figure 2.12b is the actually measured sensitivity map from the migration process (Xie and Yang, 2008). We see these kernels show excellent consistency.

Figure 2.13 compares sensitivity kernels calculated for selected image points in a five-layer/four-reflector model. Shown in the left column is sensitivity kernels calculated using

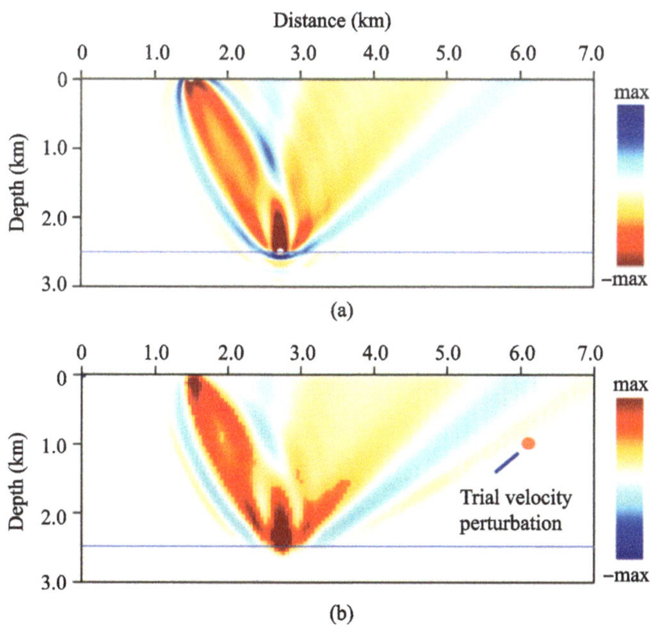

Fig. 2.12 The sensitivity kernels for a shot gather image, with (a) theoretical sensitivity kernel, and (b) the sensitivity map directly measured from migration imaging.

one-return method. As a comparison, the right column shows the actually measured sensitivity maps in the same model. The results show that the sensitivity kernels calculated using the one-return method are consistent with the measured sensitivity maps.

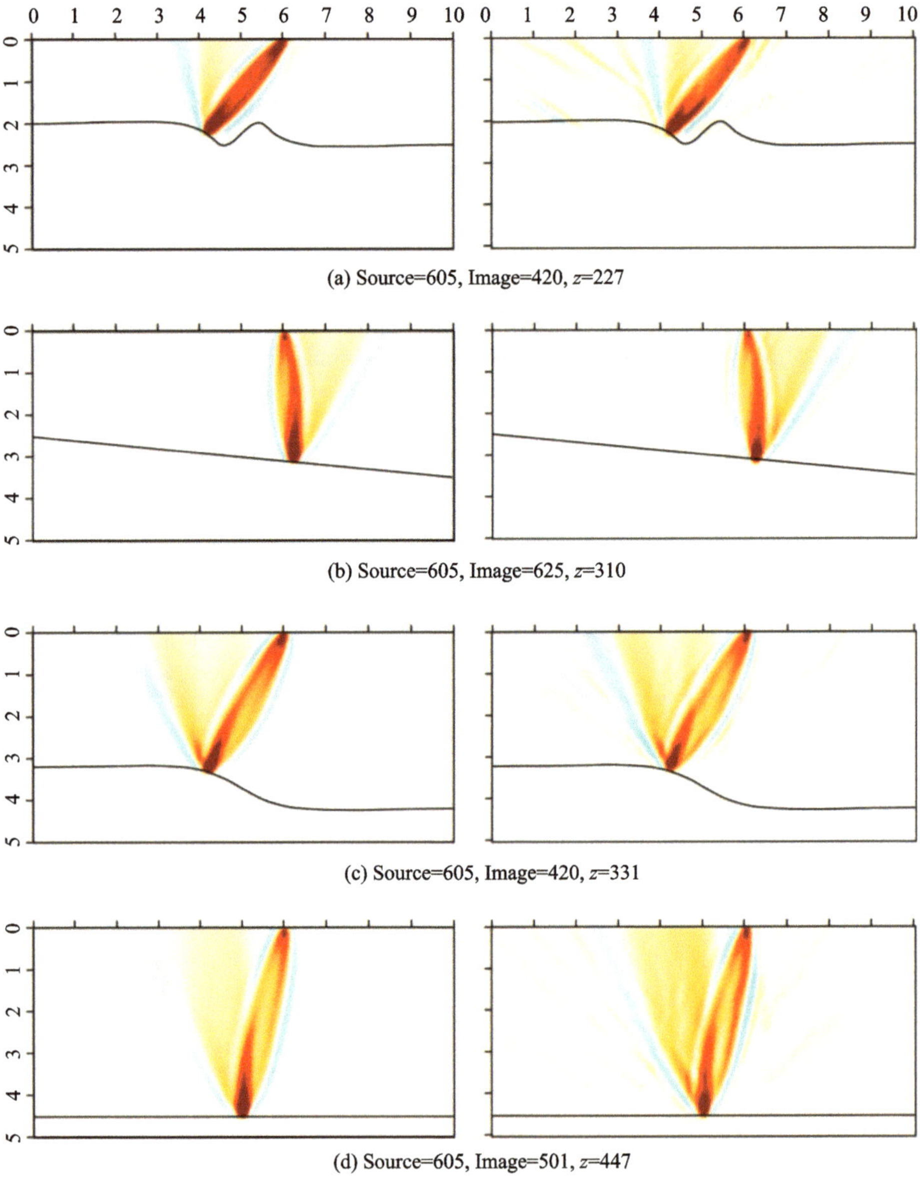

Fig. 2.13 Comparison between the theoretically calculated kernels (left column) and actually measured sensitivity maps (right column). From top to bottom are different reflectors.

2.5 Other Development of One-Return Modeling

One-way and one-return propagators have great advantages over the full-wave propagator, especially for the applications to imaging and inversion. The method is efficient and produces less artifacts compared with full-wave methods when applied to backpropagation and imaging. It has the flexibility to select different conversion modes in elastic wave propagation and imaging. It can model multiple scattering in the order of scattering multiplicity by multi-sweeps. It has great potential in applying to different inversion schemes. However, it has also some drawbacks which prevent its use in some applications. One serious drawback is the angle limitation. Even the wide-angle one-way propagators become much less accurate when the propagation angles are close to 90° (relative to the preferred direction, z-axis). Standard one-way propagators cannot go beyond 90°. Even though the one-return propagator can model the turning wave by double-sweeps, the waves propagating nearly along the horizontal direction will have less accuracy in both phase and amplitude. The research in improving accuracy of wide-angle or super-wide angle (beyond 90°) is one of the future directions of one-way, and one-return modeling. Another drawback of the one-return modeling is the convergence of the marching algorithm in strong contrast media. The one-return approximation, or the De Wolf series, is based on the perturbation series formulation. The elastic one-return methods using elastic thin-slab propagator or elastic complex-screen propagator have to limit the applications to moderate contrast media (parameter perturbations smaller than 40%) (Wu et al., 2007). Reflectivity method has been incorporated into the one-return method (Wu and Wu, 2003), but is limited only to flat layers. In the following, we summarize the progress in overcoming the two basic drawbacks described above. There are other developments in one-way, and one-return modeling, such as one-way propagator in anisotropic media (Angus et al, 2004), and improvement of the amplitude accuracy by multi-one-way modeling (Kiyashchenko et al., 2005), which will not be discussed in this chapter.

2.5.1 Super-Wide Angle One-Way Propagator

In order to improve the accuracy of wide-angle waves and overcome the fundamental angle limitation of one-way propagators, some work has been done using tilted coordinates or curved coordinate system (Shan et al., 2009; Shragge and Shan, 2010). The other approach is to use two orthogonally propagated one-way propagators to reconstruct the accurate wavefront up to nearly the full angle range (Wu and Jia, 2006; Xu and Jin, 2007; Jia and Wu, 2009a, 2009b). As shown in Figure 2.14, large-angle waves with respect to z-axis become small-angle waves to x-axis. Therefore, wavefront reconstruction method using weighted average of the two orthogonal propagated waves can keep good accuracy to super-wide angle ranges. Figure 2.15 shows the comparison of impulse responses between the regular one-way propagator (blue curve) and the superwide-angle propagator (red curve). The finite difference wavefront (green curve) is also shown as reference. The velocity of the model varies linearly in both lateral and vertical directions. We see that the wavefront from the regular one-way method has been distorted significantly at

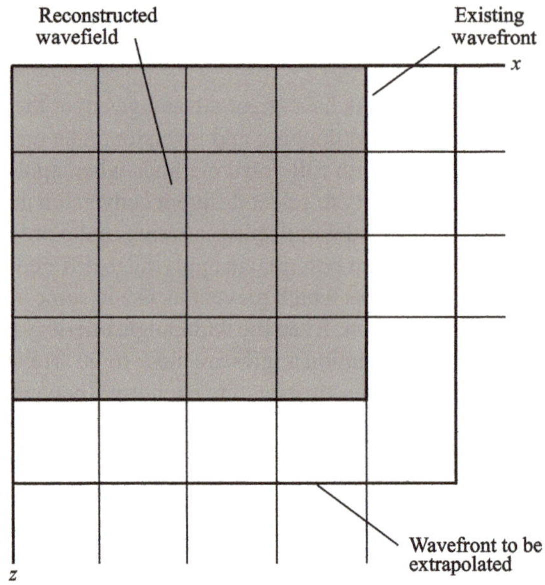

Fig. 2.14 Wavefield and wavefront reconstruction by taking a weighted average of two orthogonally propagated one-way wavefields.

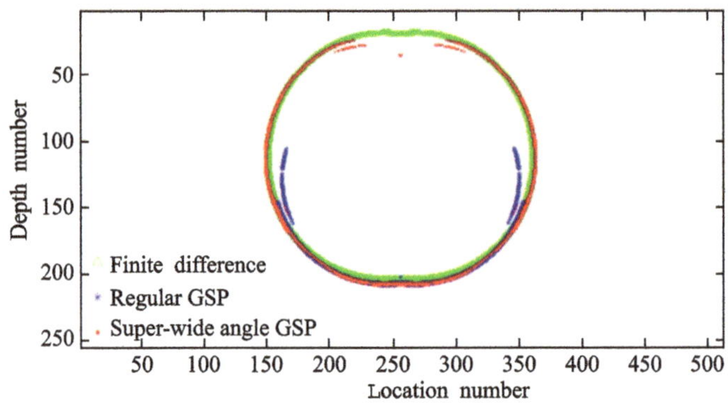

Fig. 2.15 Comparison of wavefront accuracy between regular one-way propagator (GSP) (blue curve) and superwide-angle GSP propagator (red curve). The finite difference wavefront (green curve) is shown as reference. The model has velocity gradients along $\pm x$ and z direction, $v(x,z) = v_0 + |x - 6.12| v_0/6.12 + z_0 v_0/6.12$.

large propagation angles (close to 90°). On the other hand, the new propagator models the wavefront accurately up to 135°. The performance is degenerated only for nearly backpropagation angles (close to 180°). Figure 2.16 compares the prestack images by the regular one-way propagator (left panel) and by the superwide-angle one-way propagator (right panel) for the 2D BP benchmark model. Due to the angle limitation, the regular downward one-way propagator can hardly simulate turning waves and therefore cannot image the overhanging flank. By contrast, these limitations are eliminated in the imaging

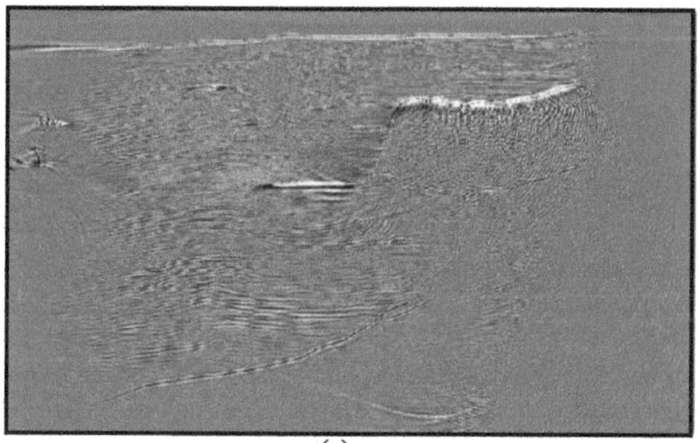

(a)

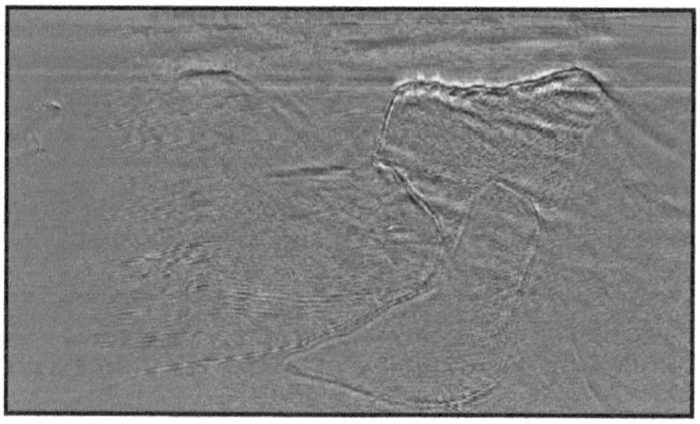

(b)

Fig. 2.16 Comparison of images obtained by the regular GSP migration (a) and by the superwide-angle GSP migration (b).

result of the superwide-angle method. The image of overhanging flank is sharp and clear and its location is accurate.

2.5.2 One-Way Boundary Element Method

As we pointed out, the one-return modeling is based on the perturbation method and has difficulty in applying to strong contrast media, such as the media with complicated salt and basalt structures. For salt models, the velocity perturbation could be more than 200%. One-way boundary-element method is proposed to partially solve the problem (He and Wu, 2009). The concepts of one-way and one-return boundary element method can be illustrated by the simple models shown in Figure 2.17a: a layered model, and 2.17b: an inclusion model. The key operation for one-way boundary-element method is to decouple multiple interactions between the upper boundary S_1 and the lower boundary S_2 and therefore eliminate the internal multiples between these two boundaries. It is quite straightforward to apply this concept to layered media. However, to extend this method to models with an inclusion, extra care must be taken. Here, we take a simple inclusion model (Figure 2.17b) as an example. The inclusion could be a salt dome. We separate the boundary of the salt dome into top part S_1 and bottom part S_2 based on the shape of

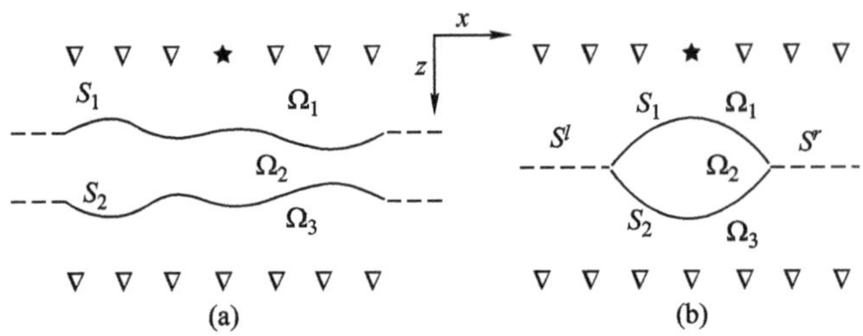

Fig. 2.17 The concepts of one-way and one-return boundary element method can be illustrated by the simple models: (a) a layered model (b) an inclusion model. The triangles indicate the receivers and the star indicates the source.

inclusion and the source-receivers configuration. We also add flat artificial interfaces S^l on the left-hand-side, and S^r on the right-hand-side, with both extending to the infinity. Now we divide the whole model into three domains: (i) Ω_1 with boundaries S_1, S^l and S^r; (ii) Ω_2 with boundaries S_1 and S_2; (iii) Ω_3 with boundaries S_2, S^l and S^r.

First, let us set up the equations to solve the full wave problem. In the first domain Ω_1, the boundary integral equation is

$$a(\mathbf{r})\mathbf{u}(\mathbf{r}) + \int_{S_1} [\mathbf{u}(\mathbf{r}')\Gamma(\mathbf{r},\mathbf{r}') - \mathbf{t}(\mathbf{r}')\mathbf{G}(\mathbf{r},\mathbf{r}')]dS(\mathbf{r}') = \mathbf{f}(\mathbf{r}^0,\omega)\mathbf{G}(\mathbf{r},\mathbf{r}^0)\mathbf{r} \in \Omega_1, \qquad (2.72)$$

where $\mathbf{u}(\mathbf{r})$ is the displacement vector, $\mathbf{t}(\mathbf{r})$ is the traction vector and $\mathbf{f}(\mathbf{r})$ is the volume source distribution. The coefficient $a(\mathbf{r})$ generally depends on the local geometry of the boundary. $\mathbf{G}(\mathbf{r},\mathbf{r}')$ and $\Gamma(\mathbf{r},\mathbf{r}')$ are the fundamental solutions (Green tensors) for displacement and traction, respectively. Similarly, in the second domain we obtain

$$\begin{aligned} a(\mathbf{r})\mathbf{u}(\mathbf{r}) &+ \int_{S_1} [\mathbf{u}(\mathbf{r}')\Gamma(\mathbf{r},\mathbf{r}') - \mathbf{t}(\mathbf{r}')\mathbf{G}(\mathbf{r},\mathbf{r}')]dS(\mathbf{r}') \\ &+ \int_{S_2} [\mathbf{u}(\mathbf{r}')\Gamma(\mathbf{r},\mathbf{r}') - \mathbf{t}(\mathbf{r}')\mathbf{G}(\mathbf{r},\mathbf{r}')]dS(\mathbf{r}') = \mathbf{0} \quad \mathbf{r} \in \Omega_2. \end{aligned} \qquad (2.73)$$

And in the third domain we have

$$a(\mathbf{r})\mathbf{u}(\mathbf{r}) + \int_{S_2} [\mathbf{u}(\mathbf{r}')\Gamma(\mathbf{r},\mathbf{r}') - \mathbf{t}(\mathbf{r}')\mathbf{G}(\mathbf{r},\mathbf{r}')]dS(\mathbf{r}') = \mathbf{0} \quad \mathbf{r} \in \Omega_3. \qquad (2.74)$$

Combined with the continuity condition of displacement and traction along the interfaces, the system of integral equations can be transformed to an equation array in matrix form.

The one-way boundary-element method can be described as follows: First, we calculate the transmitted wave through S_1 by applying the full-wave BEM to domain Ω_1 and domain Ω_2 with interface S_1, ignoring the effect of interface S_2. The second step is to propagate the output wave field to the interface S_2 as the incident wave. Finally, we obtain the transmitted wave through interface S_2 into domain Ω_3 by solving the two-domain boundary value problem involving domain Ω_2 and domain Ω_3 with interface S_2. Each time we only need to solve a much smaller matrix equation associated with the current interface rather than a much larger full rank matrix equation in full-wave BEM. However, this technique decouples the wave field interaction between interfaces S_1 and S_2, thus eliminates the multiples between S_1 and S_2, and can only obtain the primary transmitted waves. Therefore, this step (down-sweep) is called "one-way" boundary element modeling.

Next we calculate the primary reflections by one-return approximation in the up-sweep. In the same spirit of neglecting multiples, the backscattered waves at each interface are

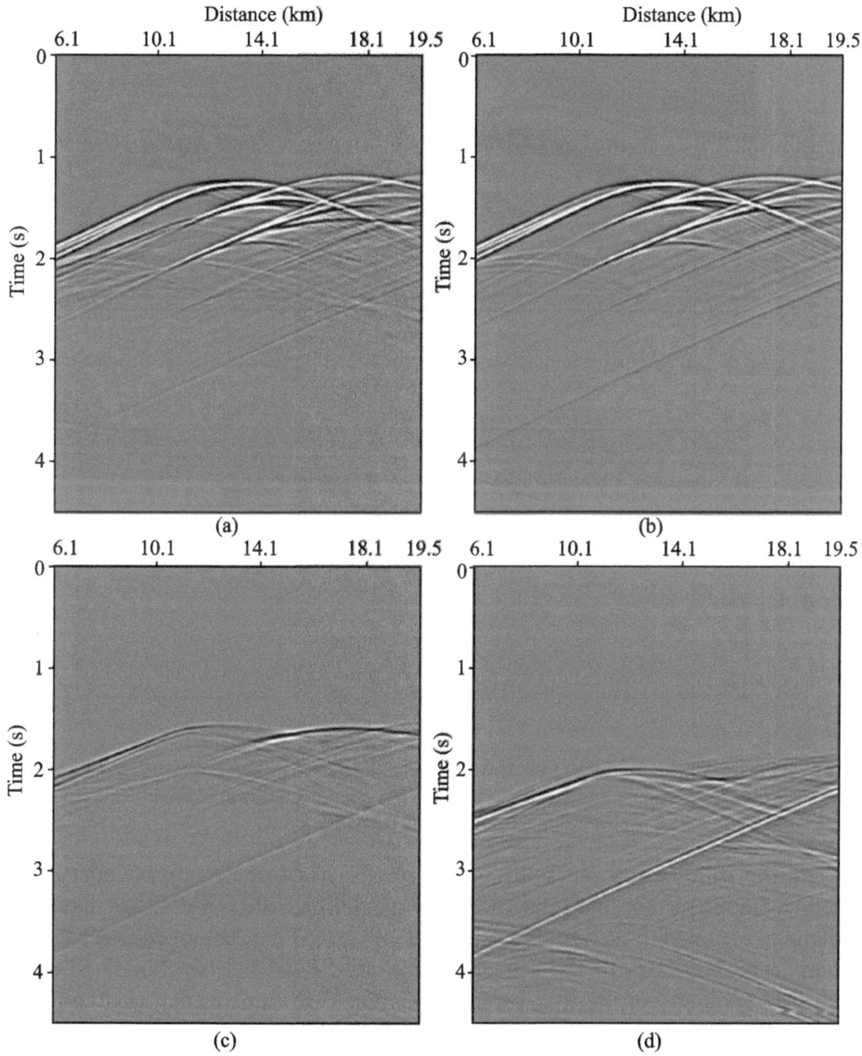

Fig. 2.18 Synthetic seismograms for the shot located at (19.5, 0.0) km for the modified SEG-EAGE model. (a) By full-wave BEM. (b) Primary reflections from the top interface. (c) Primary reflections from the bottom interface. (d) Internal multiples (amplified by a factor of ten).

picked up and propagated to the surface by the one-way boundary element propagator as formulated above. In our three-layer model, first we pick up the reflected waves on S_2(bottom interface). The reflected (backscattered) field and traction are obtained by subtraction of the incident field from the total field. Then we can calculate the transmission of backscattered fields at S_1 by solving one-way boundary-element equations. The whole process is similar to the two-sweep process of the thin-slab one-return propagator (see Figure 2.2), but the thin-slabs are replaced in this case by the thick-layers with curved interfaces.

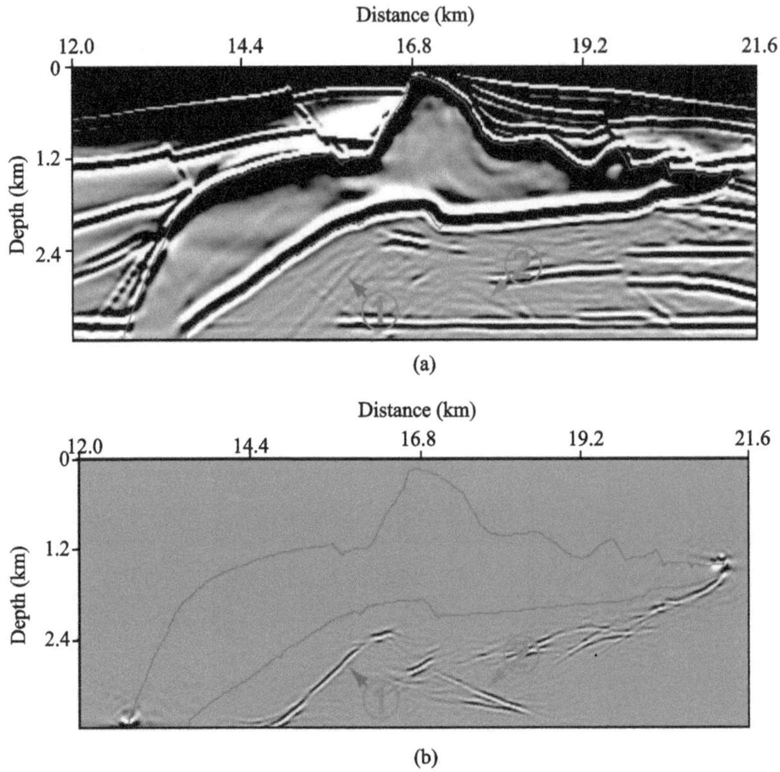

Fig. 2.19 (a) Pre-stack image of the 2D SEG-EAGE model using the full synthetic data set. (b) Pres-stack image using the data of pure internal multiples generated by one-way boundary-element method for the modified SEG-EAGE salt model (salt dome embedded in a homogeneous medium).

To test the validity and capability of this one-way and one-return boundary element method, we applied this one-return BEM to a modified SEG-EAGE salt model. The salt dome is assumed to be homogeneous and embedded in a homogeneous background medium. The boundary of the salt dome is picked from the benchmark SEG-EAGE model. The geometry of the sources and receivers is the same as that used in the SEG-EAGE model. The synthetic full acoustic arrivals, primary reflections from top interface, primary reflections from bottom interface and internal multiples are shown in Figure 2.18. Note that the internal multiples (Figure 2.18d) are amplified by a factor of ten for the comparison with the primary arrivals.

In the next example, we apply the one-way boundary-element modeling to studying the migration artifacts by salt multiples. Since internal multiple arrivals could contaminate the reflection signals from subsalt faults, migration by back-propagating these signals using one way propagators could generate artifacts on the subsalt regions. Figure 2.19a shows the pre-stack image by local cosine beamlet migration for 2D SEG-EAGE benchmark model (Cao and Wu, 2008). The image quality is generally quite good, and even some subsalt faults are clearly seen. However, there are some artifacts, which are hard to distinguish from real reflections. Applying the one-return boundary element method to the simplified salt model, we can generate the data of pure internal multiples. Then we perform migration on these data of multiples based on the same one-way propagator. By comparison of the final images, we can tell whether or not these features are due to the internal multiples. Shown in Figure 2.19b is the image obtained by migrating the data of pure internal multiples only. The features in this picture are quite consistent with some subsalt artifacts in Figure 2.19a. Further detailed study using partial data from different shot combinations can get more information about the origins of the artifacts related to salt multiples (See He and Wu, 2009 for details).

The current version of the one-way boundary element method uses Green's function in the homogeneous background media, which limits its practical applications. The challenging problem is to develop the one-return boundary element modeling using Green's function in heterogeneous media.

2.6 Conclusion

We review the one-return method for calculating seismic wave propagation in complex acoustic and elastic models. In this chapter, we give a new, intuitive derivation of the one-return approximation using a sequential thin-slab transmission/reflection operators. This derivation can reach the same formulation as the De Wolf approximation does, and in the same time provide an efficient implementation of the method. The method is based on the multiple-forescattering-single-backscattering approximation. It neglects the internal reverberations (internal multiples) but can model all forward scattering phenomena, such as focusing/defocusing, diffraction, refraction, and interference, particularly, it can handle primary reflections from heterogeneities. The method is a dual domain method. It handles wave-medium interactions in the space domain, while treating the propagation through the background in the wavenumber domain. Fast Fourier transforms are used to shuttle the wavefield between the two domains. The calculations in both domains are local, which makes this method very efficient. Two versions of the one-return method are discussed: one is the wide-angle thin-slab approximation and the other is the complex screen approximation which involves a small-angle approximation.

The one-return method has been used in seismic modeling, imaging and inversion. It can be applied to fast modeling primary reflections of elastic or acoustic waves in complex models, particularly to simulations of high-frequency waves and long propagation distances. It has been used for reservoir AVO simulations in heterogeneous visco-elastic media with moderate parameter variations (perturbations <40%). The one-return method can be used to extrapolate waves that are reflected from interfaces, including side walls. Combined with properly designed image conditions, this propagator can migrate turning wave and duplex wave (or prism wave) and therefore, image steeply dipping structure and vertical overhangs. For waveform inversion and migration velocity analysis, one-return propagator is very efficient for calculating sensitivity kernels that involve reflections from interfaces. It has been applied in wave-equation-based velocity analysis for migration velocity model updating.

Finally, we reviewed recent progress made to overcome the limitations and shortcomings of the one-way, and one-return methods. The super-wide angle one-way propagator uses two orthogonal one-way propagators to reconstruct the accurate wavefront up to nearly the full angle range. It can image the steep reflectors and overhang salt flanks. One-return boundary element method was developed to deal with strong contrast elastic media. It can generate synthetics with only primary reflections from the salt boundaries, and therefore, be used to study the artifacts caused by salt multiples in subsalt images. However, the current version uses Green's functions in homogeneous media. One-return modeling in heterogeneous, strong contrast media is still a challenging problem.

References

Aki, K. and Richards, P. (1980). Quantitative Seismology: Theory and Methods, Vol. 1 and 2. W.H. Fremman, New York.

Angus, D.A., Thomson, C.J. and Pratt, R.G. (2004). A one-way wave equation for modeling variations in seismic waveforms due to elastic anisotropy. Geophys. J. Int., 156, 595-614.

Buursink, M. L. and P. S. Routh (2007). Application of borehole-radar Fresnel volume tomography to image porosity in a sand and gravel aquifer. Expanded Abstracts, SEG 77th Annual Meeting, 668-672.

Cao, J. and Wu, R.S. (2008). Amplitude compensation of one-way wave propagators in inhomogeneous media and its application to seismic imaging. Special issue "Computational geophysics", Communications in computational physics, 3, 203-221.

Claerbout, J.F. (1970). Coarse grid calculations of waves in inhomogeneous media with applications to delineation of complicated seismic structure. Geophysics, 35, 407-418.

Claerbout, J.F. (1976). "Fundamentals of geophysical data processing". McGraw-Hill, New York.

Collins, M.D. (1989). Applications and time-domain solution of higher-order parabolic equation in underwater acoustics. J. Acoust. Soc. Am., 86, 1,097-1,102.

Collins, M.D. (1993). A two-way parabolic equation for elastic media. J. Acoust. Soc. Am., 93, 1,815-1,825.

Collins, M.D. and Westwood, E.K. (1991). A higher-order energy-conserving parabolic equation for range-dependent ocean depth, sound speed, and density. J. Acoust. Soc. Am., 89, 1,068-1,175.

Corones, J. (1975). Bremmer series that correct parabolic approximations. J. Math. Anal.

Appl., 50, 361-372.
Dahlen, F., S. Huang, and G. Nolet (2000). Fréchet kernels for finite-frequency traveltimes –I, Theory: Geophys. J. Int., 141, 157-174.
De Hoop, M., Rousseau, J. and Wu, R.S. (2000). Generalization of the phase-screen approximation for the scattering of acoustic waves. Wave Motion, 31, 43-70.
De Hoop, M.V. (1996). Generalization of the Bremmer coupling series. J. Math. Phys., 37, 3,246-3,282.
De Wolf, D.A. (1971). Electromagnetic reflection from an extended turbulent medium: cumulative forward-scatter single-backscatter approximation. IEEE trans. Ant. and Propag. AP-19, 254-262.
De Wolf, D.A.(1985). Renormalization of EM fields in application to large-angle scattering from randomly continuous media and sparse particle distributions. IEEE trans. Ant. and Propag. AP-33, 608-615.
Fishman, L. and McCoy, J.J. (1984). Derivation and application of extended parabolic wave theories II. Path integral representations. J. of Math Phys., 25, 297-308.
Fishman, L. and McCoy, J.J. (1985). A new class of propagation models based on a factorization of the Helmholtz equation. Geophys. J. Astr. Soc., 80, 439-461.
Fisk, M.D. and McCartor, G.D. (1991). The phase screen method for vector elastic waves. J. Geophys., Res., 96, 5,985-6,010.
Flatté, S.M. and Tappert, F.D.(1975). Calculation of the effect of internal waves on oceanic sound transmission, J. Acoust. Soc. Am., 58, 1,151-1,159.
Fliedner, M.M., M.P. Brown, D. Bevc, and B. Biondi (2007). Wave path tomography for subsalt velocity model building. Expanded Abstracts, SEG 77th Annual Meeting, 1,938-1,942.
French, W.S. (1974). Two-dimensional and three-dimensional migration of model-experiment reflection profiles. Geophysics, 39, 265-277.
Grimbergen, J. L. T., Dessing, F. J., and Wapenaar, K. (1998). Modal expansion of one-way operators in laterally varying media. Geophysics, 63, 995–1,005.
He, Y. and Wu, R.S., 2009, One-way/one-return boundary element method and salt internal multiples. Geophysics, submitted.
Huang, L. J., Fehler, M. C., Roberts, P. M., and Burch, C. C. (1999b). Extended local Rytov Fourier migration method. Geophysics, 64, 1,535-1,545.
Huang, L. J., Fehler, M.C., Wu, R. S. (1999a). Extended local Born Fourier migration method. Geophysics, 64, 1,524-1,534.
Hudson, A. (1980). A parabolic approximation for elastic waves. Wave Motion, 2, 207-214.
Jia, X. and Wu, R.S. (2009a). Superwide-angle one-way wave propagator and its applications in imaging steep salt flanks. Geophysics, 74, S75-83.
Jia, X. and Wu, R.S. (2009b). calculation of wave propagation angle in complex media: application to turning wave simulations. Geophys. J. Int., 178, 1,765-1,573.
Jin, S., and Wu, R.S. (1999). Common-offset pseudo-screen depth migration. Expanded Abstracts, SEG 69th Annual Meeting, 1,516-1,519.
Jin, S., Mosher, C.C., and Wu, R.S. (2002). Offset-domain pseudoscreen prestack depth migration, Geophysics. 67, 1,895-1,902.
Jin, S., Wu, R. S., and Peng, C. (1999). Seismic depth migration with screen propagators.

Computational Geosciences, 3, 321-335.
Jin, S., Wu, R.S., and Peng, C. (1998). Prestack depth migration using a hybrid pseudo-screen propagator. Expanded Abstracts, SEG 68th Annual Meeting, 1,819-1,822.
Jin, S., Xu, S., and Walraven, D. (2006). One-return wave equation migration: Imaging of duplex waves, Expanded Abstracts, SEG 76th Annual Meeting, 2,338-2,342.
Jocker, J., J. Spetzler, D. Smeulders, and J. Trampert (2006). Validation of first-order diffraction theory for the traveltimes and amplitudes of propagating waves. Geophysics, 71, T167-T177.
Kiyashchenko, D., Plessix, R.-E., Kashtan, B. and Troyan, V. (2005). Improved amplitude multi-one-way modeling method. Wave Motion, 43, 99-115.
Landers, T., and Claerbout, J. F. (1972). Numerical calculation of elastic waves in laterally inhomogeneous media. J. Geophys. Res., 77, 1,476-1,482.
Le Rousseau, J. H. and de Hoop, M. V. (2001). Scalar generalized–screen algorithms in transversely isotropic media with a vertical symmetry axis. Geophysics, 66, 1,538-1,550.
Lee, D., Shang, E-C and Buckingham, M.J. (2000). Parabolic equation development in the twentieth century. J. of Computational Acoust., 8, 527-637.
Ma, Z. (1982). Finite-difference migration with higher order approximation. Oil Geophys. Prosp. China, 1, 6-15.
McCoy, J. J. (1977). A parabolic theory of stress wave propagation through inhomogeneous linearly elastic solids. J. Appl. Mech., 44, 462-468.
McCoy, J.J. and Frazer, L.N. (1986). Propagation modeling based on wavefield factorization and invariant embedding. Geophys. J.R. astr. Soc., 86, 703-717.
Ristow, D. and Ruhl, T. (1994). Fourier finite-difference migration. Geophysics, 59, 1,882-1,893.
Sava, P.C., and B. Biondi (2004). Wave-equation migration velocity analysis - I: Theory: Geophysical Prospecting, 52, 593-606.
Shan, G., Clapp,R. and Biondi, B. (2009). 3D plane-wave migration in tilted coordinates: A field data example. Geophysics, 74: WCA199 - WCA209.
Shragge, J. and Shan, G. (2010). Inline delayed-shot migration in tilted elliptical-cylindrical coordinates. Geophysics, 75, s187-s197.
Skarsoulis, E. K. and B. D. Cornuelle (2004). Travel-time sensitivity kernels in ocean acoustic tomography, J. Acoust. Soc. Am., 116, 227-238.
Spetzler, J. and R. Snieder (2004). The Fresnel volume and transmitted waves. Geophysics, 69, 653-663.
Stoffa, P.L., Fokkema, J.T., Freire, R.M.D., and Kessinger, W.P. (1990). Split-step Fourier migration. Geophysics, 55, 410-421.
Tappert, F.D., (1977). The parabolic equation method, in "Wave Propagation and Underwater Acoustics" (ed. Keller and Papadakis). Springer, New York.
Thomson, C. J. (1999). The "gap" between seismic ray theory and "full" wavefield extrapolation. Geophys. J. Int., 137, 364–380.
Thomson, C.J. (2005). Accuracy and efficiency considerations for wide-angle wavefield extrapolators and scattering operators. Geophys. J. Int., 163, 308-323.
Van Stralen, M.J.N., de Hoop, M. and Blok, H. (1998). Generalized Bremmer series with rational approximation for the scattering of waves in inhomogeneous media. J. Acoust.

Soc. Am., 104, 1,943-1,963.

Vasco, D.W., J.E. Peterson, Jr., and E.L. Majer (1995). Beyond ray tomography: Wavepaths and Fresnel volumes. Geophysics, 60, 1,790-1,804.

Wales, S.C. (1986). A vector parabolic equation model for elastic propagation, in "Ocean Seismo-Acoustics" (ed. Akal and Berkson). Plenum, New York.

Wales, S.C., and McCoy, J.J. (1983). A comparison of parabolic wave theories for linearly elastic solids. Wave Motion, 5, 99-113.

Weston, V.H. (1989). Wave splitting and the reflection operator for the wave equation in R^3. J. Mathematics. Phys., 30, 2,545-2,562.

Wild, A.J., and Hudson, J. A. (1998). A geometrical approach to the elastic complex screen. J. Geophys. Res., 103, 707-726.

Wild, A.J., Hobbs, R.W. and Frenje, L. (2000). Modeling complex media: an introduction to the phase-screen method. Phys. of Earth and Planet. Inter., 120, 219-226.

Woodward, M. J. (1992). Wave-equation tomography. Geophysics, 57, 15-26.

Wu, R. S. (1994). Wide-angle elastic wave one-way propagator in heterogeneous media and an elastic wave complex-screen method. J. Geophys. Res., 99, 751-766.

Wu, R. S. (1996). Synthetic seismograms in heterogeneous media by one-return approximation. Pure and Applied Geophysics, 148, 155-173.

Wu, R. S., and Huang, L. J. (1995). Reflected wave modeling in heterogeneous acoustic media using De Wolf approximation. Mathematical Methods in Geophysical Imaging III, SPIE 2571, 176-193.

Wu, R. S., and Jin, S. (1997). Windowed GSP (generalized screen propagators) migration applied to SEG-EAEG salt model data. Expanded Abstracts, SEG 67th Annual Meeting, 1,746-1,749.

Wu, R. S., Jin, S., and Xie, X. B. (2000a). Energy partition and attenuation Lg waves by numerical simulations using screen propagators. Phys. Earth and Planet.Inter. 120, 227-244.

Wu, R. S., Jin, S., and Xie, X. B. (2000b). Seismic wave propagation and scattering in heterogeneous crustal waveguides using screen propagators: I SH waves. Bull. Seism. Soc. Am. 90, 401-413.

Wu, R.S. (2003). Wave propagation, scattering and imaging using dual-domain one-way and one-return propagators. Pure and Appl. Geophysics, 160, 509-539.

Wu, R.S. and Huang, L.J. (1992). Scattered field calculation in heterogeneous media using phase-screen propagator. Expanded Abstracts, SEG 62nd Annual Meeting, 1,289-1,292.

Wu, R.S. and Xie, X.B. (1994). Multi-screen backpropagator for fast 3D elastic prestack migration. Mathematical Methods in Geophysical Imaging II SPIE 2301, 181-193.

Wu, R.S., and Jia, X. (2006). Accuracy improvement for super-wide angle one-way waves by wavefront reconstruction, Expanded Abstracts, SEG 76th Annual Meeting, 2,976-2,980.

Wu, R.S., and Xie, X.B. (1993). A complex-screen method for elastic wave one-way propagation in heterogeneous media. Expanded Abstracts, 3rd International Congress of the Brazilian Geophysical Society.

Wu, R.S., Huang, L.J., and Xie, X.B. (1995). Backscattered wave calculation using the De Wolf approximation and a phase-screen propagator. Expanded Abstracts, SEG 65th

Annual Meeting, 1,293-1,296.

Wu, R.S., X.B. Xie, and X.Y. Wu (2007). One-way and one-return approximations (de Wolf approximation) for fast elastic wave modeling in complex media. *in* R.S. Wu and V. Maupin, eds., Advances in Wave Propagation in Heterogeneous Earth, Elsevier, 265-322.

Wu, X. Y., and Wu, R. S. (1999). Wide-angle thin-slab propagator with phase matching for elastic modeling. Expanded abstracts, SEG 69th Annual Meeting, 1,867-1,870.

Wu, X. Y., and Wu, R. S. (2001). Lg-wave simulation in heterogeneous crusts with surface topography using screen propagators. Geophys. J. Int., 146, 670-678.

Wu, X.Y. and Wu, R.S., (2003a). Fast modeling of 2D/3D elastic reflections using thin-slab method. Expanded Abstracts, SEG 73rd Annual Meeting, 1,865-1,868.

Wu, X.Y. and Wu, R.S., (2003b). Synthesizing AVO responses in visco-elastic media using fast one-way elastic propagators. Expanded Abstracts, SEG 73rd Annual Meeting, 208-210.

Xie, X. B., and Wu, R. S. (2001). Modeling elastic wave forward propagation and reflection using the complex-screen method. J. Acoust. Soc. Am., 109, 2,629-2,635.

Xie, X. B., and Wu, R. S., (2005). Multicomponent prestack depth migration using elastic screen method. Geophysics, 70, S30-S37.

Xie, X. B., Mosher, C. C., and Wu, R. S. (2000). The application of wide angle screen propagator to 2D and 3D depth migrations. Expanded Abstracts, SEG 70th Annual Meeting, 878-881.

Xie, X.B. and H. Yang (2007). A migration velocity updating method based on the shot index common image gather and finite-frequency sensitivity kernel. Expanded Abstracts, SEG 77th Annual Meeting, 2,767-2,771.

Xie, X.B. and Wu, R.S. (1995). A complex-screen method for modeling elastic wave reflections. Expanded Abstracts, SEG 65th Annual Meeting, 1,269-1,272.

Xie, X.B., and H. Yang (2008). The finite-frequency sensitivity kernel for migration residual moveout and its applications in migration velocity analysis. Geophysics, 73, S241-S249.

Xie, X.B., and Wu, R.S. (1998). Improve the wide angle accuracy of screen method under large contrast. Expanded abstracts, SEG 68th Annual Meeting, 1,811-1,814.

Xu, S., and Jin, S. (2006) Wave equation migration of turning waves. Expanded Abstracts, SEG 76th Annual Meeting, 2,328-2,332.

Xu, S., and Jin, S. (2007) An orthogonal one-return wave-equation migration. Expanded Abstracts, SEG 77th Annual Meeting, 2,325-2,329.

Yoon, K.H. and McMechan, G.A. (1996). 3D eighth-order elastic finite-difference modeling of refraction and strong-motion data from the Coyote Lake region, California. Bull. Seis. Soc. Am., 86, 616-626.

Zhang G. (1993). System of coupled equations for upgoing and downgoing waves, Acta Mathematica Applicata Sinica, 16, 251-263.

Zhang, Y., G. Zhang, and N. Bleistein (2005). Theory of true amplitude one-way wave equations and true amplitude common-shot migration. Geophysics, 70, E1–10.

Zhang, Y., S. Xu, and G. Zhang (2006). Imaging Complex Salt Bodies with Turning-wave One-way wave equation. SEG abstract, 2,323-2,326.

Zhang, Y., S. Xu, N. Bleistein, and G. Zhang (2007). True-amplitude, angle-domain,

common-image gathers from one-way wave equation migrations. Geophysics, 72, S49-S58.
Zhao, L., T.H. Jordan, and C.H. Chapman (2000). Three-dimensional Fréchet differential kernels for seismic delay times. Geophys. J. Int., 141, 558-576.

Author Information

Ru-Shan Wu and Xiao-Bi Xie

IGPP/CSIDE, University of California at Santa Cruz, Santa Cruz, CA 95064, USA
E-mail: rwu@ucsc.edu

Shengwen Jin

Halliburton Energy Services

Chapter 3
Fault-Zone Trapped Waves: High-Resolution Characterization of the Damage Zone of the Parkfield San Andreas Fault at Depth

Yong-Gang Li, Peter E. Malin, and Elizabeth S. Cochran

This chapter presents that highly damaged rocks within the San Andreas fault (SAF) at Parkfield form a low-velocity waveguide to trap seismic waves. The amplitudes and dispersion feature of trapped waves are sensitive to the geometry and physical properties of the fault zone due to the constructive interference conditions of these waves. We use fault-zone trapped waves (FZTWs) generated by earthquakes and explosions and recorded at a cross-fault surface array and borehole seismographs at the San Andreas Fault Observatory at Depth (SAFOD) site to document fault zone structure and rock damage at seismogenic depths with high-resolution. Observations and 3-D finite-difference simulations of these FZTWs at dominant frequencies of 2-10 Hz show the downward tapering SAF characterized by a 30-40-m wide fault core with the maximum velocity reduction up to $\sim$50% embedded in a 100-200-m wide zone with velocities reduced by 25%-35% in average from wall-rock velocities. The width and velocity reduction of the damage zone at 3 km depth delineated by FZTWs are verified by the direct measurements in SAFOD drilling and logging studies at this depth [Hickman et al., 2007]. The results indicate that the localization of severe rock damage on the SAF likely reflects pervasive cracking caused by historical earthquakes on it. The magnitude of damage varies with depth and along the fault strike due to rupture distributions and stress variations over multiple length and time scales. The damage is not symmetric across the main slip plane but extends farther on the southwest side of the main fault trace. Based on the depths of earthquakes generating prominent FZTWs, we estimate that the low-velocity damage zone along the SAF at Parkfield extends at least to depths of $\sim$ 7-8 km.

Keywords: Fault-zone trapped waves, Rock damage, Physical properties, Dynamic rupture

3.1 Introduction

Mature faults are planes of weakness in the Earth crust. They facilitate slip under the prevailing stress orientation to initiate earthquakes. Extensive field and laboratory research, and numerical simulations indicate that the fault zone undergoes high, fluctuating stress and pervasive cracking during an earthquake (e.g., Aki, 1984; Scholz, 1990; Rice, 1992; Kanamori, 1994). In order to relate present-day crustal stresses and fault motions to the geological structures formed by previous ruptures, we must understand the evolution of fault systems on many spatial and temporal scales. Rupture models involving variations in fault-zone fluid pressure over the earthquake cycle have been proposed (e.g., Dieterich, 1979; 1998; Olsen et al., 1998). Structural fault variations and rheological fault variations (e.g., Rice, 1980; Sibson et al., 1975; Angevine et al., 1982) as well as variations in strength and stress may affect the earthquake rupture (e.g., Vidale et al., 1994; Marone et al., 1995; Taira et al., 2009). On the other hand, inelastic responses of compliant fault zones to dynamic rupture of nearby earthquakes have been observed and numerically studied (e.g., Vidale and Li, 2003; Fialko, 2004; Cochran et al., 2009; Duan, 2010).

The spatial extent of fault weakness, and the loss and recovery of strength across the earthquake cycle are critical ingredients in our understanding of fault mechanics. Whereas we know slip is localized on faults because of their lower strength compared to the surrounding bedrock, critical parameters remain unknown. For example, friction laws are approximate (e.g., Richardson and Marone, 1999), and the magnitude of strength reduction and its spatial extent at seismogenic depth are still not well constrained (e.g., Hickman et al., 1995). While earthquake-related fault-zone damage and healing have been documented (e.g., Li et al., 1998, 2006; Massonnet et al., 1996; Yasuhara et al., 2004), the origin of the spatial and temporal variability in the fault zone properties still remains a mystery. Also what is unclear is the relationship between the damage magnitude and the absolute local stress level and stress drop, and the contribution of fault damage to the total earthquake energy budget.

Since the fault zone is thought to be a weak zone in the Earth crust, it facilitates slip to occur under the prevailing stress orientation. As suggested by laboratory experiments, new shear faulting is highly resisted in brittle material and strain is typically accommodated by re-activation of faults that have already accumulated considerable damage (Chester et al., 1993; Andrew, 2005). Field evidence shows that the majority of slip on a mature fault occurs in a much localized zone, often at the edge of damage zone near the contact with the intact wall rock. Assuming that this is an accurate picture of rupture propagation on the major faults, defining the fine internal structure of active fault zones is a challenging, but an important area of study for seismologists and geologists. Zones of lower seismic velocities structurally mark major crustal faults. Intense fracturing during earthquakes, brecciation, liquid-saturation and possibly high pore-fluid pressure near the fault are thought to create these low-velocity zones (LVZs) (Mooney and Ginzburg, 1986; Chester et al., 1993). Byerlee (1990) and Rice (1992) noted that pore fluids may migrate up from depth and the highly-fractured fault-zone acts as a channel due in part to its relatively high permeability. Such a high-permeability and low-strength damage zone was measured on the Nojima fault in the 1995 Kobe earthquake (Lockner et al., 2000).

Near Parkfield, California, many researchers have observed an LVZ surrounding the surface trace of the San Andreas fault (SAF) (e.g., Michelini and McEvilly, 1991; Thurber et al., 2004, 2006; Unsworth et al., 1997). This zone has been reported to span from a few hundred meters to as wide as 1 km, and has velocity reductions of 10%-30% and V_p/V_s ratios on the order of 2.3. Recent results from core sampling and well logs in the San

Andreas Fault Observatory (SAFOD) borehole located ~18 km northwest of the town of Parkfield show a ~200 m-wide zone of high porosity material, with multiple slip planes and average velocity reductions of ~30%-35% at 3 km depth (Hickman et al., 2007).

The distinct low-velocity fault zone naturally forms a waveguide to trap seismic waves as a source is located within or close to it. Since the fault-zone trapped waves (FZTWs) were first discovered at the Oroville and San Andreas fault zones in California and simulated using a simple fault-zone model with plane layers space for SH-Love trapped waves (Li et al., 1990; Li and Leary, 1990; see Appendix in this chapter), these waves have been shown to have ability to reveal detailed information of the fine structure at the heart of fault zones and its lateral variation (e.g., Li et al., 1994a, 1994b, 1998, 2002; Malin et al., 1996; Ben-Zion et al., 2003). Because FZTWs arise from constructive interference of multiple reflections at the boundaries between the low-velocity fault zone and high velocity surrounding rocks, the feature of trapped waves (including amplitudes and frequency contents) is strongly dependent on the fault-zone geometry and physical properties (Li and Vidale, 1996; Ben-Zion, 1998; Igel et al., 2002). We can resolve fault-zone width from tens to several hundreds of meters using the records of FZTWs.

The LVZ on the Parkfield segment of SAF has been studied using FZTWs (e.g., Li et al., 1990, 1997, 2004; Korneev et al., 2003; Li and Malin, 2008, 2010; Wu et al., 2010). The damage zone is estimated to be ~200-m-wide and extends to depths of at least 7 km, with seismic velocities reduced by 25%-50% and seismic attenuation Q values of ~10-50. This zone experienced coseismic damage and post-seismic healing on a logarithmic time scale during and after the 28 September, 2004 $M6$ Parkfield Earthquake (Li et al., 2006, 2007). In the present chapter, we use the data recorded at the SAFOD borehole seismographs and the dense linear arrays across the San Andreas fault for a large number of aftershocks of the 2004 M6 Parkfield Earthquake and local microearthquakes to determine the volume and magnitude of the low-velocity anomalies on the SAF at seismogenic depths. The SAFOD borehole seismograms include many hundreds of examples of FZTWs recorded at the seismograph installed at the depth of ~3 km where the SAFOD is deviated from main hole drilled into the core of the SAF. We analyze the surface and borehole data to examine the amplitude-frequency contents of FZTWs, and the manner which is related to the physical properties of the fault damage zone, including vertical and lateral velocity variations and damage zone widths across the SAF at the SAFOD site. The SAFOD borehole data have high signal-to-noise ratios and are less contaminated by the near-surface geological complexity, giving us better constraints on the damage structure of the SAF with high-resolution at seismogenic depth using FZTWs. We also examine the variations of rock damage magnitude along the fault line using the data recorded at seismic stations near the town of Parkfield.

3.2 Fault-Zone Trapped Waves at the SAFOD Site

We systematically analyzed the data recorded at the dense linear surface array that was deployed across and along the San Andreas fault near the SAFOD site in the fall of 2003 for ~120 local Earthquakes. We then combine these results with the data of ~350 aftershocks of the 2004 *M*6 Parkfield Earthquake recorded at the SAFOD borehole seismographs for fault zone trapped waves. Figure 3.1 and Figure 3.2 show the location of SAFOD site, approximately 14.5 km northwest of Parkfield, California and 1.7 km southwest of the surface trace of the San Andreas Fault (SAF). At this location, the SAF separates Pacific

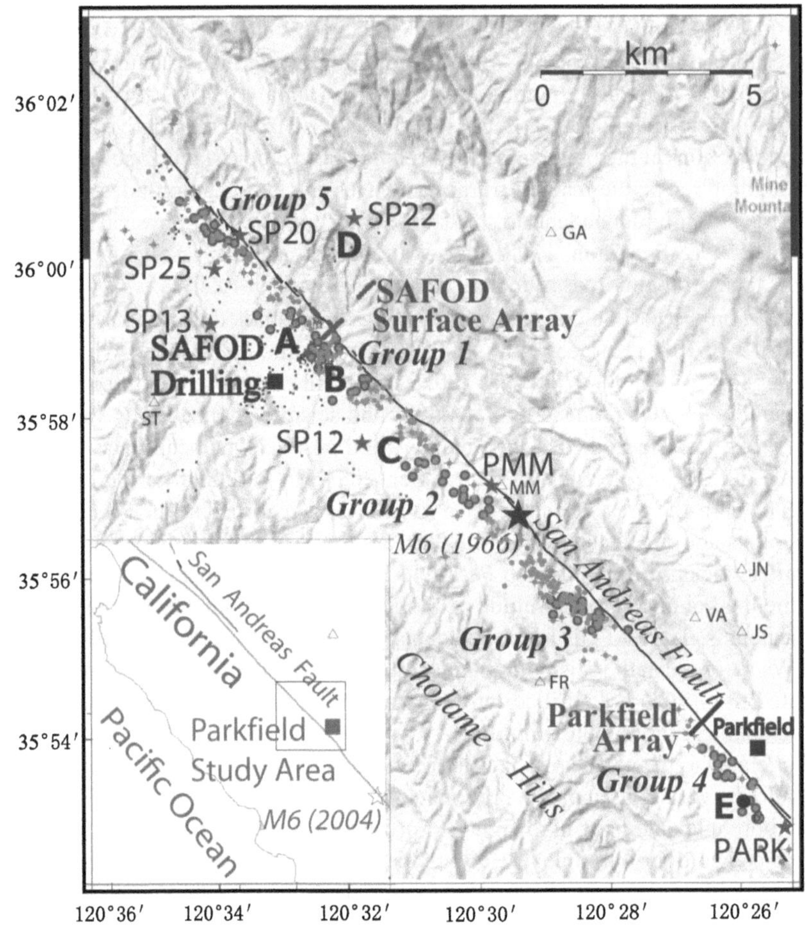

Fig. 3.1 Location of the study area (box in inset map). Circles denote ~350 aftershocks of the 2004 *M*6 Parkfield Earthquake, signals from which were recorded at the SAFOD borehole seismographs. The aftershocks marked by black outlines occurred within the fault zone and are divided into 5 groups based on epicentral distances from the SAFOD site. Grey dots are 120 microearthquakes and stars are 5 explosions SP1-SP5, signals from which were recorded at the surface seismic array deployed near the SAFOD site in 2003. Event A, the SAFOD drilling target, occurred at ~3 km depth while Events B, C and D are deep events occurring within and away from the fault zone recorded at the SAFOD surface array in 2003. Two explosions PMM and PARK detonated within the fault zone, signals from which were recorded at the cross-fault array deployed ~1.5 km northwest of the town of Parkfield in the experiment in 2002 (Li et al., 2004). Event E is an aftershock recorded at the Parkfield surface array in 2004, seismograms from which are shown in Figure 3.18a. Triangles are Parkfield borehole network stations; first observed fault-zone trapped waves at station MM are shown in Figure 3A-2.

Plate Salinian rocks from North American Plate Franciscan mélange (Fig. 3.2a). Vertical seismic profiling (VSP) in the SAFOD Pilot-Hole from nearby micro-earthquakes contains signals scattered by the local geologic structure, the strength of which suggests sharp contrasts in material properties, as in cracks and fluids of the SAF at depth (Chavarria et al., 2003; Malin et al., 2006). Additionally, it is the transition between the creeping and locked segments of the SAF (Nadeau and McEvilly, 1997). SAFOD's goal is to investigate small, persistently repeating earthquakes ("target" earthquakes), as well as the structure and mechanics of this unique section of fault.

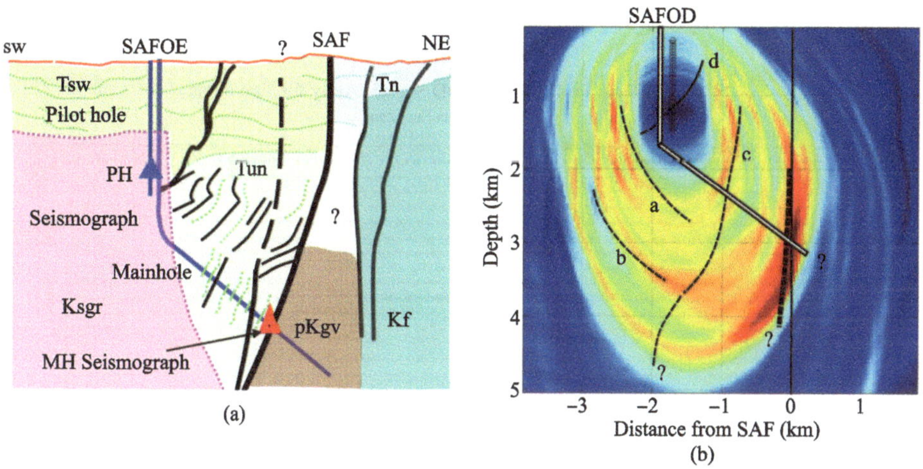

Fig. 3.2 (a) SW-NE cross-section along the SAFOD Main Hole towards the surface trace of the SAF. The borehole seismographs installed in the Main Hole (MH) and Pilot Hole (PH) (denoted by red and blue triangles). The geological interpretation is based on the results of a Drill Bit Seismic reflection profile gathered during the 2004 Phase 1 drilling (Ellsworth and Malin, 2006). Ksgr = Gabilan Granite; pKgv = pre-Cretaceous Great Valley formation; Kf = Franciscan mélange; Tsw and Tn = Tertiary cover SW and NE of the SAF surface trace; Tun = unnamed tilted Tertiary encountered along MH drilling. (b) A cross section through cells of migrated microearthquake VSP signals assuming P-to-S scattered waves recorded at the array of 32-levels of seismographs (shown by squares) installed in the SAFOD PH (Chavarria et al., 2003; Malin et al., 2006). The sections intersect a grid of 141*124 cells, each 40 m on a side. The solid line at $x = 0$ km is the extension of the surface trace of the SAF to depth. Microearthquakes located > 4.5 km away and underneath the PH were used in the migration. The colors indicate the amplitude of scattered energy. The highest energy scattering cells (black dash line) correlate is interpreted as the downdip extension of the San Andreas fault. Interpreted faults from other scattering modes (grey dash line) are included.

3.2.1 The SAFOD Surface Array

In the fall of 2003, as part of a seismic characterization program, we deployed a dense linear seismic array of 31 PASSCAL RT130 seismometers across the San Andreas Fault near the SAFOD site ~15 km NW of Parkfield (Fig. 3.1). The array recorded fault-zone trapped waves (FZTWs) to provide site characterization prior to the start of SAFOD drilling. Station spacing was 25 m for central part of the array near the main fault trace at the surface, and increased to 50 m, 100 m or more for stations located farther away from the fault. A three-component 2 Hz L22 sensor was buried at each station site. Station ST0, the center of the linear array, was located at the main trace of the SAF at the surface. The three-components of the sensor were oriented vertically, parallel and perpendicularly to the fault strike. Seismometers were powered by deep-cycle batteries charged by solar panels. Locations and internal clocks of seismometers were synchronized by GPS. We recorded ~120 local earthquakes with hypocentral depths between 2 and 12 km over roughly two months and 5 explosions in a fan-geometry detonated by the USGS. The FZTWs generated by the near-surface explosions are used to delineate the low-velocity

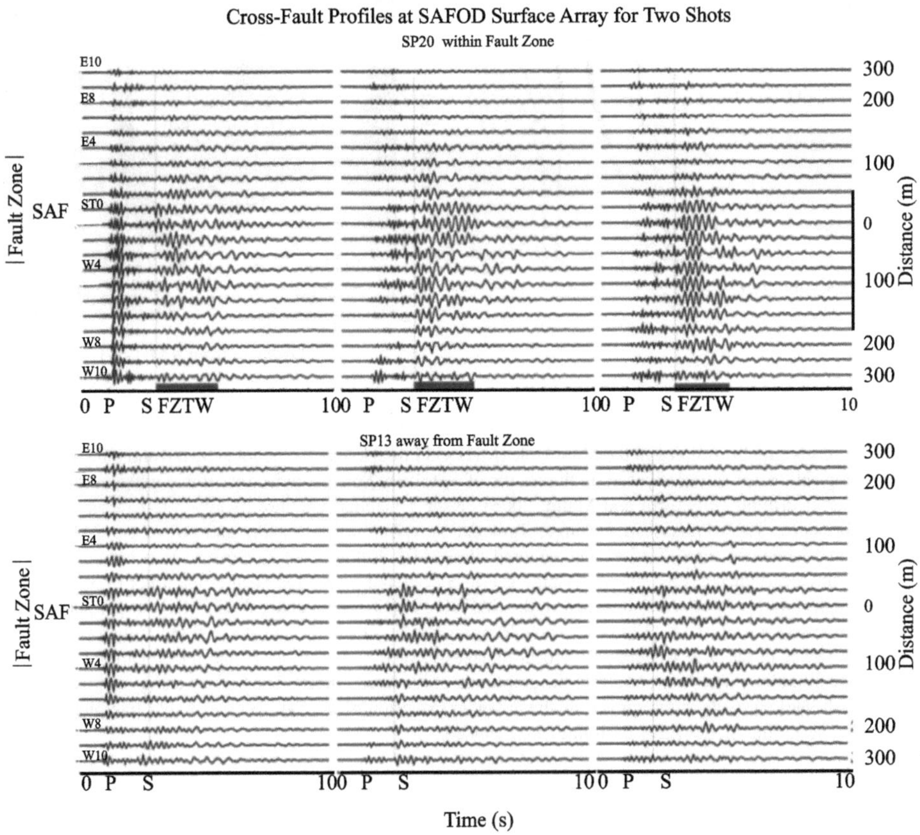

Fig. 3.3 Three-component seismograms for shots SP20 and SP13 detonated within and away from the fault zone recorded at the surface array (see Fig. 3.1). Station ST0 of the array was located on the SAF main fault trace. Seismograms have been low pass filtered at 3 Hz and are plotted using a fixed amplitude scale for each event. The recordings have been deconvolved by the sensor and instrument responses. Prominent fault-zone trapped waves (FZTWs) with large amplitudes and long wavetrains (marked by brackets) following *S*-arrivals are observed on seismograms for shot SP20 at stations located close to the fault trace, within ~200 m, as marked by the vertical grey bar. By contrast, body waves with brief wavetrains are dominant in seismograms for shots SP22 and SP13 although some scattering of seismic energy is shown at stations located within the fault zone.

structure on the SAF at shallow depth and then microearthquakes occurring at greater depths first are used to examine depth extent of the damage zone.

Figure 3.3 shows seismograms of 3 shots detonated within and outside the fault zone ~3 km NW of SAFOD and recorded at the surface array deployed across the SAF main trace. We observed prominent FZTWs with large amplitudes for shot SP20 detonated within the fault zone, at stations located within a 150-200-m wide zone near the main trace, but FZTWs are much weaker for shot SP13 detonated ~2 km away from the fault zone. The FZTWs are characterized by relatively high amplitude, low frequency, long-duration wavetrains with slight dispersion following the S-arrivals and result from constructive interference of critically reflected waves (Li and Leary, 1990; Li et al., 1990). Their presence indicates a velocity reduction within the fault zone where seismic energy is trapped, known to be confined to faults with low-velocity cores. Our observations of FZTWs generated by the shot detonated within the fault zone indicate that a distinct low-velocity zone exists along the SAF, at least for shallow depths. We note that the low-velocity damage zone is asymmetric with the main fault trace, with the LVZ extending farther on the southwest side. This observation is consistent with our previous study near the town of Parkfield, ~15 km southeast of the SAFOD site (Li et al., 2004). The asymmetry may imply that the fault zone has a significant cumulative damage due to previous large earthquakes on the SAF. When a fault ruptures, it may preferentially damage the already weakened rocks in the zone, even though those rocks are not symmetrically distributed on either side of the main slip plane (Chester et al., 1993).

Seismograms for ~35 events in our records in 2003 show clear fault-zone trapped wave energy. Figure 3.4a shows 3-component seismograms for 3 micro-earthquakes (events A, C and D in Fig. 3.1 and Fig. 3.5). These events occur over a range of source and-receiver paths, which can be used to determine the velocity structure of the SAF at SAFOD. The first observation is that events A and C whose source-and-receiver points are closest to the projected fault trace generate the most prominent, longest lasting, fault-zone trapped waves. The second observation is that the time duration of these signals is a strong function of event depth. Seismic envelopes recorded within the fault zone for these 3 earthquakes show a correlation of longer duration FZTWs with increasing event depth (Fig. 3.4b). Taken together, these two characteristics imply that the FZTWs result from a relatively continuous low velocity zone between the surface and events at seismogenic depths along these source-and-receiver paths. Again, we note that the damage zone is not symmetric, but is broader on the southwest side of the main fault trace. In addition, some seismic energy appears to be trapped within a secondary fault that passes near stations W12 and W14 located ~850-1,050 m southwest of the SAF main trace (Fig. 3.4a).

It would seem that the newly discovered fault must be connected to the main fault section to the northeast, but below the SAFOD target earthquake at ~3 km depth. This is likely the result of the flower structure of the Middle Mountain segment of the SAF, characterized by divergent fault splays and a synclinal formation, typical of strike-slip fault zones. It is not surprising that this branch has not been identified until now, given that any obvious surface expression would be hidden; rapid uplift has left many of Middle Mountain's flanks covered with landslides [provided by M. Rymer, unpublished data, 2005].

In order to examine the depth extension of the low-velocity zone on the SAF, we used

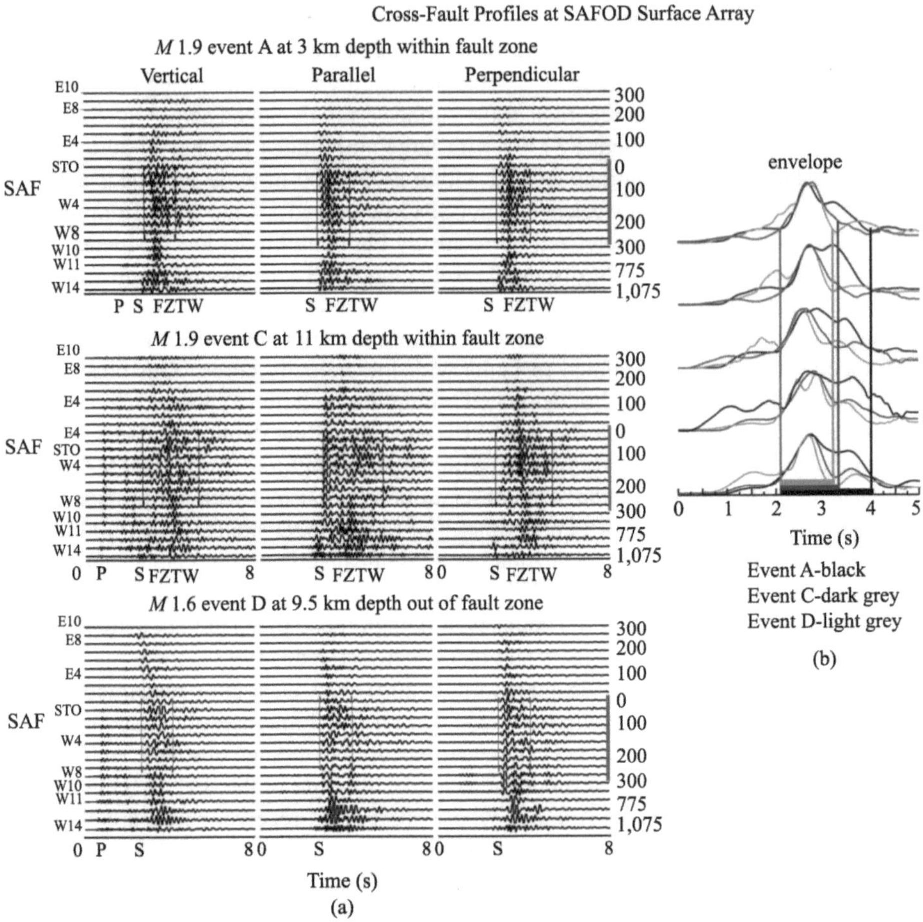

Fig. 3.4 (a) Three-component seismograms for 3 micro-earthquakes A, B and C (locations shown in Fig. 3.1 and Fig. 3.5) recorded at the surface seismic array. Seismograms have been low pass filtered at 6 Hz and plotted using a fixed amplitude scale. Prominent fault-zone trapped waves (FZTW) with large amplitudes and long wavetrains are observed following the *S*-arrivals for on-fault events A and C. FZTWs are observed in a ~200-m wide zone across the SAF between stations E6 through W4. Vertical arrows indicate FZTW arrivals. The FZTW durations for event C, occurring at ~11 km depth, are ~2.2 s long, longer than the FZTW durations for event A (1.2 s) that occurred at 3 km depth. The wavetrain for the off-fault event D is much lower amplitude and shorter in duration. (b) Normalized envelopes of vertical-component seismograms for 5 stations (ST0-W4) located within the fault zone. Longer duration FZTWs energy is observed for the deeper event (black lines) other than for the shallower (dark grey lines) and the off-fault events (light grey lines). Envelopes are aligned on the *S*-arrivals. Horizontal bars denote FZTW durations for FZTW amplitudes twice greater than background noise amplitude.

the data from 33 local earthquakes located within the fault zone at different depths with the raypath incidence angles to the array smaller than 30° from vertical (Fig. 3.5a). Figure 3.5b illustrates seismograms and envelopes at station ST0 located on the main fault trace for 11 on-fault events at different depths near SAFOD site, showing a move-out of trapped

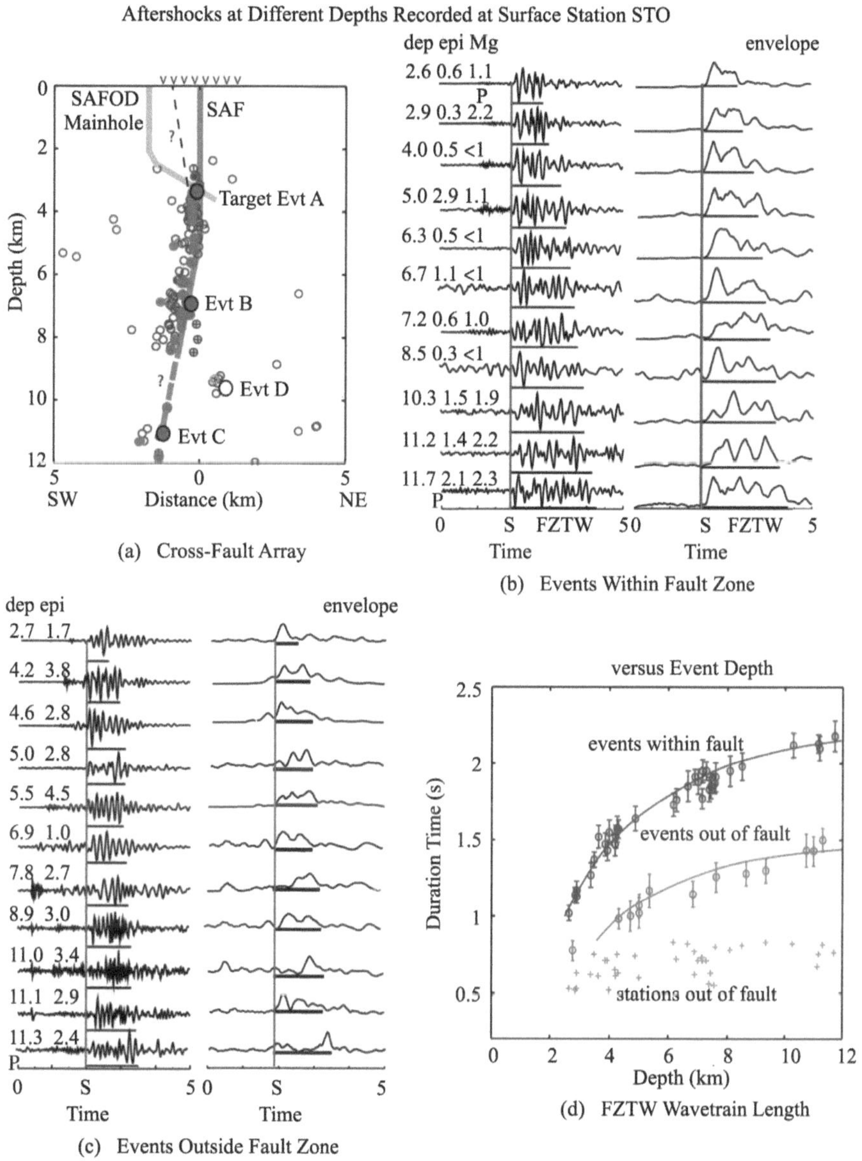

Fig. 3.5 (a) Vertical section across the SAF near the SAFOD main-hole (grey line) shows locations of micro-earthquakes (circles) recorded at our seismic array in 2003. Solid (open) circles denote events showing (without) FZTWs with long wavetrains. Waveforms of events A-D are shown in Figure 3.4. The branch fault (thin dash lines) may connect to the main fault at depth. (b) Vertical-component seismograms and envelopes at station ST0 on the SAF main trace for 11 on-fault earthquakes show an increase in wavetrain length (marked by solid horizontal bars) of FZTWs as event depths increase. *S*-arrivals for these events are aligned at the same time. Seismograms have been < 6 Hz filtered. (c) The same plot but for 11 off-fault events in the similar range of depths. (d) The measured FZTW wavetrain lengths versus focal depths for on-fault (black circles) and off-fault (grey circles) events recorded at stations within and out of the fault zone. Each data point is averaged from measurements at 4 stations for the event. Error bars are standard deviations. Curves are polynomial fits to the data. Grey crosses denote the data at off-fault stations for all these events.

energy in traveltime with travel distance (Li and Malin, 2008). The time duration of fault-zone trapped wavetrains following S-arrivals progressively increases from $\sim$1.2 s to $\sim$2.4 s as the event depths increase from 2.6 km to 11.7 km. By contrast, seismograms and envelopes recorded at the same station for 11 other events located away from the fault zone in the similar depth range show much shorter wavetrains ($<$ 1.5 s) after S-arrivals and without move-out (Fig. 3.5c).

Figure 3.5d shows the measured wavetrain lengths of FZTWs registered at stations within the fault zone for 33 on-fault events and 13 off-fault events at depths between 2 km and 12 km. The lengths (time durations) of FZTWs for on-fault events increase from $\sim$ 1.0 s to $\sim$ 2.2 s as the depth increases from $\sim$ 2 km to $\sim$ 12 km, but shorter wavetrains with flat depth-dependent changes are measured at the same stations for off-fault events. Stations located out of the fault zone registered much short wavetrains after S-arrivals for all the events. These observations indicate that the low-velocity waveguide formed by damaged rocks on the SAF extends across seismogenic depths to at least $\sim$ 7 km although the velocity reduction (damage magnitude) within the zone becomes smaller with decreasing depth due to the larger confined stress at greater depths.

3.2.2 The SAFOD Borehole Seismographs

In the following section, we examine aftershock data from the 2004 Parkfield Earthquake recorded at the SAFOD borehole seismographs. The SAFOD observatory currently consists of two boreholes (Fig. 3.2a). The vertical 2.1-km-deep pilot-hole (PH) was drilled in summer 2002 and operated as both a seismic monitoring station and instrument test bed. The deviated main-hole (MH) was drilled vertically for about 2 km, then the drillstring was deviated by $\sim$ 55° from vertical and steered toward the target earthquakes, which were thought to lie near the vertical down-dip extension of the SAF surface trace. In the fall of 2004, the bottom of the MH was $\sim$ 0.7 km west of the surface trace, passing a highly fractured lithological contact seen in the recovered drill core (Solum et al., 2007). The MH reached a depth of 3.1 km in the summer of 2005.

Following completion of the PH, a multi-level array (consisting of 3-component seismographs spaced 40 meters apart) was installed in the deepest portion of hole. SAFOD investigators have studied the PH data extensively in an effort to locate the drilling program's "target" earthquakes (Thurber et al., 2004). The PH data from microearthquakes has been used to produce a 2-D tomographic P-wave velocity cross-section through the drill site and SAF. The tomographic model detail was improved (Chavarria et al., 2003; Malin et al., 2006) using results from a 2003 refraction line along this same section. These data show that a zone of low seismic velocities exists between the drill pad and the SAF with its southwest edge being $\sim$ 0.7 km west of the surface trace (Fig. 3.2b). These data imply that the structure seen in the migration image and drill core is a fault zone that enables trapping of seismic waves from the aftershocks occurring on the main SAF at depths beneath the MH seismograph.

In December of 2004, a 3-component, 4.5 Hz seismograph was installed in the SAFOD MH at $\sim$ 3 km depth where the highly fractured and low velocity zone of the SAF was found in the SAFOD drilling and well logs (Hickman et al., 2007). The MH seismograph and a smaller group, consisting of the seven levels of the original array in PH, recorded ~ 350 aftershocks of the 2004 *M6* Parkfield Earthquake that occurred over a range of hypocentral distances and depths (Fig. 3.1 and Fig. 3.6a). Of these data, ~ 80 aftershocks

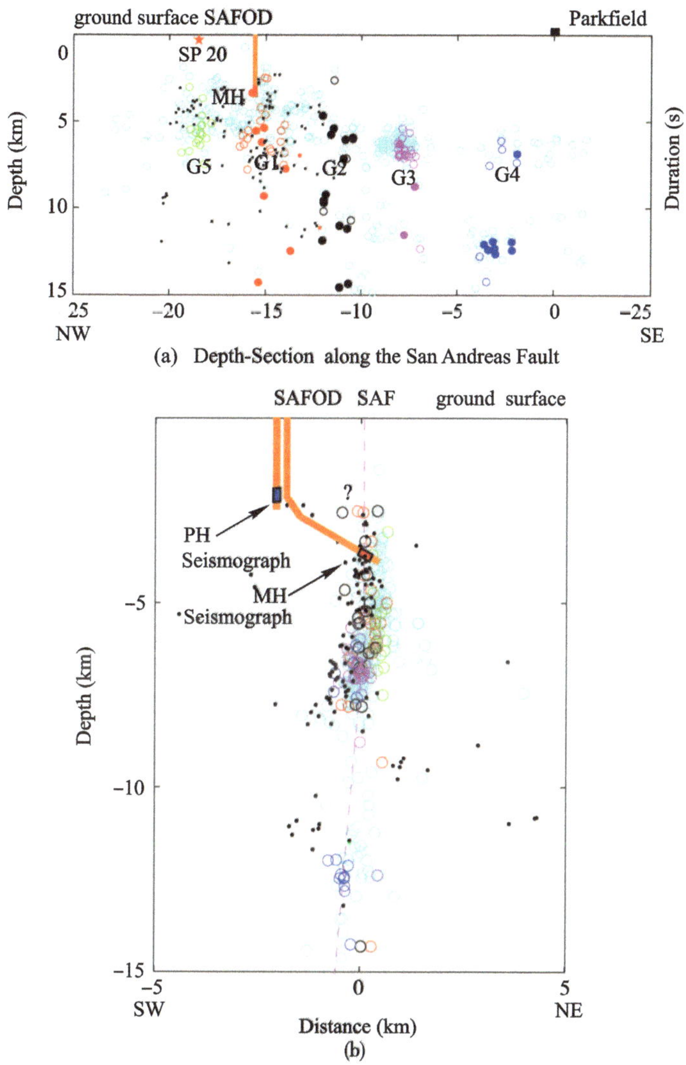

Fig. 3.6 (a) Cross-section parallel to SAF strike showing the locations of ~ 350 aftershocks (circles) of the 2004 *M6* Parkfield Earthquake recorded at the SAFOD MH seismograph (red square) located at ~ 3 km depth on the SAF. The aftershocks in five groups G1-G5 (denoted by red, black, pink, blue and green colors, respectively) used in the waveform analysis are also shown. Other aftershocks are denoted by cyan color. Small black dots denote the micro-earthquakes recorded at the surface array across the SAF near the SOFOD site in 2003. (b) Cross-section across the SAF showing the location of ~ 350 aftershocks (circles) recorded at SAFOD borehole seismographs. The red, black, pink, blue and green circles denote the on-fault aftershocks for Groups 1-5. Other notations are the same as in (a).

contain clear fault-zone trapped waves (Fig. 3.6b). In order to relate the amplitude and dispersive feature of FZTWs to travel distance, we sorted those aftershocks showing clear FZTWs into 5 groups at the epicentral distance ranges of 1-2 km (G1), 4-6 km (G2), 8-10 km (G3) and 14-16 km (G4) southeast of the SAFOD site and 3-5 km (G5) northwest of the site (Fig. 3.6). These aftershock data are used in this study to investigate the detailed internal structure and rock damage of the SAF at seismogenic depths. Accurate aftershock locations were made available in the Northern California Seismic Network Catalog.

For example, Figure 3.7 shows seismograms recorded at the SAFOD borehole seismographs for clustered aftershocks in group G2 occurring within the fault zone at ~ 4.4 km

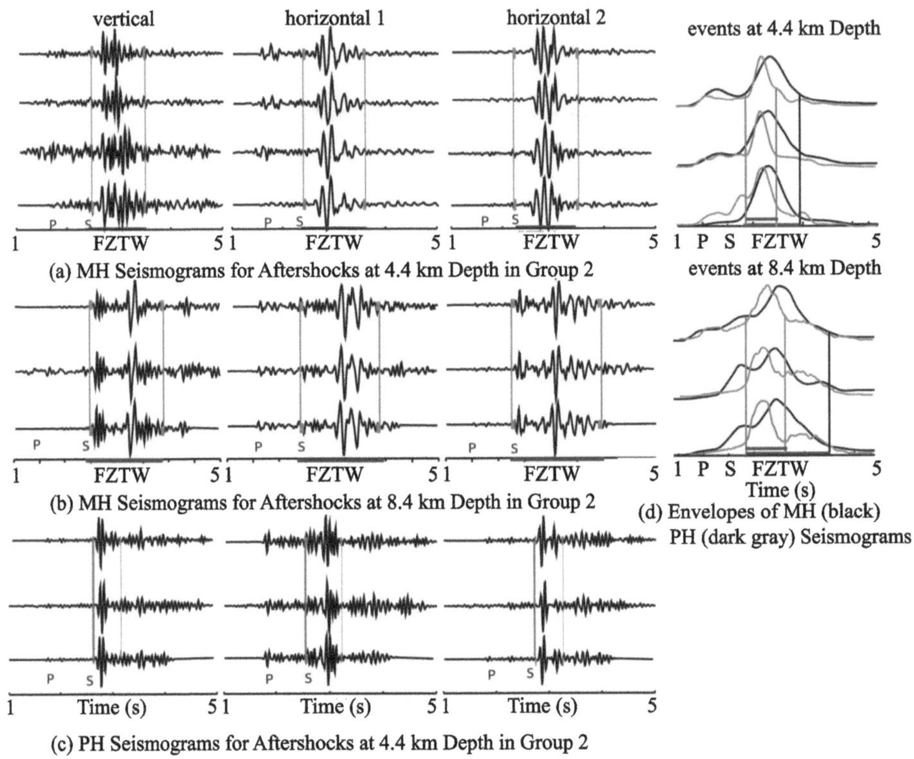

Fig. 3.7 (a) 3-component MH seismograms for 4 clustered on-fault aftershocks at ~ 4.4 km depth in group G2 with epicentral distances of 4-6 km from the SAFOD site show prominent FZTWs. Seismograms have been < 10 Hz filtered and are trace-normalized in plot. The recordings have been deconvolved by the sensor responses. (b) The same as in (a) but for 3 clustered on-fault aftershocks at ~ 8.4 km depth in Group 2 showing longer time durations of FZTWs after *S*-arrivals than those for shallower events at ~ 4.4-km depth, indicating a continuous low-velocity zone formed by damage rocks on the SAF extending to the depth at least below the aftershock at 4.4-km depth. (c) PH seismograms show high-frequency body waves at high frequencies for the same aftershocks in (b). (d) Normalized envelopes of amplitudes derived from MH (black line) and PH (dark grey line) seismograms for these aftershocks. The horizontal bars denote the time duration of FZTW wavetrains after *S* waves, within which the amplitudes of FZTWs are at least twice higher than the amplitude of later coda waves.

depth and 8.4 km depth. We observed prominent FZTWs with large-amplitudes and long wavetrains at low frequencies following *S*-arrivals at the MH seismograph. By contrast, the PH seismograph registered brief body waves with short wavetrains at high frequencies for the same events. The amplitude envelopes of MH seismograms for aftershocks at the 8.4 km depth show longer time duration ($\sim$ 1.5 s) of FZTW wavetrains than those ($\sim$ 1 s) for events at the 4.4 km depth, indicating that a continuous low-velocity waveguide formed by damage rocks on the SAF extends to the depth at least below $\sim$ 5 km.

To more clearly show the variation in FZTW duration with depth, Figure 3.8a shows MH and PH seismograms for 6 on-fault aftershocks in group G5 occurring at different depths northwest of the SAFOD site. Prominent FZTWs with long wavetrains after *S*-arrivals are observed in the MH seismograms. The time duration of FZTWs increases from $\sim$ 0.7 s to 1.8 s as the event depth increases from 2.7 km to 8.2 km. However, the PH seismograms show high-frequency body waves with short wavetrains and flat changes in duration for the same aftershocks. The resulting observations indicate that a low-velocity waveguide formed by damage rocks on the SAF connects the MH seismograph and these on-fault events. The low-velocity waveguide likely extends to the depth of at least $\sim$ 7 km near the SAFOD site.

We compared the SAFOD borehole data recorded for on-fault and off-fault aftershocks. For example, Figure 3.8b shows the seismograms for two *M*1.3 events occurring at the similar depths of $\sim$ 5 km and similar epicentral distances of $\sim$ 4 km northwest of the SAFOD site. One aftershock was located within the fault zone while the other aftershock occurred $\sim$ 3 km away from the fault. The MH seismographs show prominent FZTWs with large amplitudes and long wavetrains ($\sim$ 1.6 s time duration) after *S*-arrivals for the on-fault event while the body waves with brief wavetrains are dominant in PH seismograms for the same event. By contrast, both MH and PH seismograms show short wavetrains after *S*-arrivals for the off-fault event. These observations confirm the existence of a low-velocity waveguide formed by damage rocks on the SAF that connects the SAFOD MH seismograph at $\sim$ 3 km depth and the on-fault aftershocks occurring beneath it.

We also compared the waveforms recorded at the SAFOD MH and PH seismographs for aftershocks in Groups G1-G4 (Fig. 3.9). Figure 3.9a and Figure 3.9b illustrates seismograms for four on-fault aftershocks in groups G1 and G2, respectively, near the SAFOD site to show the variation in wavetrain durations of FZTWs with epicentral distance and depth. Prominent FZTWs with large amplitudes and long wavetrains appear after *S*-arrivals in the MH seismograms while the PH seismograms are dominated by body waves with brief wavetrains for the same events. The time duration of FZTW wavetrains increases as the travel distance between the MH seismograph and the on-fault event increases. FZTWs from deeper aftershocks in groups G1 and G2 occurring at 7-8 km depths exhibit longer duration ($\sim$ 1.2 s) of wavetrains after *S*-arrivals than those (0.5-1 s) from shallower events at 4-5 km depths and the deepest events at 13-14 km depths exhibit the longest duration ($\sim$ 1.5-2 s) of wavetrains. By contrast, PH seismograms show much shorter wavetrains (less than 0.5 s) with flatter changes for events at different depths, indicating that a low-velocity waveguide (damage zone) on the SAF likely extends to the depth of at least 7-8 km.

Figure 3.9c and Figure 3.9d shows the data recorded for aftershocks in groups G3 and

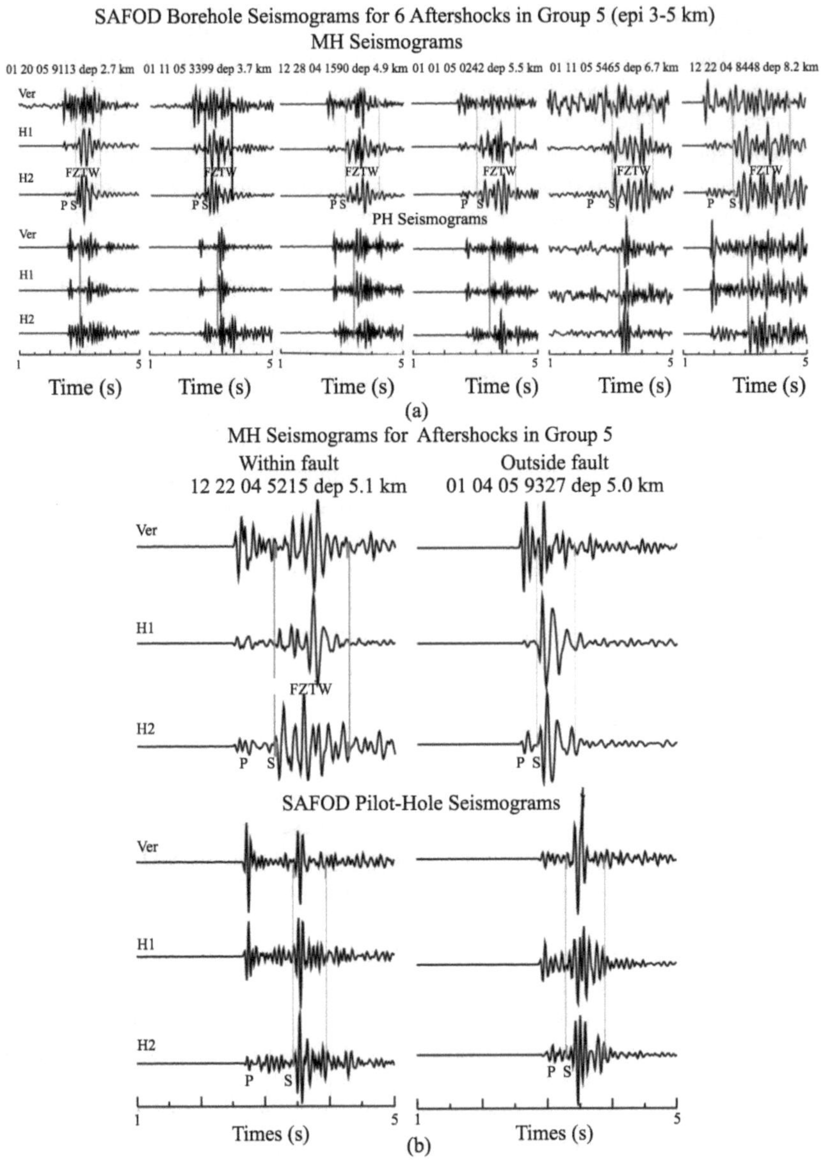

Fig. 3.8 (a) 3-component seismograms recorded at the SAFOD MH and PH seismographs for 6 on-fault aftershocks in group G5 located at 3-5 km northwest of the SAFOD site and at different depths show prominent FZTWs in the MH seismograms with long wavetrains after *S*-arrivals while high-frequency body waves with brief wavetrains are dominant in the PH seismograms. H1 and H2 denote two horizontal components of seismograms. The event date, index and focal depth of these aftershocks are plotted at the top of seismograms. Seismograms have been < 10 Hz filtered and are plotted in trace normalized. FZTWs travelling longer distances from the deeper events show longer durations (marked by two vertical bars) than those from the shallower events, indicating a low-velocity waveguide formed by damage zones on the SAF extending to deep seismogenic depths. (b) 3-component seismograms recorded at the SAFOD MH and PH seismographs for on-fault (left) and off-fault (right) aftershocks at $\sim$ 5-km depths. The waves from the on-fault aftershock in Group G5 show prominent FZTWs with long wavetrains after *S*-arrivals in MH seismograms while waves from the off-fault event occurring 3 km away from the fault zone show much shorter wavetrains after *S*-arrivals. Seismograms have been <10 Hz filtered and amplitudes are normalized.

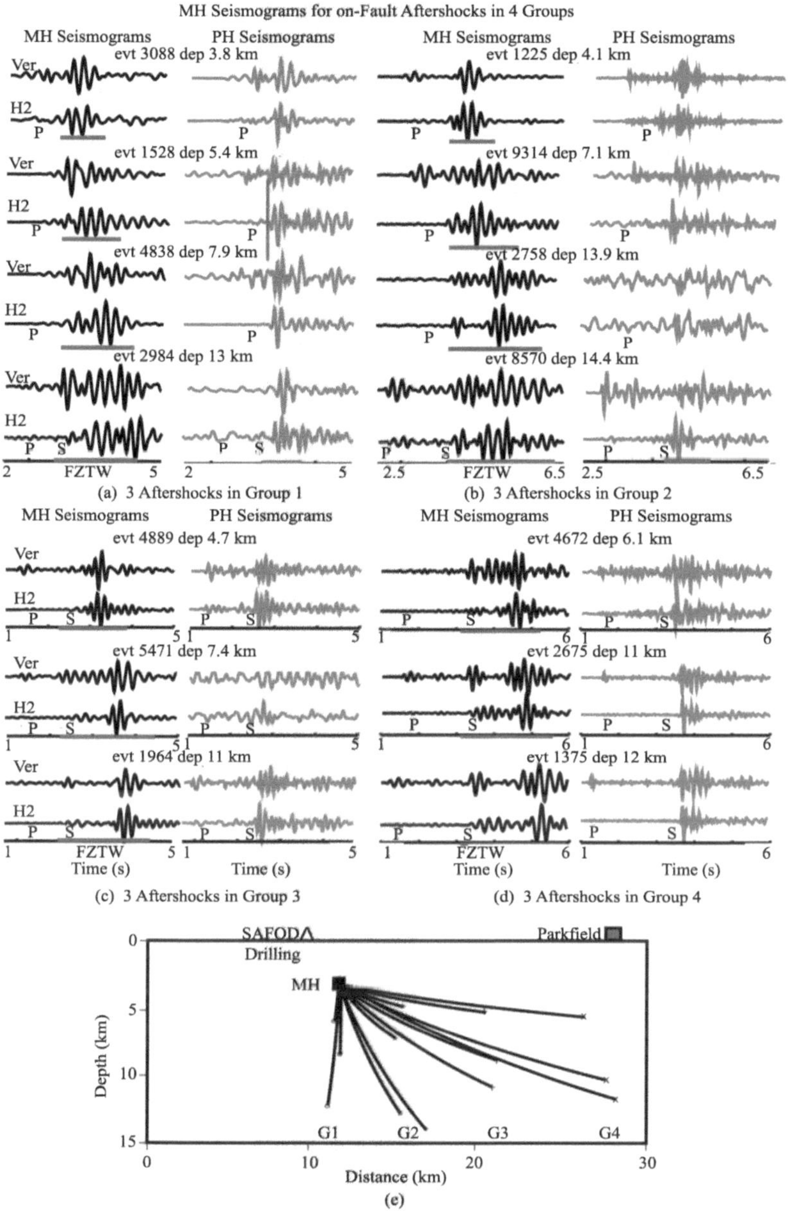

Fig. 3.9 Vertical and horizontal seismograms recorded at the MH and PH for aftershocks occurring within the fault zone at different depths in 4 groups with different epicentral distance ranges. (a) group G1 located 1-2 km SE of the array, (b) group G2 located 4-6 km SE of the array, (c) group G3 located 8-10 km SE of the array, (d) group G4 located 14-16 km SE of the array. Earthquakes are ordered by source depth and aligned on the *S*-arrival. Event index and focal depths are listed above the seismograms. Seismograms have been < 6 Hz filtered and are normalized by amplitude. Prominent FZTWs with large amplitudes and long wavetrains are observed after *S*-arrivals in MH seismograms while the high-frequency body waves with brief wavetrains are dominant in PH seismograms. (e) Raypaths from aftershocks in 4 groups shown in (a)-(d) to SAFOD MH seismograph.

G4 with epicentral distances between 8-10 km and 14-16 km southeast of the SAFOD site, respectively, to illustrate prominent FZTWs in the MH seismograms. The wavetrains of FZTWs show 1.8 s to 2.6 s duration after *S*-arrivals, longer than those for aftershocks in groups G1 and G2 located at shorter epicentral distances from the array site. Note that the FZTW wavetrains for the events in G4 are even longer than those for the events in G3 at similar depths. These observations indicate that the low-velocity waveguide formed by damage rocks on the SAF at seismogenic depths extends for at least $\sim$ 20 km along strike of the SAF passing through the town of Parkfield and the SAFOD site (Langbein et al., 2005). Also note that the duration of FZTW wavetrains for the deeper aftershocks is longer than that for shallower events. This trend is not as obvious for G4 because the ray paths from the shallower events to the MH seismograph are sub-horizontal and mostly travel in the top part of the fault zone with lower velocities. This difference suggests that the velocity structure on the SAF is depth-dependent.

In order to further examine the depth extension of the low-velocity zone on the SAF, we used the SAFOD borehole data for aftershocks occurring at different depths within the fault zone and with short epicentral distances. The ray-path incidence angles from these events to the seismic array smaller than 30° from the vertical. Figure 3.10a shows seismograms recorded at MH and PH seismographs for 9 on-fault aftershocks of the 2004 *M*6 Parkfield Earthquake, occurring at depths between 3.8 km and 14.4 km. We measured time duration of FZTW wavetrains after *S*-arrivals from MH seismograph, in which wave amplitudes are more than twice the background level in later coda (Fig. 3.10b). The measured durations of these wavetrains progressively increase from $\sim$ 0.5 s to $\sim$ 1.9 s as the event depths increase, showing a move-out of trapped wave propagation in traveltime with event depth. By contrast, much shorter wavetrains (less than 1 s) with flat changes in duration time after *S*-arrivals are registered at the PH seismograph for these same events at different depths. These observations indicate that the low-velocity waveguide formed by the damaged rock on the SAF likely extends from the ground surface across the seismogenic depths. We note that FZTWs recorded at the surface station (Fig. 3.5) show longer wavetrains than those recorded at the SAFOD main-hole seismograph for the events located at the similar depths and epicentral distances, indicating that the low-velocity waveguide formed by damage rocks on the SAF is more prominent at the shallow depth.

In summary, we measured time durations of FZTWs for $\sim$ 80 Parkfield aftershocks in 5 groups G1 – G5 occurring within the fault zone at different depths and epicentral distances and recorded at the SAFOD MH seismograph. Figure 3.10c illustrates the measured time durations of FZTW wavetrains versus event distance from the SAFOD borehole, showing that the duration of FZTWs scales with epicentral distance. These observations indicate that a roughly continuous low-velocity waveguide formed by the severely damaged rocks exists along the SAF over the distance of at least $\sim$ 20 km in the Parkfield area. This low-velocity waveguide extends from the surface to depths of at least 7-8 km, likely extending across the entire seismogenic zone. We note that the trend of increasing length of FZTW wavetrains with increasing travel distance is more marked for aftershocks at shallower depths than that for deeper aftershocks, suggesting that greater velocity contrasts and lower seismic velocities occur within the shallowest portion of the fault zone. Therefore, it implies that the fault-zone rocks at the shallowest depths experienced greater damage

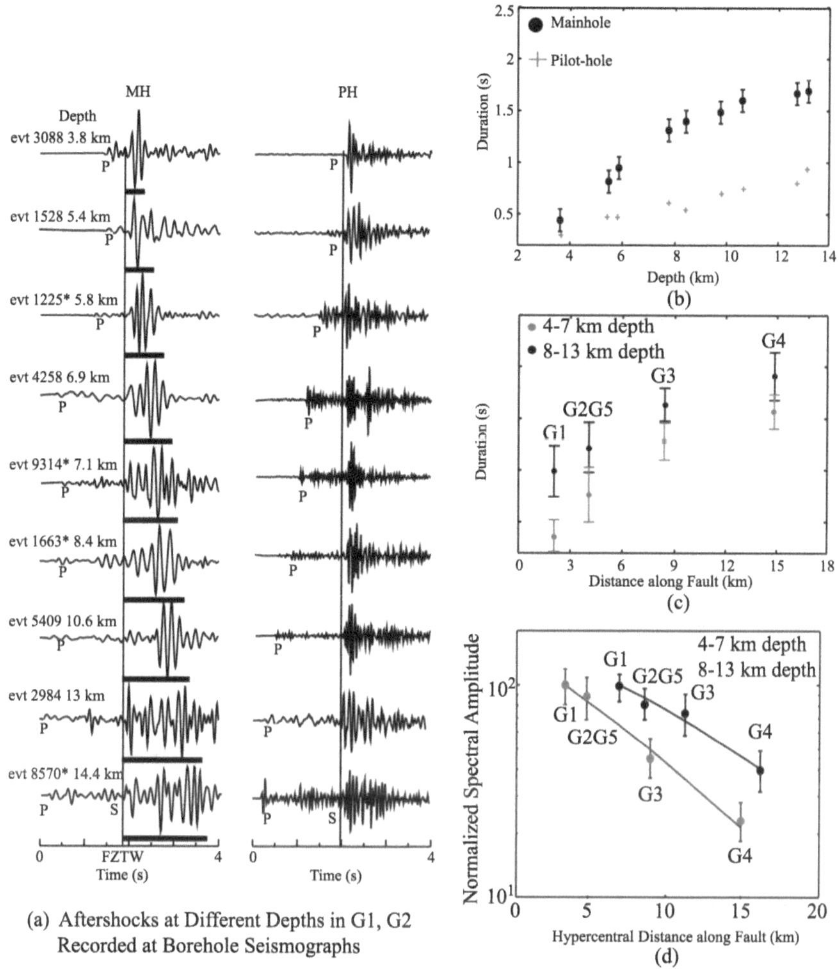

Fig. 3.10 (a) Vertical-component seismograms recorded at the SAFOD MH and PH seismographs for 9 on-fault aftershocks at different depths in groups G1 and G2. The incidence angles of ray paths from these events to the MH seismograph are smaller than 30° to the vertical and traces are aligned on *S*-arrivals. Seismograms have been < 6 Hz filtered. Horizontal bars denote FZTW durations for FZTW amplitudes twice greater than background noise amplitude. (b) Measured FZTW durations from the MH (black dots) and PH (grey crosses) seismograms plotted versus focal depths for 9 on-fault aftershocks in groups G1 and G2. Each data point is the average FZTW duration measurements on the 3 component earthquake data. Error bars indicate standard deviations of the measured durations. (c) Measured FZTW durations from MH seismograph for $\sim$ 80 aftershocks in groups G1 - G5 versus epicentral distance from the SAFOD site. Each data point is the averaged for all on-fault aftershocks in each group. Violet and red dots denote the measurements for aftershocks occurring at depths of 4-7 km and 8-13 km, respectively. Error bars indicate standard deviations of the measured durations in each group. (d) Normalized spectral amplitudes of FZTWs versus hypocentral distances for aftershocks in 5 groups G1 - G5. Each point denotes the mean of coda-normalized spectral amplitude peaks at 4-8 Hz at the SAFOD MH seismograph. We fit the measurements using the formula $\ln(A_1/A_i) = \pi f(r_i - r_1)/QV_s$. Spectral amplitudes of trapped waves are multiplied by a factor of $(r_i/r_1)^{1/2}$, $i = 1, \cdots, n$, to correct geometrical spreading. The light line fits the data for shallow events at depths of 4-7 km using Q of 30 while the black line fits the data for deep events at depths of 8-13 km using Q of 60.

during earthquake ruptures. The smaller velocity reduction within the deeper portion of the fault zone is probably due to the closure of cracks under the larger confining stress at greater depths.

In order to evaluate Q values of the fault rock in the rupture zone, we calculated coda-normalized spectral amplitudes of FZTWs recorded at the MH seismograph versus hypocentral distances for aftershocks in 5 groups. We used the same method to estimate Q values of fault-zone rock at the rupture zones of the 1999 *M*7.2 Hector Mince Earthquake (Li et al., 2002). Normalized spectral amplitudes of FZTWs decrease with the hypocentral distances of these events along the rupture zone. Each point in Figure 3.10d denotes the mean of spectral amplitude peaks between 4 Hz and 8 Hz for aftershocks in each group. The error bar at each point denotes the standard deviation of measurements. The large standard deviation may be due to the contamination of other phases, such as direct S-waves in seismograms, and the heterogeneity of the fault zone. We divided the measurements of normalized spectral amplitudes into two groups for aftershocks at shallow depths of 4-7 km and at deep depths of 8-13 km. We fit the data using the formula ln $(A_1/A_i) = \pi f(r_i - r_1)/QV_s$, where A_i is the normalized spectral amplitude for the event located at distance r_i to the array. The amplitude A_i has been multiplied by a factor $1/\sqrt{r_i}$ to correct geometrical spreading for trapped waves. For shallow events, shear velocity V_s is assumed to be 1.5 km/s and frequency f is 5 Hz. For deep events, V_s is assumed to be 2.0 km/s and f is 7 Hz. We obtained the best fit to the data using a Q of 30 for shallow events and a Q of 60 for deep events, showing that fault zone Q increases with depth. The measured Q values have been used as a constraint on modeling of FZTWs.

3.2.3 Finite-Difference Simulation of Fault-Zone Trapped Waves at SAFOD Site

Based on our observations of fault-zone trapped waves at the SAFOD borehole seismographs and the cross-fault surface array and compared with our previous velocity models at Parkfield (Li et al., 2004), we construct a velocity and Q model of the SAF near the SAFOD site as shown in Figure 3.11a. Model parameters are listed in Table 3.1. The width and velocities of the fault zone at 3 km depth in this model are constrained by the logging measurements in the SAFOD mainhole (Fig. 3.11b) (Hickman et al., 2007). The wall-rock velocities are constrained by the velocity contours from travel-time tomography at Parkfield (Fig. 3.11c) (Thurber et al., 2004). Using a 3-D finite-difference code (Graves, 1996), we simulate FZTWs generated by explosions and recorded at the surface array to determine the shallow 1-2 km fault zone structure. Then, we simulate trapped waves generated by earthquakes at different depths and epicentral distances that are recorded at the borehole seismograph and surface array to obtain a model of the SAF with depth-variable structure at seismogenic depths.

The 3-D finite difference computer code is second order in time and fourth order in space. It propagates the complete wave field through elastic media with a free-surface boundary and spatially variable anelastic damping (an approximate Q). The grid spacing is 12.5 m. The grid volume was changed based on the distance and depth of the events to receivers to minimize computer run time and memory. The fault zone waveguide is sand-

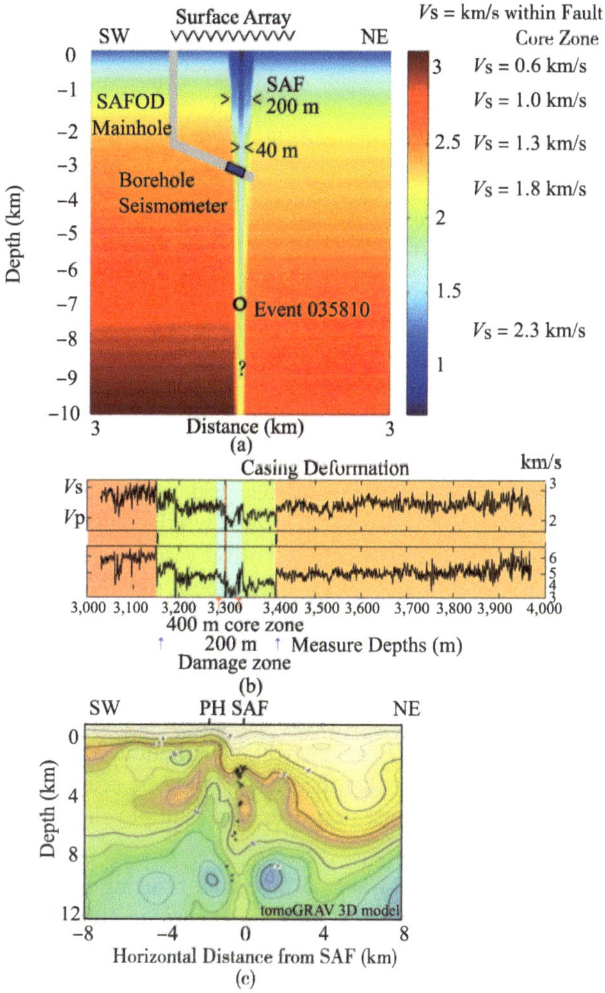

Fig. 3.11 (a) Cross-section of the *S*-wave velocity model across the SAF used to compute synthetic fault-zone trapped waves (FZTWs). The velocities within the 100-200-m wide waveguide on the SAF and surrounding rocks were found by 3-D finite-difference fits to the FZTWs generated by explosions and aftershocks. Model parameters are listed in Table 3.1. (b) SAFOD drilling log data showing a 40-m fault core surrounded by a 200 m low velocity zones (Hickman et al., 2007). The red line indicates the location where fault creep is deforming the borehole casing. (c) The cross-section through the SAFOD site from the tomography DD 3D velocity model (Thurber et al., 2004). Earthquakes within 1 km of the section are shown (filled circles), and the positions of the Pilot Hole (PH) and SAF trace (SAF) are indicated. Depths are relative to sea level. The 0.2 contour of the diagonal element of the model resolution matrix is shown in the result (dash line).

wiched between wall rocks, and placed in the center of the grid. We use a double-couple source for aftershocks and an explosion source for shots. The source was placed within or close to the rupture zone. To find model parameters that best fit observed trapped waves, we tested various values for the fault zone width, velocity and Q, the wall rock velocity

Table 3.1 Parameters for the SAF near Parkfield

Parameters Layer No.	1	2	3	4	5
Depth of the layer, km	0.5	1.0	2.0	6.0	12.0
Waveguide width, m (Damage zone/core)	200/40	200/40	150/40	125/25	100/25
Waveguide *V*s, km/s	0.6	1.0	1.3	1.8	2.3
Waveguide *V*p, km/s	1.5	2.0	2.8	3.8	4.5
Waveguide	10	25	30	50	80
NE wall-rock *V*s, km/s	1.1	1.8	2.1	2.8	3.2
NE wall-rock *V*p, km/s	2.2	3.5	4.2	5.5	6.0
SW wall-rock *V*s, km/s	1.3	2.0	2.5	3.1	3.5
SW wall-rock *V*p, km/s	2.5	4.0	5.0	6.0	6.3
Wall-rock *Q*	30	50	60	100	200

and *Q*, the layer depths, and the source location. To test the various model parameters, we changed the waveguide width by a step of one grid, velocity by 0.2 km/s, and *Q* by 5 in the test ranges, respectively. When the fault zone width varies two grids, or *S* velocity varies 0.2 km/s, or *Q* value varies 10, or source offset varies 2 grids from the fault, or the fault zone depth is only a few kilometers for aftershocks at deeper levels, the amplitudes and dispersion of trapped waves change observably.

To better match *P* and *S* arrival time, we allow the hypocentral distances to float by up to 0.5-1.0 km from catalog locations to account for possible location error and also the lateral heterogeneity along the fault zone. We also note that synthetic *P* waves show smaller amplitudes than recorded *P* waves, while synthetic *S* and trapped waves match observations quite well, indicating that the waves might be scattered by near-fault heterogeneities (e.g., asperities, barriers, step-over, and multiple slip planes), which are not included in our model.

The SAFOD fault-zone trapped waves can be well fit using a tapered, 35-40 m wide fault core with velocities reduced by 40%-50% inside a wider, 100-200 m zone with velocities reduced by ~20%-30%. The relatively intact wall rocks surrounding this composite damage zone have different velocities on the east and west sides the SAF to match the geologic material velocities determined from tomographic modeling. Further, by matching the observed increase in FZTW duration with event depth, it would appear that the main low velocity zone, and probably its interior core extend downward to at least 7 km depth. This is the depth which is determined using the most clearly resolved and the fit test FZTWs seen in the SAFOD borehole seismograms modeled in this study.

For example, Figure 3.12a shows 3-D finite-difference synthetic waveforms using the model in Figure 3.11a to fit seismograms recorded at the SAFOD MH seismograph for 4 on-fault aftershocks in group G1 at depths of 3.8 km, 5.4 km, 7.9 km and 13 km, respectively. The ray paths from these aftershocks to the MH seismograph are nearly vertical. The synthetic FZTWs fit the amplitudes and wavetrains of observed ones quite well. The fault zone in the model extends to 13-km depth. A double-couple source is used for aftershocks at certain depths within the fault zone. The increase in time duration of FZTW wavetrains after *S*-arrivals versus event depth for these on-fault aftershocks shows that the low-velocity damage zone on the SAF likely extends across the entire

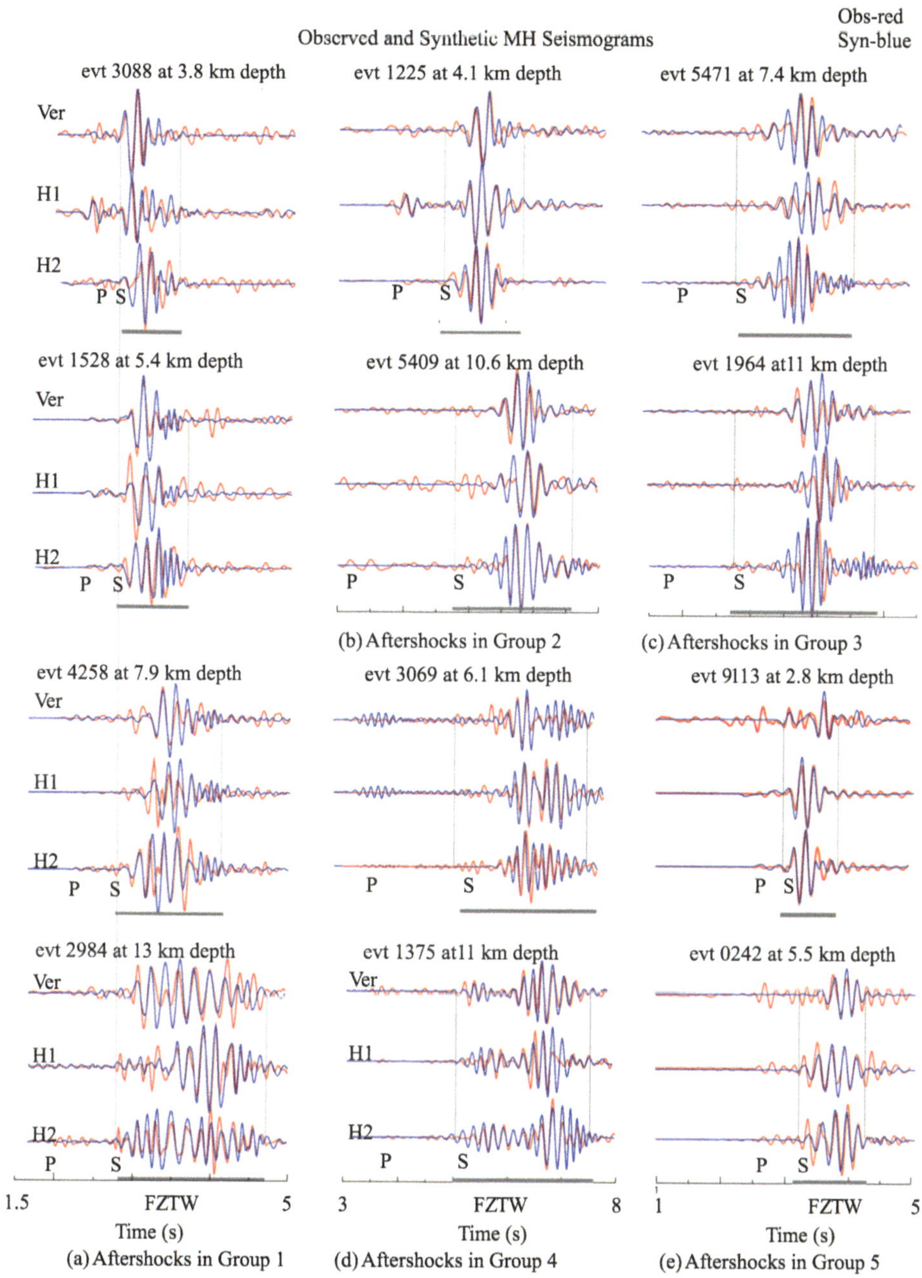

Fig. 3.12 Comparison of observed and synthetic seismograms from the SAFOD MH for Groups 1-5. The observed seismograms are shown in red and synthetic seismograms are shown in blue. Synthetic seismograms are computed using 3-D finite difference techniques, as described in the text, and the velocity model shown in Figure 3.11a and listed in Table 3.1. Observations and synthetics are shown for a range of source depths from 2.8 to 13 km. The model adequately reproduces the observed FZTW amplitudes and durations. Data have been low pass filtered at 6 Hz.

seismogenic depths although the velocity reduction within the deeper portion of the fault zone is smaller than those at the shallower depth.

Figure 3.12b to Figure 3.12e shows observed and the fit test synthetic MH seismograms using the model in Figure 3.11a for on-fault aftershocks at different depths in groups G2-G5 with different epicentral distances to the SAFOD site, including 2 events at depths of 4.1 km and 10.6 km in group G2, 2 on-fault aftershocks at depths of 7.4 km and 11 km in group G3, and 2 events at depths of 7.9 km and 13 km in group G4 located southeast of the SAFOD site as well as 2 events at depths of 2.8 km and 5.5 km in group G5 located northwest of the SAFOD site. The synthetic waveforms fit observed FZTWs with large amplitudes and long wavetrains which elongate either as the event depths or epicentral distances increase.

We further tested the effect of the fault-zone depth on the feature of FZTWs. Figure 3.13 shows 3-D finite-difference MH seismograms for the 2 on-fault aftershocks occurring at depths of 8.4 km and 10.5 km in group G2, using a 12-km deep low-velocity fault zone. The synthetic seismograms are agreeable to observed FZTWs with long-duration wavetrains after *S*-arrivals. However, as the fault zone in the model is truncated at the depth of 4 km, the computed seismograms show short wavetrains after *S*-arrivals, which cannot match the long wavetrain of observed FZTWs. This test manifests that the low-velocity waveguide formed by damage rocks on the SAF at Parkfield likely extends to the depth of ~ 8 km or more.

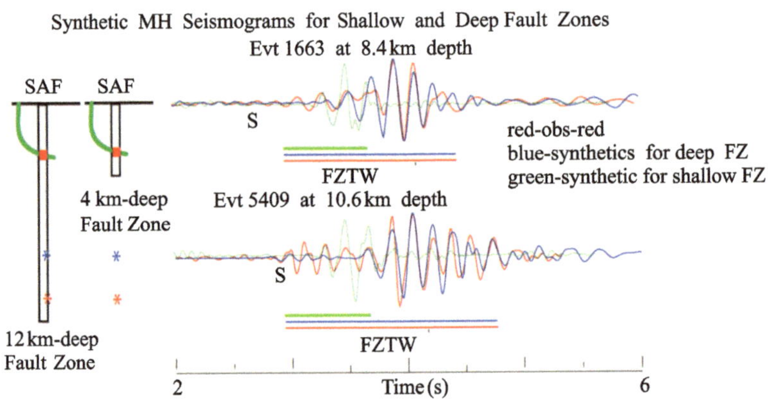

Fig. 3.13 Comparison of synthetics computed using 12 km and 4 km deep fault zones for an earthquake that occurs at 8.4 km depth. The observed seismogram is shown in red. Synthetic seismograms computed using a 12 km deep LVZ are shown in blue and those computed using a 4 km deep LVZ are shown in green. The 12 km deep fault zone provides a better match to the observed data.

We then simulated seismograms at the SAFOD MH seismograph for 9 Parkfield aftershocks occurring within the fault zone at depths between 3.8 km and 14.4 km (Fig. 3.14a), and seismograms at station ST0 of the SAFOD surface array for 11 on-fault microearthquakes at depths between 3.6 km and 11.2 km (Fig. 3.14b), using the model with a deep fault zone in Figure 3.11a. Synthetic seismograms show that wavetrain lengths of FZTWs increase with focal depths, matching the observed move-out of trapped waves in

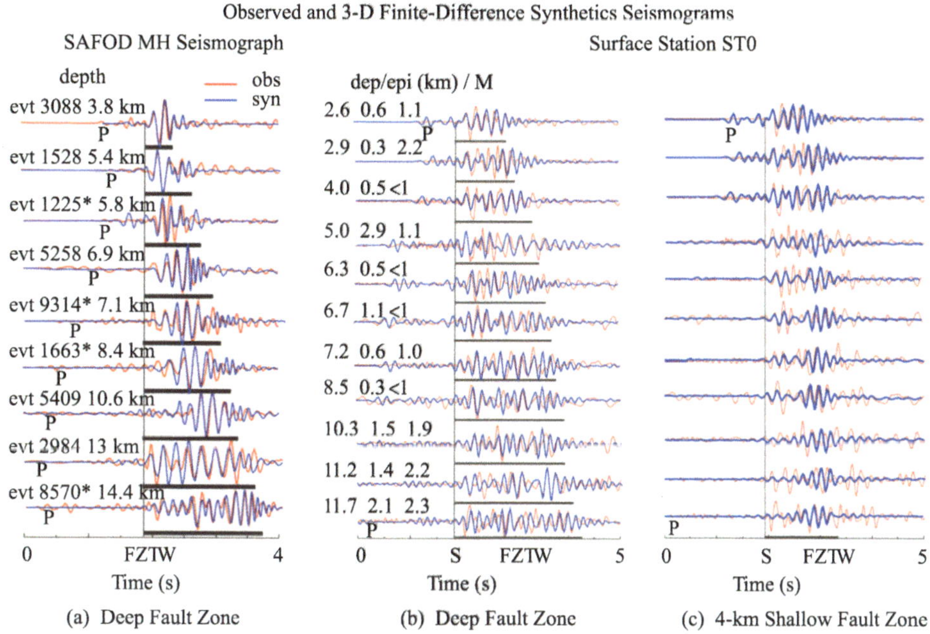

Fig. 3.14 (a) Observed (red lines) and synthetic (blue lines) vertical-component seismograms at the SAFOD MH seismograph at ~ 3 km depth for 9 on-fault aftershocks in groups G1 and G2 (with stars) at depths between 3.8 km and 14.4 km. The *S*-arrivals for these events are aligned at the same time. The focal depth is plotted for each event. The finite-difference synthetic seismograms were computed using the model in Figure 3.11a. Seismograms have been low pass filtered below 8 Hz and are plotted for trace normalized. Bars denote the post-*S* wave durations, in which amplitude envelopes of guided waves are more than twice the level of the background noise coda. (b) Observed and synthetic vertical-component seismograms at station ST0 of the cross-fault surface array for 11 on-fault microearthquakes with magnitudes of *M*0.5-2.3 at depths between 2.6 km and 11.7 km. The focal depth and epicentral distance from ST0 are plotted for each event. (c) Synthetic seismograms at ST0 for 11 on-fault events using a 4-km shallow fault zone for comparison with observations. The synthetic waveforms with short wavetrains for events at depths below 4 km cannot match the long wavetrain of observed FZTWS.

traveltime with event depth. By contrast, the synthetic seismograms for the same events but using a shallow fault zone truncated at 4 km depth show shorter wavetrains after *S*-arrivals and flat change in their wavetrain length with event depth larger than 4 km, mismatching observations (Fig. 3.14c). These modeling results further indicate that the low-velocity waveguide formed by damage rocks on the SAF at the SAFOD site likely extends across entire seismogenic depths although the magnitude of rock damage is below 7-8 km depth due to greater confining stress.

Finally, Figure 3.15 illustrates observed and synthetic seismograms at the SAFOD surface array for the near-surface explosion SP20 detonated within the fault zone at ~ 3 km northwest of the SAFOD site, an SAFOD drilling target event occurring at ~ 3 km depth, and a deep microearthquake at 7-km depth near the array, using the model in Figure 3.11a. Note that FZTWs generated by explosion SP20 travel slower and show longer wavetrains than those generated by earthquakes at depth, indicating the lower seismic velocity and

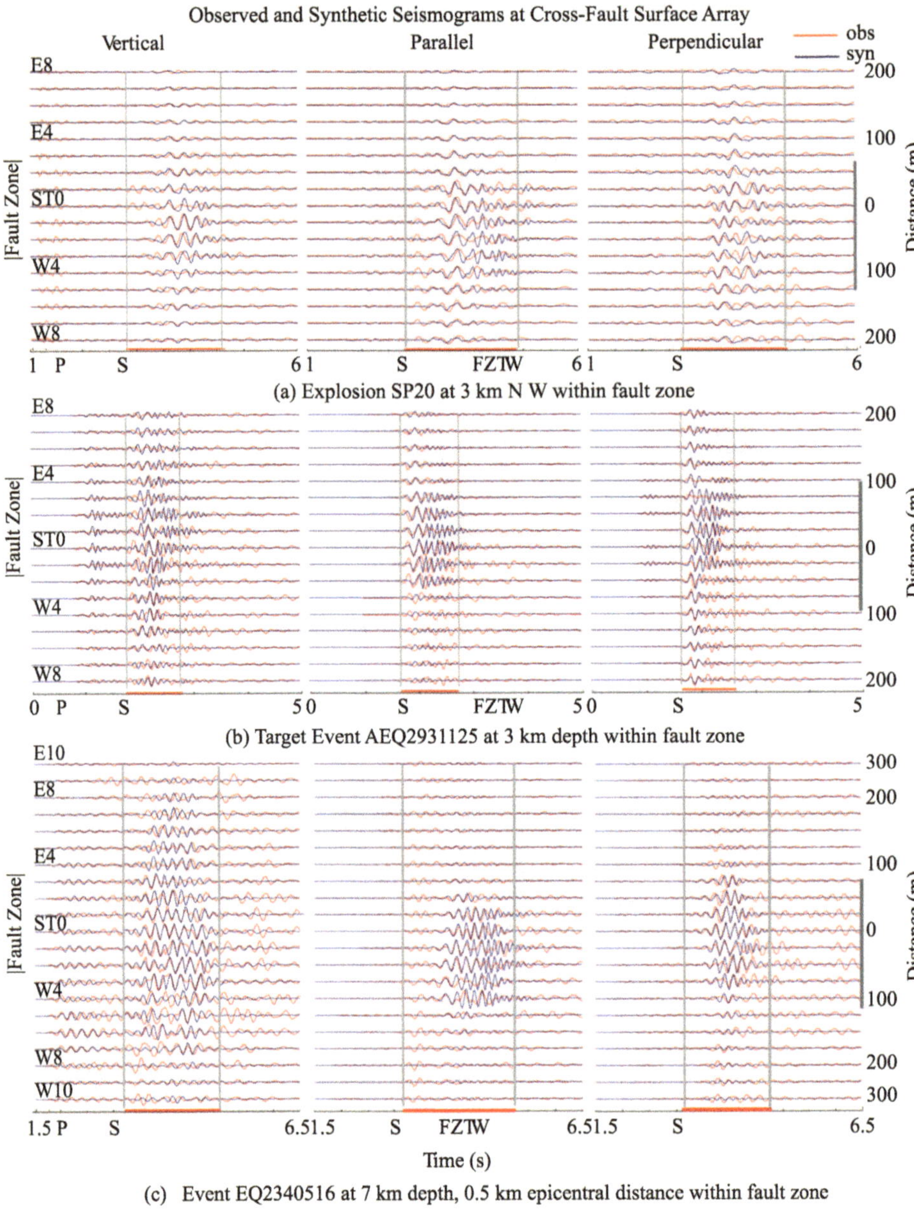

Fig. 3.15 Comparison of observed (red lines) and synthetic (blue lines) seismograms at the surface array for (a) shot SP20, (b) the SAFOD drilling target event at ~ 3 km depth and (c) a micro-earthquake occurring at 7 km depth and 0.5 km epicentral distance from the surface array. An explosive source is used for the shot and a double-couple source is used for the earthquakes. Seismograms have been low pass filtered at 6 Hz and are plotted using a fixed amplitude scale.

larger velocity reduction within the shallower portion of the fault zone formed by more fractured rocks. Figure 3.16 shows observed and synthetic seismograms at the SAFOD surface array for other two deep microearthquakes at depths of 6.9 km and 11 km (events B and C in Fig. 3.1 and Fig. 3.5). In modeling, we computed seismograms using a shallow fault zone truncated at 4-km depth for the deep event C occurring at 11-km depth, showing much shorter wavetrains after *S*-arrivals than observed long-duration FZTWs. This test illuminates a deep low-velocity zone formed by the damage rocks on the SAF at the SAFOD site.

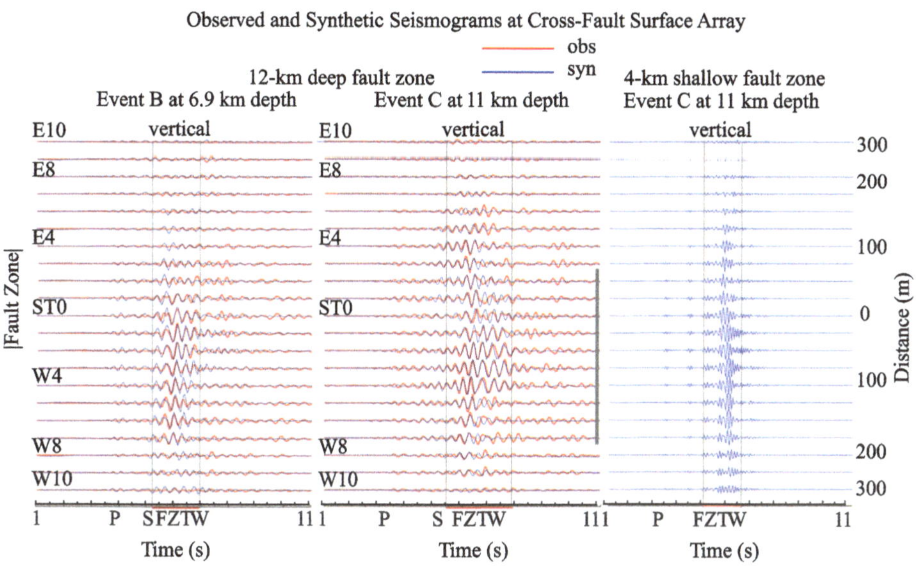

Fig. 3.16 Observed (red lines) and synthetic (blue lines) seismograms at the surface array for 2 deep microearthquakes at depths of 6.9 km with 1 km epicentral distance and 11 km with 3.5 km epicentral distance (events B and C in Fig. 3.1 and Fig. 3.5). Synthetics were computed using the velocity model shown in Figure 3.11a. Other notations are the same as in Figure 3.15. The fault zone extends to 12-km depth. A double-couple source is used for earthquakes located within the fault zone at the focal depth. By contrast, the computed seismograms using a shallow fault zone truncated at 4-km depth for event C show much shorter wavetrains of FZTWs than those in recorded seismograms for this earthquake.

The velocity model in Fig. 3.11a derived from FZTWs recorded at the SAFOD surface array and MH seismograph represents a gross average of the fault-zone damage structure of the San Andreas fault near the SAFOD site at seismogenic depths. The actual structure in 3-D will certainly be more complicated. The damage magnitude and extent may vary along fault strike and with depth due to rupture distributions and stress variations over multiple length and time scales. For instance, the model in Figure 3.11a does not include velocity variations along the fault strike, which have been implied in observed variations of fault slip and rock damage along the SAF at Parkfield (Langbein et al., 2005; Li et al., 2006, 2007; Lewis et al., 2010). The along-fault heterogeneity is discussed in the section below.

3.3 Fault-Zone Trapped Waves at the Surface Array near Parkfield Town

In our previous experiment conducted in 2002, we deployed linear seismic arrays across and along the SAF near the town of Parkfield (called the Parkfield array), ~ 15 km southeast of the SAFOD site and recorded FZTWs generated by explosions detonated within the fault zone and the on-fault microearthquake (Li et al., 2004). 3-D finite-difference simulations of the FZTWs recorded at the Parkfield array delineate a distinct low-velocity waveguide formed by damaged rocks along the Parkfield SAF segment at depths. The best-fit model parameters used in simulations of these FZTWs are shown in Table 3.2.

Table 3.2 Parameters for the SAF near Parkfield

Parameters Layer No.	1	2	3	4
Main fault NW/SE of Parkfield array:				
Depth of the layer, km	0.25	1.0	2.0	5.0
Waveguide width, m	175/150	175/150	150/125	125/100
Waveguide *V*s, km/s	0.5/0.35	0.65/0.55	1.0/0.9	1.7/1.4
Waveguide *V*p, km/s	1.3	1.8	2.3/2.1	3.5/3.0
Waveguide *Q*	10	25	30	50
NE wall-rock *V*s, km/s	0.8/0.6	1.0/0.9	1.5/1.4	2.3/2.0
NE wall-rock *V*p, km/s	2.0/1.5	2.5/2.2	3.3/3.0	5.0/4.2
SW wall-rock *V*s, km/s	0.8/0.6	1.0/0.9	1.6/1.5	2.5/2.2
SW wall-rock *V*p, km/s	2.0/1.5	2.5/2.2	3.5/3.2	5.2/4.5
Wall-rock *Q*	20	50	60	100

It is noticed that seismic velocities of the waveguide southeast of the Parkfield array are remarkably lower by ~ 20% in average than waveguide velocities northwest of the array, indicating the heterogeneity in rock damage magnitude on the SAF near the SAFOD site. The along-fault variations are evident in the FZTWs recorded at the Parkfield array for two explosions PMM and PARK detonated within the fault zone at distances of 7 km northwest and 4.2 km southeast from the array, respectively, in 2002 (Fig. 3.17). The distance from the array to shot PARK is much shorter than the distance to shot PMM, but the FZTWs for these two shots show nearly the same time durations (~ 5-6 s) after *S*-arrivals, indicating the lower velocities and higher damage magnitude of fault rocks within the SAF segment between the array and shot PARK where the highest surface slip was found in the 2004 *M*6 Parkfield Earthquake (Langbein et al., 2005; Ammon et al., 2005). It is noted that seismograms in Figure 3.3 show only ~ 2 s duration of FZTWs after *S*-arrivals for short SP20 detonated ~ 3 km northwest of the SAFOD surface array, indicating that the velocities are lower and damage magnitude of fault rocks near Parkfield Town is higher than those near the SAFOD site.

Immediately after the *M*6 Parkfield Earthquake on September 28, 2004, we re-deployed the Parkfield array to record aftershocks and repeated shot PMM at the same places in our experiment at Parkfield in 2002 for study of the co-seismic damage and after-mainshock healing of the SAF (Li et al., 2006, 2007). Waveform cross-correlations of repeated shots and microearthquakes recorded at the Parkfield array deployed in 2002 and 2004 show a peak of an approximately 2.5% decrease in seismic velocity at stations within the fault

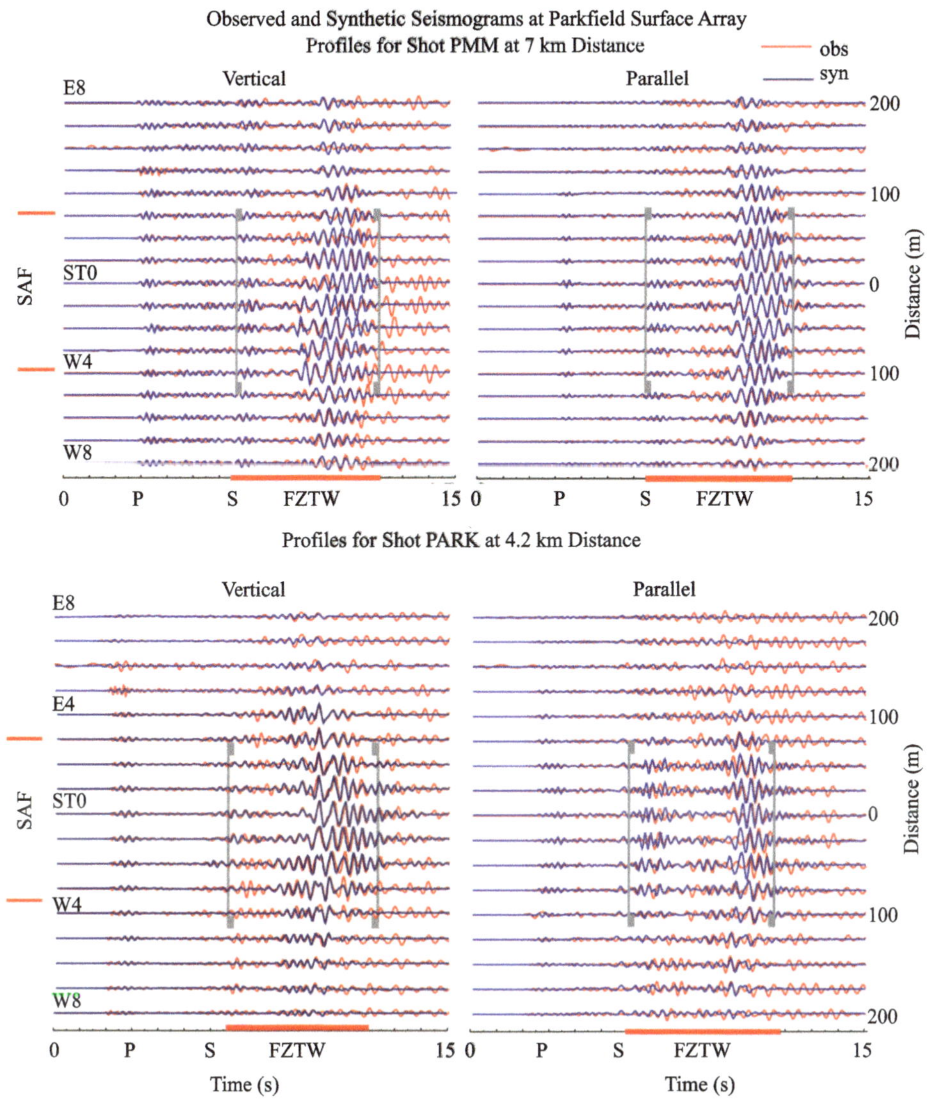

Fig. 3.17 Observed and 3-D finite-difference simulations of seismograms for shots PMM and PARK detonated within the fault zone at distance 7 km NW and 4.2 km SE from Parkfield surface array across the SAF ~ 1.5 km NW of Parkfield Town in 2002 (see Fig. 3.1). Stations ST0 of the array was located on the SAF main trace (SAFm). Seismograms have been <3 Hz filtered and are plotted using a fixed amplitude scale for each event. Prominent fault-zone trapped waves (FZTWs) with large amplitudes and long wavetrains (marked by brackets) following *S*-arrivals are observed at stations located close to the fault trace, within ~ 200 m. Other notations are the same as in Figure 3.1 and Figure 3.14.

zone, most likely due to the co-seismic damage of fault-zone rocks during dynamic rupture of the 2004 *M*6 Parkfield Earthquake. The width of the damage zone characterized by marked velocity changes is consistent with the width of the low-velocity waveguide along the Parkfield SAF delineated by FZTWs recorded at the Parkfield surface array (Table 3.2). The data from repeated aftershocks show that seismic velocities within the fault zone increased by ~ 1.2% in the following 3-4 months after the mainshock, indicating the recovery (or healing) of damaged rocks with time. We also observed that the magnitude of fault damage and healing varied across and along the rupture zone, showing that the greater damage was inflicted and thus greater healing is observed in regions with larger slip in the mainshock.

Figure 3.18a illustrates FZTWs with large amplitudes and long wavetrains after *S*-arrivals recorded at the Parkfield array for an on-fault aftershock (ID 325003619) occur-

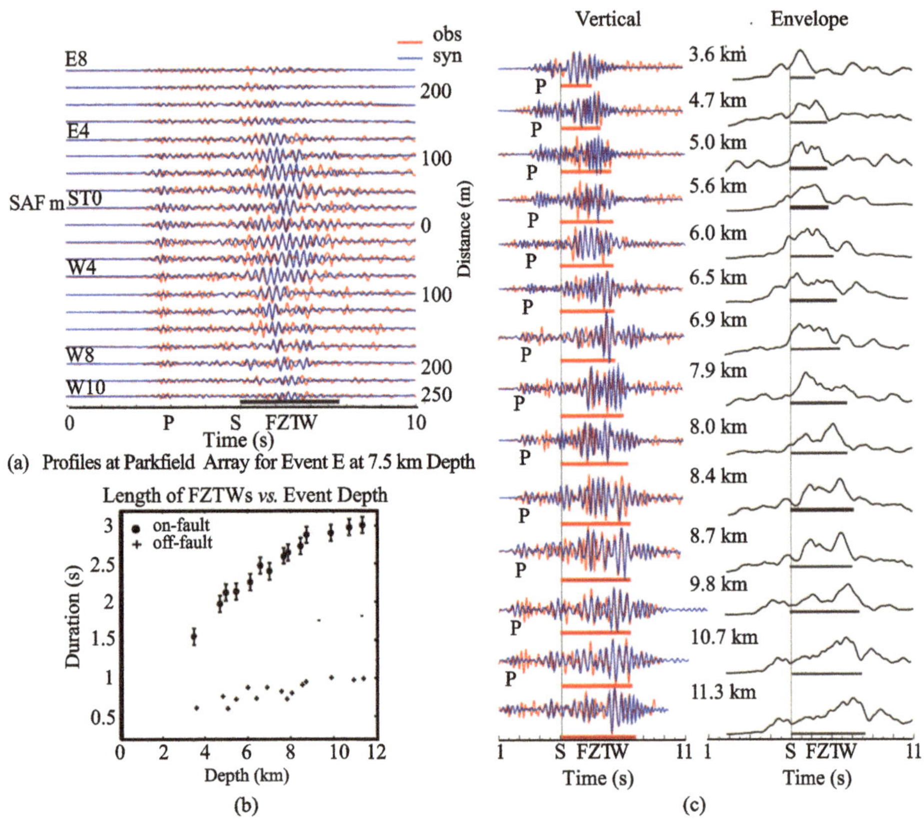

Fig. 3.18 (a) Observed and synthetic vertical-component seismograms recorded at the array near Parkfield for an on-fault aftershock (event E in Fig. 3.1), showing fault-zone guided waves with large amplitudes and long-duration wavetrains after *S*-arrivals at stations close to the main fault trace. Seismograms have been < 6 Hz filtered. (b) The measured FZTW wavetrain lengths versus focal depths for these 14 aftershocks recorded at on-fault (black circles) and off-fault (grey crosses) stations. Each data point is averaged from measurements at 4 stations within or away from the fault zone for each aftershock. Error bars are standard deviations. (c) Observed and synthetic vertical-component seismograms and envelopes at the on-fault station ST0 of the Parkfield array for 14 on-fault aftershocks near the array at different depths show an increase in wavetrain length (marked by solid horizontal bars) of FZTWs as event depths increase. Other notations are the same as in Figure 3.5b.

ring at 7.5 km depth and 5 km southeast of the array on October 23, 2004, showing a remarkable low-velocity fault-zone waveguide southeast of Parkfield Town. In order to examine the depth extension of the damage zone on the Parkfield SAF segment, we measured FZTW wavetrain lengths after *S*-arrivals recorded at the on-fault station ST0 of the Parkfield array for 14 on-fault aftershocks at Parkfield occurring at different depths and with epicenters less than 3 km from the array (Fig. 3.18b). The measured FZTW wavetrain durations increase from 1.5 s to $\sim$ 3.2 s with an obvious move-out as event depths increase from 3.6 km to 11.3 km, showing that the low-velocity damage zone on the SAF near Parkfield likely extends across the seismogenic depths although the magnitude of rock damage in the deeper portion of the fault zone is smaller due to the higher confined stress. It is also noted that FZTWs recorded at the Parkfield array show wavetrains longer than those recorded at the SAFOD surface array (Fig. 3.5) for the events at similar depths, indicating the lower velocities and greater damage magnitude on the SAF Parkfield segment where the larger slip and anomalous high peak acceleration are documented at Turkey Trail road $\sim$2 km SE of Parkfield (Shakal et al., 2005).

Finally, we simulated FZTWs using the model parameters in Table 2 to fit seismograms recorded at the Parkfield array for shots PMM and PARK (Fig. 3.17) and the aftershocks (Fig. 3.18). The best-fit model parameters in Table 3.1 and Table 3.2 illuminate the heterogeneity of fault zone properties with an increase trend in rock damage magnitude along the SAF from the SAFOD site southeastward to Parkfield Town.

3.4 Conclusion and Discussion

Observations and modeling of fault-zone trapped waves recorded at the SAFOD surface array and down-hole seismograph installed in the SAFOD mainhole (MH) show the existence of a distinct low-velocity damage zone along the SAF near the SAFOD site, consistent with our previous results from FZTWs recorded at the Parkfield array near Parkfield Town (Li et al., 2004; Li and Malin, 2008). The borehole data, recorded at 3-km depth in the SAFOD MH, provide high-quality records of earthquakes with a wide range of magnitudes that occur deep on the SAF, within the damage zone. The MH seismograph allows us to avoid some of the complexity of the near surface geology, and wave scattering due to near-surface effects. With these data, we obtain a high-resolution image of the detailed internal structure of the SAF at seismogenic depths near the SAFOD drilling site.

In the modeling procedure, we use the 3-D finite-difference method to simulate FZTWs recorded in the borehole and at surface. The model parameters for the SAF at shallow depth are determined by simulation of FZTWs generated by explosions and recorded at the surface array and shallow earthquakes. The deep structure of the SAF is determined from FZTWs generated by deep earthquakes and recorded at the SAFOD MH seismograph and the surface array. The structure of the narrow fault core zone with highly damaged rocks is mainly constrained by FZTWs recorded at the MH, which show frequencies higher than those recorded at the surface array. To best fit the amplitude, frequency, and travel-time characteristics of observed FZTWs, the SAF is modeled as a downward tapering, 30-40-m wide fault-core embedded in a 100-200-m wide low velocity zone. The fault core is thought to be a zone in which material has been severely damaged due to high on- and near-fault strains produced during dynamic rupture of major earthquakes at Parkfield. Compared with the wall-rock velocities, the seismic velocities are reduced by ~40%-50% in the fault core and by ~25%-35% in the surrounding damage zone. The width and velocities of the SAF damage zone at 3 km depth in our model resolved by FZTWs (Fig. 3.11a) have been verified by the SAFOD drilling and logging data (Fig. 3.11b) which show that a ~ 200 m-wide zone of high porosity material, with multiple slip planes and average velocity reductions of ~ 25%-35%, is at ~ 3 km depth (Hickman et al., 2007). It is plausible for downward tapering of the damage zone with high porosity rocks because the confining stress increases with depth.

This distinct low-velocity zone is interpreted as an accumulation of damage caused by intense fracturing during successive earthquakes. Alternatively, the low velocity zone may reflect high pore-fluid pressure near the fault, giving the observations of fluid in the SAFOD well. However, fluids are a less-favored explanation as permeability variation in faults is complex with higher permeability observed within the near-fault damage zone but very low permeability across the main slip plane (Lockner et al., 2000). We expect a combination of *in-situ* seismic observations and stress measurements with laboratory studies on fault rock, fluids and hydrologic conditions to provide more detailed information about the physical and chemical processes controlling faulting and earthquake generation. Furthermore, the damage zone on the SAF at Parkfield imaged by FZTWs recorded at the across-fault surface array shows asymmetry, extending farther on the southwest side of the main fault trace. The asymmetry may reflect the different strength of the material on each side of the SAF or a preferred rupture direction. When a fault ruptures, it may preferentially damage the more compliant rocks in the zone; increased compliance may be the result of different geologic materials juxtaposed across the fault or weakness resulting from accumulated damage by previous events (Chester et al., 1993). Alternately, preferred rupture direction would result in asymmetric damage, with greater damage occurring in the extensional quadrant of the earthquake rupture (Andrews, 2005).

We note that the subsurface structure of the fault-zone simply obtained from forward modeling of FZTWs would not be unique due to a trade-off in the model parameters (e.g., the fault zone width, velocity contrast between the fault wall rocks, Q value, source location within the fault zone, and travel distance along the fault zone). This non-uniqueness and other 3-D effects have been discussed in previous numerical studies of fault zone structure using trapped waves (e.g., Li and Leary, 1990; Li and Vidale, 1996; Igel et al., 1997; Ben-Zion, 1998). For instance, either increasing the waveguide width or decreasing the waveguide velocity in modeling will lower the dominant frequency of trapped waves. Low values of Q also affect the dominant frequency and duration of trapped waves. Reducing Q causes a shorter duration of trapped waves at lower frequencies. Smaller velocity contrasts between the waveguide and surrounding rocks also reduce the duration of trapped waves. Moving the source from the middle to the edge of the waveguide will reduce the amplitudes of trapped waves with respect to the P and S waves. However, the trade-offs among the model parameters could be reduced by constraints on multiple observations with various methods. At Parkfield, we use the velocities and width of the SAF directly measured by well log in the SAFOD mainhole at 3-km depth (Hickman et al., 2007) and the velocity structure in the Parkfield region determined using P-wave tomography (e.g., Thurber et al., 2006) as constraints on our FZTWs modeling.

The model parameters in Table 3.1 and Table 3.2 reflect variations in velocity and damage magnitude along the SAF from the SAFOD site southeastward to Parkfield Town where the fault zone has slower seismic velocities. For instance, seismograms recorded at the Parkfield array for shot PARK detonated within the SAF at 4.2 km southeast of the array show remarkable low velocities (Fig. 3.17). Accordingly, anomalous large slip and higher peak acceleration were documented at Turkey Trail road $\sim$ 2 km SE of Parkfield Town (Shakal et al., 2005). The along-fault variations in fault zone properties are also indicated in our observations of fault rock damage and healing associated with the 2004 *M*6 Parkfield Earthquake observed (Li et al., 2006, 2007), and observations and modeling of head-waves at Parkfield (Lewis and Ben-Zion, 2010).

Although a shallow low-velocity zone along the SAF at SAFOD was well known from surface observations of FZTWs, its extension into the seismogenic zone ($>$ 3 km) has been controversial. Some researchers argue that the low-velocity damage zone on faults is a near-surface feature that reaches only down to the top of seismogenic zone at $\sim$ 3 km (e.g., Ben-Zion et al. 2003; Lewis et al. 2010). Some argue that it extends across the seismogenic zone at depths from 3 km to 10 km (e.g., Korneev et al. 1993; Li & Malin 2008; Wu et al., 2010). The differing conclusions illustrate the non-uniqueness of FZTWs interpretations based on surface measurements. Recently, Ellsworth and Malin (2010) document a profound zone of rock damage on the SAF downwards to at least half way ($>$ 5-6 km) through the seismogenic crust at Parkfield using both Rayleigh-type and Love-type of FZTWs recorded at the SAFOD MH seismograph. Wu et al. (2010) also show that the low-velocity waveguide on the SAF at Parkfield extends to the depth $>$ 10 km using SAFOD borehole data. A comprehensive analysis of FZTWs recorded at Parkfield surface and borehole stations in our present study shows that the low-velocity zone on the SAF at Parkfield extends to the depth of at least $\sim$ 7 km or deeper although the velocity reduction within the damage zone decreases with depth due to the increasing confining pressures.

The 100-200 m width of the low-velocity damage along the SAF at Parkfield delineated by the FZTWs is similar to the width of the damage zone observed along the 1992 *M*7.4 Landers and 1999 *M*7.1 Hector Mine Earthquake ruptures (Li et al., 2000; 2002). However, a 1.2-1.5 km wide low-velocity zone with greater velocity reduction in a 400-500-m wide fault zone was documented at the Calico fault (Fialko, 2004; Cochran et al., 2009). The difference in width of low-velocity fault-zone might reflect the evolution of fault zones over the seismic cycle. The studies at Parkfield, Landers and Hector Mine were initiated following a large mainshock rather than late in the interseismic period like on the Calico fault. The wider low-velocity zone observed at the Calico fault might be attributed to complex fault geometry where strain is distributed across several fault strands. Alternatively, it might imply the width variations of damage zones on the different faults, and the variations along the fault strike.

Our observations and modeling of FZTWs at multiple arrays deployed at different places along the SAF show seismic velocity variations along the fault line suggestively due to rupture distributions and stress variations over multiple length and time scales. Detailed mapping of near-field fault structure and strain accumulation will provide further insights into the ubiquity and spatial variability of permanent damage around active faults. Our present study contributes essential information towards further understanding of faulting mechanics and earthquake hazards along mature faults such as the San Andreas fault.

Acknowledgements

This article is dedicated to the late Professor Keiiti Aki for his invaluable contributions to seismology and pioneering research at the San Andreas Fault, Parkfield. This study was supported by EarthScope Grant EAR0342277, USGS Grant NEHRP20060160, and NSF Grant EAR-0910911. Special thanks to S. Hickman, W. Ellsworth, and M. Zoback, SAFOD PIs, for their coordination of our experiments to seismically characterize the SAFOD drill site. We acknowledge the IRIS for the use of PASSCAL instruments. We are grateful to J. Vidale, P. Leary, P. Chen, C. Thurber, S. Roecker, M. Rymer, R. Catchings, A. Snyder, R. Russell, L. Powell, B. Nadeau, N. Boness, D. McPhee, and F. Niu for their collaboration in our research at Parkfield. This research was supported by the Southern California Earthquake Center (contribution number 1321). This article was written partly during the lead author's visits to Fudan University and Ningbo Nottingham University in China.

References

Aki, K. (1984). Asperities, barriers, characteristic earthquakes, and strong motion prediction. J. Geophys. Res., 89, 5867-5872.

Andrews, D. J. (2005). Rupture dynamics with energy loss outside the slip zone. J. Geophys. Res., 110, B01307, dio: 0.1029.

Angevine, C. L., D. L., Turcotte, and M. D. Furnish (1982). Pressure solution lithification as a mechanism for the stick-slip behavior of faults, Tectonics. 1, 151-160.

Ben-Zion, Y. (1998). Properties of seismic fault zone waves and their utility for imaging low velocity structures. J. Geophys. Res., 103, 12567-12585.

Ben-Zion, Y., Z. Peng, D. Okaya, L. Seeber, J. G. Armbruster, N. Ozer, A. J. Michael, S. Barris, and M. Aktar (2003). A shallow fault zone structure illuminated by trapped waves in the Karadere-Dusce branch of the North Anatolian Fault, Western Turkey. Geophys. J. Int., 152, 699-717.

Byerlee, J. (1990). Friction, overpressure and fault-normal compression. Geophys. Res. Lett., 17, 2109-2112.

Chavarria, J. A., P. E. Malin, E. Shalev, and R. D. Catchings (2003). A look inside the San Andreas Fault at Parkfield through Vertical Seismic Profiling. Science, 302, 1746-1748.

Chen, Q., and J. Freymueller (2002). Geodetic evidence for a near-fault compliant zone along the San Andreas fault in the San Francisco bay area. Seismological Society of America Bulletin, 92, 656-671, doi:10.1785/0120010110.

Chester, F. M., J. P. Evans, and R. L. Biegel (1993). Internal structure and weakening mechanisms of the San Andreas fault. J. Geophys. Res., 98, 771-786.

Cochran, E. S., Y. G Li, P. M. Shearer, S. Barbot, Y. Fialko, and J. E. Vidale (2009). Seismic and geodetic evidence for extensive, long-lived fault damage zones. Geology, 37(4): 315-318. doi:10.1130/G25306A.1; Data Repository Item 2009082.

Dieterich, J. H. (1997). Modeling of rock friction: 1. Experimental results and constitutive equations. J. Geophys. Res., 84, 2161-2168.

Duan, B. C. (2010). Inelastic response of compliant fault zones to nearby earthquakes. Geophys. Res. Lett., L16303, doi:10.1029/2010GL044150.

Ellsworth W. L. and P. E. Malin (2010). Deep rock damage in the San Andreas fault revealed by *P*- and *S*-type fault zone guided waves, Sibson's volume, New Zealand, in press.

Ellsworth W. L., P. E. Malin, K. Imanishi, S. W. Roecker, R. Nadeau, V. Oye, C. H. Thurber, F. Waldhauser, N. L. Boness, S. H. Hickman and M. D. Zoback (2007). Scientific Drilling, Part 4: The Physics of Earthquake Rupture, Special Issue No. 1, doi:10.2204/iodp.sd.s01.04.2007.

Fialko, Y. (2004). Probing the mechanical properties of seismically active crust with space geodesy: Study of the coseismic deformation due to the 1992 *M*w 7.3 Landers (southern California) earthquake. J. Geophys. Res., 109, B03307, doi: 10.1029/2003JB002756.

Graves, R. W. (1996). Simulating seismic wave propagation in 3D elastic media using staggered-grid finite differences. Bull. Seismol. Soc. Am., 86, 1091-1106.

Hickman, S. H., M. D. Zoback, W. L. Ellsworth, N. Boness, P. Malin, S. Roecker and C. Thurber (2007). Structure and properties of the San Andreas Fault in Central California: recent results from the SAFOD experiment, Scientific Drilling, Special Issue, No.1, doi:10.2204 /iodp.sd.s01.39.2007, 29-32.

Hickman, S., R., Sibson, and R. Bruhn (1995). Introduction to special section: mechanical involvement of fluids in faulting. J. Geophys. Res., 100, 12831-12840.

Igel, H., Y. Ben-Zion and P. Leary (1997). Simulation of SH and P-SV wave propagation in fault zones. Geophys. J. Int., 128, 533-546.

Kanamori, H. (1994). Mechanics of earthquakes. Ann. Rev. Earth Planet., Sci., 22, 207-237.

Korneev, V. A., R. M. Nadeau, and T. V. McEvilly (2003). Seismological studies at Parkfield IX: Fault-zone imaging using guided wave attenuation. Bull. Seism. Soc. Am. 80, 1245-1271.

Langbein, J., R. Bocherdt, D. Dreger, J. Fletcher, J. L. Hardbeck, M. Hellweg, C. Ji, M. Johnston, J. R. Murray, R. Nadeau, M. J. Rymer, and J. A. Trieman (2005). Preliminary report on the 28 September 2004, *M* 6.0 Parkfield, California earthquake. Seism. Res. Lett. **76**(1): 10-26.

Lewis, M. A. and Y. Ben-Zion (2010). Diversity of fault zone damage and trapping structures in the Parkfield section of the San Andreas Fault from comprehensive analysis of near fault seismograms. Geophys. J. Int., doi: 10.1111/j.1365-246X.2010.04816.x.

Li, Y. G. and P. E. Malin (2008). San Andreas Fault damage at SAFOD viewed with fault-guided waves. Geophys. Res. Lett., 35, L08304. doi: 10.1029/2007GL032924.

Li, Y. G., and P. C. Leary (1990). Fault-zone trapped seismic waves. Bull. Seism. Soc.

Am., 80, 1245-1271.

Li, Y. G., J. E. Vidale, K. Aki, and F. Xu (2000). Depth-dependent structure of the Landers fault zone from trapped waves generated by aftershocks. J. Geophys. Res., 105, 6237-6254.

Li, Y. G., J. E. Vidale, K. Aki, F. Xu, and T. Burdette (1998). Evidence of shallow fault zone strengthening after the 1992 *M*7.5 Landers, California, earthquake. Science, 279, 217-219.

Li, Y. G., J. E. Vidale, S. M. Day and D. Oglesby (2002). Study of the *M*7.1 Hector Mine, California, earthquake fault plan by fault-zone trapped waves. Hector Mine Earthquake Special Issue, Bull. Seism. Soc. Am., 92, 1318-1332.

Li, Y. G., J. E., Vidale, and S. E. Cochran (2004). Low-velocity damaged structure of the San Andreas fault at Parkfield from fault-zone trapped waves. Geophy. Res. Lett., 31, L12S06.

Li, Y. G., K. Aki, J. E. Vidale, and M. G. Alvarez (1998). A delineation of the Nojima fault ruptured in the *M*7.2 Kobe, Japan, earthquake of 1995 using fault zone trapped waves. J. Geophys. Res., 103, 7247-7263.

Li, Y. G., P. C. Leary, K. Aki, and P. E. Malin (1990). Seismic trapped modes in Oroville and San Andreas fault zones. Science, 249, 763-766.

Li, Y. G., P. Chen, E. S. Cochran, and J. E. Vidale (2007). Seismic velocity variations on the San Andreas Fault caused by the 2004 *M*6 Parkfield earthquake and their implications. Earth Planets and Space, 59, 21-31.

Li, Y. G., P. Chen, E. S. Cochran, J. E., Vidale, and T. Burdette (2006). Seismic evidence for rock damage and healing on the San Andreas fault associated with the 2004 *M*6 Parkfield earthquake. Special Issue for Parkfield *M*6 earthquake, Bull. Seism. Soc. Am., 96, No. 4, S1-15, doi:10.1785/0120050803.

Li, Y. G., W. L. Ellsworth, C. H. Thurber, P. E. Malin, and K. Aki (1997). Observations of fault-zone trapped waves excited by explosions at the San Andreas fault, central California. Bull. Seism. Soc. Am., 87, 210-221.

Li, Y.-G, K. Aki, D. Adams, A. Hasemi, and W. H. K. Lee (1994a). Seismic guided waves trapped in the fault zone of the Landers, California, earthquake of 1992. J. Geophys. Res. 99, 11705-11725.

Li, Y.-G., J. E. Vidale, K. Aki, C. Marone, and W. H. K. Lee (1994b). Fine structure of the Landers fault zone; segmentation and the rupture process. Science 256, 367-370.

Lockner, D. A., H. Naka, H. Tanaka, H. Ito, and R. Ikeda (2000). Permeability and strength of core samples from the Nojima fault of the 1995 Kobe earthquake, in Proceedings of the International Workshop on the Nojima Fault Core and Borehole Data Analysis, Tsukuba, Japan, Nov 22-23, 1999, USGS Open file Report 00-129, edited by H. Ito, K. Fujimoto, H. Tanaka, and D. A. Lockner, 147-152.

Malin, P. E., M. Lou and J. A. Rial (1996). FR waves: a second fault-guided mode with implications for fault property studies. Geophysical Research Letters, 23, 3547-3550.

Malin, P., E. Shalev, H. Balven, and C. Lewis-Kenedi (2006). Structure of the San Andreas Fault at SAFOD from P-wave tomography and fault-guided wave mapping. Geophys. Res. Lett., 33, L13314. doi:10.1029/2006GL025973.

Marone, C. (1998). The effect of loading rate on static friction and the rate of fault healing during the earthquake cycle. Nature, 391, 69-72.

Massonnet, D, W. Thatcher, and H. Vadon (1996). Detection of postseismic fault-zone collapse following the Landers earthquake. Nature, 382, 612-616.
Michelini, A. and T. V. McEvilly (1991). Seismological studies at Parkfield. I. Simultaneous inversion for velocity structure and hypocenters using cubic B-splines parameterization. Bull. Seism. Soc. Am. 81, 524-552.
Mooney, W. D., and A. Ginzburg (1986). Seismic measurements of the internal properties of fault zones. Pure Appl. Geophys., 124, 141-157.
Nadeau, R. M. and T. V. McEvilly (1997). Seismological Studies at Parkfield V: Characteristic microearthquake sequences as fault-zone drilling targets. Bull. Seism. Soc. Am., 87, 1463-1472.
Olsen, M., C. H. Scholz, and A. Leger (1998). Healing and sealing of a simulated fault gouge under hydrothermal conditions for fault healing. J. Geophys. Res., 103, 7421-7430.
Rice, J. R. (1992). Fault stress states, pore pressure distributions, and the weakness of the San Andreas fault, in Fault Mechanics and Transport Properties of Rocks, edited by B. Evans and T.-F. Wong, 475-503, Academic, San Diego, Calif.
Rice, J. R., The mechanics of earthquake rupture, in Physics of the Earth's Interior, edited by A. M. Dziewonski and E. Boschi, pp. 555-649, North-Holland, Amsterdam, 1980.
Richardson, E., and C. Marone (1999). Effects of normal stress vibrations on frictional healing. J. Geophys. Res. 104, 28859-28878.
Scholz, C. H. (1990). The Mechanics of Earthquakes and Faulting. Cambridge Univ. Press, New York.
Shakal, A. F., V. G. Graizer, M. Huang, R. Borcherdt, H. R., Haddadi, K. W. Lin, C. Stephens, and P. Roffers (2005). Preliminary analysis of strong-motion recordings from the 28 September 2004 Parkfield, California earthquake. Seismo. Res. Lett. 76(1): 27-40.
Sibson, R. H., J. M. Moore and A. H. Rankin (1975). Seismic pumping – A hydrothermal fluid transport mechanism. Geological Society of London Journal, 131, 653-659.
Solum, J.G., Hickman, S., Lockner, D.A., Tembe, S., Evans, J.P., Draper, S.D., Barton, D.C., Kirschner, D.L., Chester, J.S., Chester, F.M., van der Pluijm, B.A., Schleicher, A.M., Moore, D.E., Morrow, C., Bradbury, K., Calvin, W.M., and Wong, T.-F. (2007). San Andreas fault zone mineralogy, geochemistry and physical properties from SAFOD cuttings and core. Scientific Drilling Special Issue, No. 1, doi:10.2204 /iodp.sd.s01.34.2007, 64-67.
Taira T., Paul G. Silver, F. Niu, and R. M. Nadeau (2009). Remote triggering of fault-strength changes on the Parkfield. Nature, 461, 636-639. dio:10.1038/nge:10.1038/nature08395.
Thurber, C., H. Zhang, F. Waldhauser, J. Hardbeck, A. Michael and D., Eberhart-Phillips (2006), Three-dimensional compressional wavespeed model, earthquake relocations, and focal mechanisms for the Parkfield, California, region. Bull. Seism. Soc., 96, S38–S49. dio:10.1785/0120050825.
Thurber, C., S. Roecker, H. Zhang, S. Baher and W. Ellsworth (2004). Fine-scale structure of the San Andreas fault zone and location of the SAFOD target earthquakes. Geophys. Res. Letter, 31, L12S02. doi:10.1029/2003GL019398.
Unsworth, M., P. Malin, G. Egbert, and J. Booker (1997). Internal structure of the San

Andreas fault at Parkfield, CA. Geology, 356-362.
Vidale, J. E. and Li, Y. G. (2003), Damage to the shallow Landers fault from the nearby Hector Mine earthquake. Nature, 421, 524-526.
Vidale, J. E., W. L. Ellsworth, A. Cole, and C. Marone (1994), Rupture variation with recurrence interval in eighteen cycles of a small earthquake. Nature, 368, 624-626.
Wu, J., J. A. Hole, and J. A. Snoke, (2010), Fault-zone structure at depth from differential dispersion of seismic guided waves: evidence for a deep waveguide on the San Andreas fault. Geophys. J. Int., 182, 343-354, doi: 10.111.j.1365-246X.2010.04612.x.
Yasuhara, H., C. Marone, and D. Elsworth (2005). Fault zone restrengthening and frictional healing: The role of pressure solution. J. Geophys. Res., 110, B06310, doi: 10.1029/2004JB003327.

Author Information

Yong-Gang Li

Department of Earth Sciences, University of Southern California, Los Angeles California 90089, USA

E-mail: ygli@usc.edu

Peter E. Malin

Institute of Earth Science and Engineering, University of Auckland, Auckland 1142, New Zealand

Elizabeth S. Cochran

Department of Earth Sciences, University of California-Riverside, Riverside California 92521, USA

Appendix: Modeling Fault-Zone Trapped *SH-Love* Waves

In our early study of fault-zone trapped waves observed at the San Andreas fault near Parkfield, we employed a modified propagator matrix method to compute the vertical wavenumbers that lead to propagating *SH-Love* energy at a specified frequency for planar multilayer velocity models appropriate to test the utility of simple modeling of trapped energy (Li et al., 1990; Li and Leary, 1990). Since the vertical wavenumber controls the amplitude decay of the trapped wave motion as observed by sensors away from the fault zone and is related to the phase velocity of the trapped wave, we can compare simulated *SH-Love* amplitudes and phase velocities with observations, giving an estimate of the reliability of a simple modeling approach to the complicated *in-situ* data. The details in computing the trapped *SH/Love*-wave dispersion curves, wave amplitudes, amplitude spectra, and simulated waveforms are described in the following sections.

In order to reproduce the observed *SH-Love* trapped wave behavior, we synthesized the modes trapped in a simple stratified velocity model containing a major population of cracks aligned parallel to the vertical fault zone which is embedded in a full space (Fig. 3A-1). The layers numbered $j = 0$, 4 are the two different half spaces of intact high speed

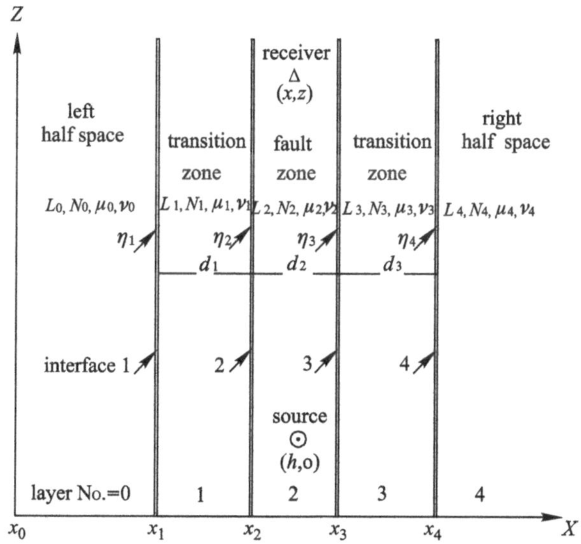

Fig. 3A-1 The simple fault zone model with multiple vertical planar layers in between two half spaces. The fault zone is sandwiched by two transition zones. The horizontal variable increases from $j = 0,1,2,3,4$. Layers numbered $j = 0,4$ are the two half spaces of intact high speed crystalline rocks, $j = 1,3$ are transition zones of moderate fracture density, $j = 2$ is a fault zone of high fracture density. The interface x coordinates are x_1, x_2, x_3, and x_4. Thicknesses of two transition zones and fault zone are d_1, d_2 and d_3. The shear rigidity μ_j for isotropic medium, L_j and N_j for transversely isotropic medium, and vertical wavenumber ν_j are assigned to each layer. The phase shift η_j is assigned to each interface. An *SH* source with motion along y-axis is located at coordinates $(h,0)$ within the fault zone. A receiver is located at position (x,z) in the model.

crystalline rocks, $j = 1, 3$ are transition zones with moderate fracture density, and $j = 2$ is a fault core of high fracture density. The horizontal variable increases from the $j = 0$ layer to the $j = 4$ layer. The x coordinates of interfaces are x_1, x_2, x_3, and x_4. We give here the formalism for computing the relationship between wave frequency and vertical wavenumber necessary for propagating phases, for constructing seismograms from the phase and group velocity information. We introduce the known transverse isotropy of thin aligned fractures parallel to the fault zone into the two-dimensional scalar wave equation for *SH-Love* displacements in an axially symmetric medium.

The equation of Love-type motion in the jth layer may be written as

$$L_j \frac{\partial^2 U_j}{\partial x^2} + N_j \frac{\partial^2 U_j}{\partial z^2} = \rho_j \frac{\partial^2 U_j}{\partial t^2}, \tag{3A1}$$

where U_j is the y-component displacement in the jth layer. L_j and N_j are rigidities of the transverse (parallel to the fault zone) isotropic medium with material density ρ_j.

$$L_j = (C_{44})_j = (\rho V_{sh}^{\perp 2})_j, \quad N_j = (C_{66})_j = (\rho V_{sh}^{\parallel 2})_j,$$

where $V_{sh}^{\parallel}$ and $V_{sh}^{\perp}$ are the velocities of *SH* waves propagating parallel and perpendicular to the fault zone, respectively. The displacement $U_j(x,z,t)$ as a function of time t and position in the $x-z$ plane at $y=0$ is expressed as

$$U_j(x,z,t) = A_j(x)\exp\left[ik(z-ct)\right], \tag{3A2}$$

where $A_j(x)$ is the x-dependent amplitude term, k is the wavenumber and c is the phase velocity.

Putting (3A2) into the equation of motion (3A1), we obtain

$$\frac{\partial^2 A_j(x)}{\partial x^2} = -k_j^2\left(\frac{c^2\rho_j}{L_j} - \frac{N_j}{L_j}\right)A_j(x) = -\nu_j^2 A_j(x), \tag{3A3}$$

where ν_j is the vertical wavenumber (normal to the fault zone) for the jth transversely isotropic layer. It can be written as

$$\nu_j = k\sqrt{\left(\frac{c^2}{V_{sh}^{\perp 2}} - \frac{V_{sh}^{\parallel 2}}{V_{sh}^{\perp 2}}\right)_j} = \xi_j k\sqrt{\left(\frac{c^2}{V_{sh}^{\parallel 2}} - 1\right)_j} = \xi_j k \nu_j^{\circ}, \tag{3A4}$$

where

$$\nu_j^{\circ} = \sqrt{\left(\frac{c}{V_{sh}^{\parallel}}\right)^2 - 1}, \quad \text{and} \quad \xi_j = \left(\frac{V_{sh}^{\parallel}}{V_{sh}^{\perp}}\right).$$

ν° is the expression of the vertical wavenumber for isotropic medium. ξ is called anisotropy factor. We define ν_j as a pseudo-vertical wavenumber for the jth layer. Replacing ν_j° by ν_j, all formulations for the isotropic medium can be used for the transversely isotropic medium with respect to a vertical fault zone and vice versa.

Introducing the right-going and left-going plane waves with wavenumber ν_j and amplitudes $\grave{S}_j$ and $\acute{S}_j$, respectively, into the differential equation (3A3), the total displacement amplitude $Aj(x)$ of the *SH-Love* wavefield in the jth layer is defined as

$$A_j(x) = \grave{S}_j \exp[-i\nu_j(x-x_j)] + \acute{S}_j \exp[+i\nu_j(x-x_j)]. \tag{3A5}$$

The right-ward and left-ward plane waves may yield a constructive interference in the jth layer if the difference between their phases are $2n\pi$ $(n = 0, 1, \cdots, n)$; n corresponds to the order of trapped modes. Following Rader et al. (1985), we introduce the complex phase shift η_j at the jth interface between the jth layer and the $j+1$th layer to relate amplitude $\grave{S}_j$ of the right-going wave to amplitude $\acute{S}_j$ of the left-going wave by

$$\acute{S}_j = \grave{S}_j \exp[2i\eta_j]. \tag{3A6}$$

Then (3A5) may be written as

$$\begin{aligned} A_j(x) &= \grave{S}_j \exp[i\eta_j]\{\exp[-i(\nu_j(x-x_j)+\eta_j)] + \exp[i(\nu_j(x-x_j)+\eta_j)]\} \\ &= B_j \cos[\nu_j(x-x_j)+\eta_j], \end{aligned} \tag{3A7}$$

where

$$B_j = 2\grave{S}_j \exp[i\eta_j]. \tag{3A8}$$

The *SH-Love* type trapped waves radiate in the two half spaces with the amplitudes exponentially decaying

$$A_0(x) = B_0 \exp[i\nu_0(x-x_1)] \tag{3A9}$$

$$A_4(x) = B_3 \exp[i\nu_3(x-x_4)]. \tag{3A10}$$

Because the displacement and stress must be continuous at the interface, the continuity of displacement $A_j(x)$and stress $L_j\partial_x A_j$ across the interface between layers j and $j+1$ can be expressed by the condition

$$\begin{pmatrix} \cos(\nu_j d_j + \eta_j) & \cos(\eta_{j+1}) \\ L_j\nu_j \sin(\nu_{j+1} d_j + \eta_j) & -L_{j+1}\nu_j \sin(\eta_{j+1}) \end{pmatrix} \begin{pmatrix} B_j \\ B_j \end{pmatrix} = 0, \tag{3A11}$$

where d_j is the thickness of the jth layer. The first matrix on the left side of this homogeneous equation is called the layer matrix. It forms a standard eigenvalue problem. For a non-trivial solution, the displacement of the layer matrix must be equal to zero, yielding

$$L_{j+1}\nu_{j+1} \tan[\eta_{j+1}] = L_j\nu_j \tan[\eta_j + \nu_j(d_j)]. \tag{3A12}$$

Instead of the standard Thomson-Haskell propagator matrix method, we used a rapid phase-shift recursion algorithm to compute *SH-Love* trapped wave dispersion and amplitude response in the model [Li and Leary, 1990]. From the recursion equation (3A10), a set of unknown phase shifts η_j at successive interfaces $(j = 1, 2, 3, 4)$ can be recursively computed with the starting phase shift η_1 at the interface between the half space and transition zone left of the fault zone,

$$\eta_1 = \arctan\left(-i\frac{L_0 v_0}{L_2 v_2}\right). \tag{3A13}$$

The phase shift at the interface between the transition zone and half space right of the fault zone is computed by

$$\eta_4 = \arctan\left(-i\frac{L_4 v_4}{L_3 v_3}\right) = \eta_3 + v_3 d_3. \tag{3A14}$$

The first term η_3 on the right-hand side of (3A14) is the phase shift at the 3rd interface between the fault zone and east transition zone. The second term $v_3 d_3$ is the phase delay due to the traveltime in the 3rd layer (the transition zone to right). For a given frequency ω, the phase velocities for different modes may be computed by recursively solving (3A12), (3A13) and (3A14) as a root-finding condition. Because the recursion algorithm operation requires only the inverse tangent of complex phase shift with logarithm operation, it works fast and accurately. (3A12) permits the procedure to be numerically stable for a wide range of frequencies and wavenumbers. After knowing a set of phase shift values at interfaces and phase velocities in multiple layers and the two half spaces for the specified mode n, the amplitudes of *SH-Love* trapped waves can be computed as a function of the location in the model using equations (3A7) to (3A10).

With phase velocities c at each frequency for each layer known via the above procedure, group velocities of *SH-Love* type trapped waves can be stably computed from integral relations given by Aki and Richards [1980],

$$V_g = \frac{1}{c}\frac{I_2}{I_1}, \quad I_1 = \frac{1}{2}\int_{-\infty}^{\infty} \rho(x)A^2(x)dx, \quad I_2 = \frac{1}{2}\int_{-\infty}^{\infty} L(x)A^2(x)dx. \tag{3A15}$$

To compare model and observed wave motion, the simulated seismograms for particle motion at receiver position (x,z) due to a source at position $(h,0)$ may be computed using the *Love*-wave Green's function as given by Aki and Richards [1980]

$$G_{yy}(x,z;h,0;\omega) = \sum\nolimits_N \frac{A(\omega,v_n,h)A(\omega,v_n,x)}{4v_n I_2}\exp\left[i(v_n z + \pi/2)\right], \tag{3A16}$$

where ω and v_n, are respectively the frequency and the vertical wavenumber, and n is the order of the mode of trapped waves, $n = 0$ for fundamental mode, $n = 1$ for the first-high mode and so on. z is the distance between the source and receiver along the vertical fault zone. $A(\omega,v_n,h)$ and $A(\omega,v_n,x)$ are displacement contribution terms dependent on the source location $(h,0)$ and the receiver location (x,z) in the model for a given frequency ω. They can be computed by equations (3A7) to (3A10). The energy integral I_2 is obtained from equation (3A15).

A wave packet of the *SH-Love* trapped waves for a single mode is then computed by the integration of the spectral density $|F(\omega)|$ given by (3A16) in the interesting frequency range,

$$f(z,t) = \frac{1}{2\pi}\int_{-}^{+} |F(\omega)|\exp\left[-i\omega t + iv_n z + i\phi(\omega)\right]d\omega. \tag{3A17}$$

where $\phi(\omega)$ is the initial phase at the source position.

The number of layers can be expanded arbitrarily as required for the purpose. Replacing L and N by μ, all formulations for the transversely isotropic medium can be used for the isotropic medium. In the isotropic medium, $\xi_j = \left(\frac{V_{sh}^{\parallel}}{V_{sh}^{\perp}}\right) = 1$. The above equations are rewritten as

$$\mu_j \frac{\partial^2 U_j}{\partial x^2} + \mu_j \frac{\partial^2 U_j}{\partial z^2} = \rho_j \frac{\partial^2 U_j}{\partial t^2}, \tag{3A1'}$$

$$L_j = N_j = \mu_j = (\rho V_{sh}^2)_j.$$

$$\frac{\partial^2 A_j(x)}{\partial x^2} = -k_j^2 \left(\frac{c^2 \rho_j}{\mu_j} - 1\right) A_j(x) = -v_j^2 A_j(x), \tag{3A3'}$$

$$v_j = k\sqrt{\left(\frac{c^2}{V_{sh}^2} - 1\right)_j} = k v_j^{\circ}, \tag{3A4'}$$

where

$$v_j^o = \sqrt{\left(\frac{c}{V_{sh}^2}\right)^2 - 1}, \quad \text{and} \quad \xi_j = \left(\frac{V_{sh}^{\parallel}}{V_{sh}^{\perp}}\right) = 1.$$

$$\begin{pmatrix} \cos(v_j d_j + \eta_j) & \cos(\eta_{j+1}) \\ \mu_j v_j \sin(v_{j+1} d_j + \eta_j) & -\mu_{j+1} v_j \sin(\eta_{j+1}) \end{pmatrix} \begin{pmatrix} B_j \\ B_j \end{pmatrix} = 0, \tag{3A11'}$$

$$\mu_{j+1} v_{j+1} \tan[\eta_{j+1}] = \mu_j v_j \tan[\eta_j + v_j(d_j)], \tag{3A12'}$$

$$\eta_1 = \arctan\left(-i\frac{v_0}{\mu_2 v_1}\right), \tag{3A13'}$$

$$\eta_4 = \arctan\left(-i\frac{\mu_4 v_4}{\mu_3 v_3}\right) = \eta_3 + v_3 d_3. \tag{3A14'}$$

$$V = \frac{1}{c}\frac{I_2}{I_1} \quad I_1 = \frac{1}{2}\int_{-} \rho(x) A^2(x) dx \quad I_2 = \frac{1}{2}\int_{-} \mu(x) A^2(x) dx. \tag{3A15'}$$

Li et al. (1990) first discovered fault-zone trapped waves generated by earthquakes when they examined seismograms recorded at station MM of the Parkfield ten-station borehole seismic network of three-component seismographs installed at depths of 200-300 m and operated by the U.S. Geological Survey and University of California. Station MM is located at about 200 m northeast of the surface trace of the San Andreas fault (SAF) in the Middle Mountain, ~ 8 km northwest of Parkfield Town while other borehole network stations are far away from the SAF (Fig. 3.1). The SAF at Parkfield separates two crustal blocks. The Franciscan block lies to the northeast, which contains fluid-rich metamorphic oceanic sedimentary rocks. The southwest crustal block is primarily harder granite, in which the seismic velocities are faster. Seismic tomography shows the asymmetry in seismic velocity variations across the SAF and suggests the abruption at the southwest granitic block and graduation at the northeast Franciscan block (Thurber et al., 2004).

A suite of Parkfield borehole seismic network data was inspected for evidence of fault-zone trapped waves (FZTWs) characterized by large-amplitude dispersive wavetrains following S-waves (Li et al., 1990; Li and Leary, 1990). For example, Figure 3A-2a shows seismograms recorded at two borehole network stations MM and JN. We recognized prominent FZTWs at station MM close to the SAF, but not at station JN located ~ 3 km away from the SAF for a microearthquake occurring within the fault zone at depths of 4 km and 11 km northwest of station MM. Figure 3A-2b shows seismograms recorded at station MM for two microearthquakes occurring within and away from the fault zone. Prominent FZTWs with large-amplitude dispersive wavetrain appear in seismograms for the on-fault event. By contrast, body waves are dominant in seismograms for the off-fault event. These observations indicate the existence of a low-velocity waveguide on the SAF at Parkfield and show the sensitivity of FZTW features to positions of the source and receiver relative to the fault-zone waveguide.

Fault-zone trapped waves recorded at station MM were first modeled with the simplest possible physical theory (*SH-Love* trapped waves), the simplest component of motion (vertical), and the simplest waveguide model (infinite, uniform, parallel planar layers, non-attenuation media) to obtain an initial estimate of the fault-zone waveguide properties (Li et al., 1990, Li and Leary, 1990). The model used for simulations of fault-zone trapped

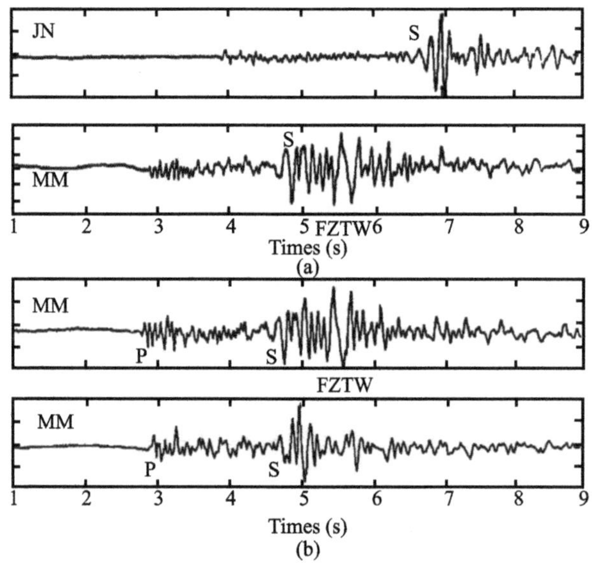

Fig. 3A-2 (a) Vertical-component seismograms recorded at borehole stations MM and JN of Parkfield borehole seismic network for a microearthquake located within the San Andreas fault zone at depths of 4 km and 11 km northwest of station MM (adapted from Li at el., 1990). Fault-zone trapped waves (FZTW) with large amplitudes and long dispersive wavetrains following *S*-arrivals appear in seismograms recorded at near-fault station MM while *P* and *S* waves are dominant in seismograms recorded at the away-fault station JN. (b) Vertical-component seismograms recorded at station MM show prominent FZTWs for the on-fault event, but not for the event at ~ 1.5 km away from the fault zone although the two events occurred at the similar depths of ~ 4.2 km and epicentral distances of ~ 11 km NW of station MM.

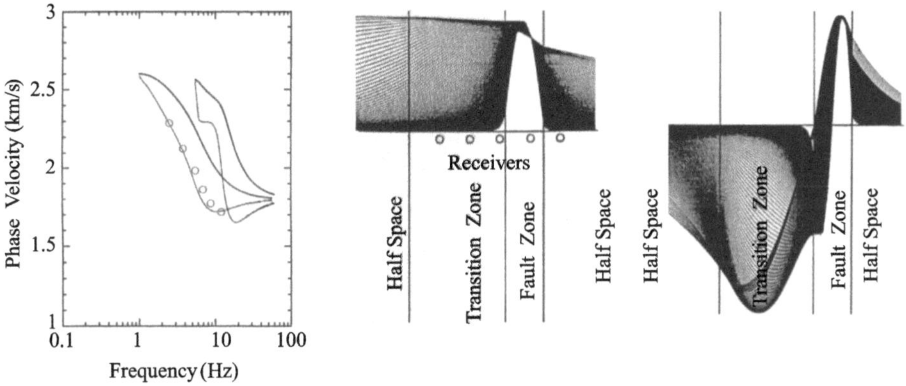

(a) Dispersion of SH/LOVE Trapped Modes
(b) Spectral Amplitudes Fundamental Mode
(c) Spectral Amplitudes First-Higher Mode

Fig. 3A-3 (a) Computed dispersion curves of the fundamental and first-higher modes of *SH-Love* type fault-zone trapped waves in the simplest fault-zone model of isotropic medium containing a 160-m-wide vertical fault zone and a 400-m-wide transition zone in between two half spaces plotted in (b) and (c). (b) Spectral amplitudes for the fundamental mode in frequency range of 3.5-29.1 Hz as a function of position normal to the fault zone. (c) Spectral amplitudes for the first-higher mode in frequency range of 5.1-30.7 Hz as a function of position normal to the fault zone. The shear velocities in the model are 1.8 km/s within the fault zone, 2.4 km/s in the transition zone, 2.75 km/s for the west half space, and 2.6 km/s for the east half space. Circles in (a) are group velocities measured from recorded dispersive waveforms of fault-zone trapped waves at 6 specific frequencies using a narrow-band filter with 1 Hz band, which are comparable to the synthetic dispersion curve of fundamental mode. 5 circles in (b) denote locations of receivers for computation of trapped waves shown in Figure 3A-4. Figures are adapted from Li and Leary (1990).

waves is of a four-part vertical velocity structure, containing a fault zone with lowest seismic velocity abutting a high-velocity half-space on the southwest side and a transition zone with mild low velocity on the northeast side, which abuts a half-space with the lower velocity other than the SW half-space. Model parameters include velocities of multi-layers, widths of the fault zone and transition zone, and locations of source and receivers. For example, Figure 3A-3 illustrates computed dispersion curves of the fundamental and first-higher modes and wave component amplitudes as a function of distance normal to the fault zone for a series of frequencies, using formulas for the isotropic medium given in the sections above and specific model parameters. The dispersion curves show slower phase and group velocities at higher frequencies, and the Airy-phase with the minimum group velocity. The maximum spectral amplitude for the fundamental mode appears at the center of the low-velocity fault-zone waveguide while the first-higher mode has highly diagnostic amplitude extinction in the center of the fault zone due to phase reversal.

Figure 3A-4 shows the simulated *SH/Love* trapped waveforms at 5 receiver locations (Fig. 3A-4b) and a source located within the fault zone at distance of $\sim$ 12 km from receivers for comparison with observations. The main features of observed fault-zone trapped waves are in general matched by the simulated waveforms in frequency content

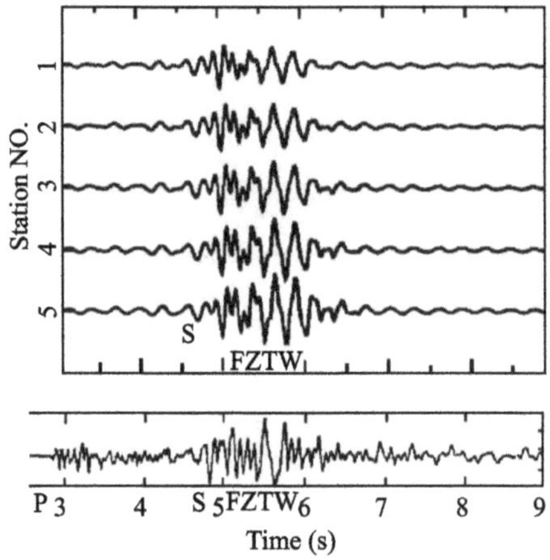

Fig. 3A-4 Comparison of observed and synthetic seismograms for the Parkfield microearthquake occurring within the San Andreas fault as shown in Figure 3A-2 b. Top panel shows *SH-Lo*ve trapped waves at 5 receivers located in the model shown in Figure 3A-3. The station spacing is 100 m. A source is located within the fault zone at the distance of ~ 12 km from the receivers. The computed wave components for the fundamental and first-higher modes are overlapped. Synthetic trapped waves at stations within and close to the fault zone are comparable with observations at station MM for the microearthquake as shown in Figure 3A-2.

and dispersive wavetrain at receivers located within and close to the low-velocity fault-zone waveguide. In the trial and error forward modeling, the model parameters are perturbed. The most sensitive parameters are the width and velocity of the fault zone, and there is a trade-off between them with regard to the effect on amplitudes and dispersion of trapped waves. In order to examine the trade-off, 150 pairs of these two parameters in the range of 50 to 200 m of fault-zone width at 10-m intervals and of 1.0 to 2.0 km/s of velocity at 0.1 km/s intervals have been computed to obtain synthetic seismograms fit test to observed fault-zone trapped waves.

In these examples, we illuminate that the constructive interference conditions of seismic trapped waves are able to offer information on the physical properties of the low-velocity structure of the fault zone, and show that a basic structural control of trapped *SH-Love* waves propagation can be obtained by an analytic solution using rapid phase recursion computation for the simplest model with plane layers space. However, the structure of a realistic fault zone is expected to be much complicated because the increasing pressure with increasing depth will strongly affect the crack density, fluid pressure, and amount of fluids, as well as the rate of healing of damage caused by earthquakes (Sibson, 1975; Byerlee, 1990; Rice, 1992). It may also influence the development of fault gouge (Scholz, 1990; Marone, 1998). For all these reasons, realistic fault zones are not uniform with depth and along the fault strike. Therefore, we need to use more sophisticated 3-D finite difference code with attenuation factor to document their detailed internal structure and physical properties at seismogenic depths (see main part of this chapter).

Chapter 4
Fault-Zone Trapped Waves at a Dip Fault: Documentation of Rock Damage on the Thrusting Longmen-Shan Fault Ruptured in the 2008 *M*8 Wenchuan Earthquake

Yong-Gang Li, Jin-Rong Su, and Tian-Chang Chen

This chapter presents observations and 3-D finite-difference simulations of fault-zone trapped waves (FZTWs) recorded at the south Longmen-Shan fault (LSF) with varying dip angles, which was ruptured in the 2008 *M*8 Wenchuan earthquake in Sichuan, China. Results of the FZTWs show a distinct low-velocity zone (LVZ) composed by severely damaged rocks at seismogenic depths. Through numerical investigations of trapping efficiency for a dip fault, we imaged a damaged zone several hundred meters wide along the thrusting LSF, within which seismic velocities are reduced by $\sim$ 30%-60% from wall-rock velocities with the maximum velocity reduction in the fault core at shallow depth. We interpret this remarkable LVZ as a break-down zone accumulating damages caused by dynamic rupture in historical major earthquakes, mainly in the 2008 *M*8 Wenchuan earthquake, which eventually forms a distinct low-velocity waveguide to trap seismic waves. Because the amplitude and dispersion features of FZTWs are sensitive to the source location with respect to the waveguide, these waves allow us to delineate the geometry of fault-zone damage along with the principal slip of the Wenchuan mainshock at seismogenic depth based on locations of those aftershocks generating prominent FZTWS. By examining the changes in the dispersion features of FZTWS recorded at the same station for similar earthquakes occurring before and after the 2008 Wenchuan earthquake, we estimate that seismic velocities within the LVZ along the south LSF was reduced by $\sim$10%-15% likely due to the co-seismic damage of fault rocks (with rigidity weakening) during the 2008 *M*8 mainshock. This value is greater than the damage magnitude of fault rocks caused by the 1992 *M*7.4 Landers, 1999 *M*7.1 Hector Mine and 2004 *M*6 Parkfield earthquakes in California (Li et al., 1998, 2003, 2006, 2007; Vidale and Li, 2003), probably due to the different sizes of slip and stress drop, and faulting mechanisms in these earthquakes.

Keywords: Dip fault-zone structure, Trapped waves, Co-seismic rock damage

4.1 Geological Setting and Scientific Significance

The tectonic setting of western China is dominated by the boundary between the Indian and Eurasian plates, the largest continental collision zone in the world (Burchfield et al., 1995). Indian plate began moving northward about 200 million years ago at the rate of ∼90 mm/year. When it ultimately collided with the Eurasian plate about 50 million years ago, its moving rate slowed down, but the northeastward motion continues at the rate of ∼40 mm/year. This Indo-Asia collision uplifted the Tibetan Plateau, created the still growing Himalayas and Tien-Shan mountains, and resulted in many large thrust and strike-slip faults in western China. The deformation associated with the collision is expressed in a north-south shortening of the Himalayas and an eastward movement of crustal materials. Most, if not all, of the deformation in continental China is attributed to this monumental collision (Zhang et al., 2004). The NNE-trending Longmen-Shan fault located at the east margin of the Tibetan Plateau is featured by this active tectonics (Densmore et al., 2007; Burchfield et al., 2008).

The Longmen-Shan fault (LSF) is the southern part of the so-called "North-to-South Seismic Belt Zone" in China, which has been evaluated as a highly risky region in China's seismic hazard model given by Applied Insurance Research (AIR). AIR natural catastrophe models simulate the physical characteristics of natural hazards, including earthquakes in the Asia-Pacific region. The hazard component draws upon China's extensive historical earthquakes experience dating back to 780 B.C., as well as detailed paleo-seismic, geodetic and GPS information, to depict the locations, magnitudes and occurrence rates of potentially damaging future earthquakes. The May 12, 2008 *M*8 Wenchuan earthquake ruptured in total length of 300 km along the LSF with the maximum slip of ∼9 m and most part of ruptures emerged at the ground surface (Xu et al., 2008; Ji, 2008; Lin et al., 2009). Its length, slip and magnitude are comparable with those in the 1857 *M*8 Tejon Pass earthquake and *M*7.9 San Francisco earthquake in 1906. This devastating earthquake, causing ∼90,000 lives dead, 340,000 people injured, and 1,500,000 people homeless, is the worst natural disaster in China after the *M*7.8 Tangshan earthquake in 1996.

Results of moment-tensor inversion of the data recorded at Global Broad-Band Seismic Network show multiple events in this big earthquake (Chen et al., 2008). The source mechanisms from waveform inversion show the reverse-thrusting at the first stage after initiation of this earthquake and then become strike-slipping gradually, indicating the complexity in faulting mechanism and stress heterogeneity along the LSF. More than 30,000 aftershocks were recorded at Sichuan Seismological Network in the first 2 years. The southern ruptures show bifurcation, consistent with the branching structure of the Longmen-Shan fault zone system in geological map (Fig. 4.1a). The main fault dips NWW at a high angle near the surface and becomes low angles at seismogenic depths. The aftershock zone extends for ∼300-km along the LSF and to the depth of ∼30 km (Fig. 4.1b, c). In order to document the subsurface extent and magnitude of rock damage on the Longmen-Shan fault caused by the *M*8 Wenchuan earthquake, we used fault-zone trapped waves (FZTWs) recorded at seismic stations of Sichuan Seismic Network (SSN) and a dense linear seismic array deployed across the south LSF for aftershocks occurring in the 2008 *M*8 Wenchuan earthquake source region (Fig. 4.1d, e).

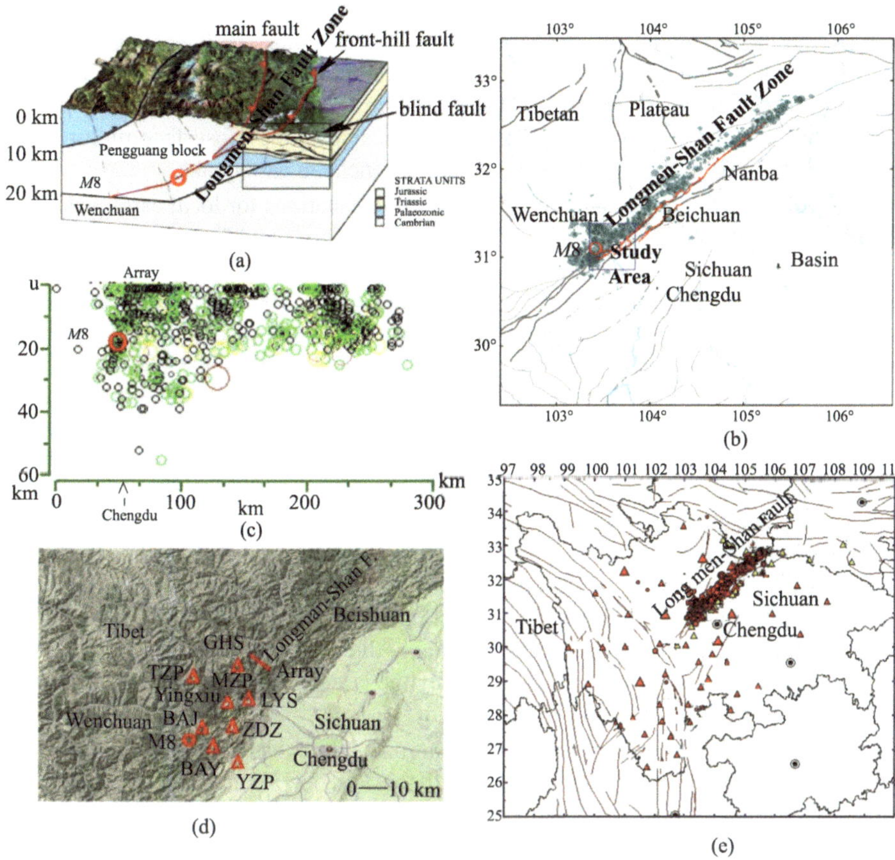

Fig. 4.1 (a) 3-D geological structure model of the south Longmen-Shan fault (LSF) shows the thrust-faulting and bifurcation of ruptures of the 2008 *M*8 Wenchuan earthquake. The map is adapted from "The Investigation Report of Surface Ruptures in the Wenchuan Earthquake, China Seismological Administration Special Report of Wenchuan Earthquake (The Emergency Science and Research Team of China Seismological Administration, 2008)". The main rupture dips to NWW at a high angle at shallow depth and becomes lower angles at the deeper portion. The branch rupture along front-hill fault dips at the lowest angle. (b) and (c) Map view and the depth section along the LSF show epicenters and hypocenters of part of ~6,000 aftershocks in the first year after Wenchuan earthquake recorded at 8 stations of Sichuan Seismological Network (SSN) located in the study area (blue square). (d) Locations of 8 SSN stations (red triangles) labeled by names, the data from which are used in this study. The red line denotes the dense linear seismic array of 16 stations deployed across the main rupture on the LSF after the 2008 *M*8 Wenchuan earthquake. (e) Triangles denote entire Sichuan Seismic Network stations used for locating earthquakes in the region.

First, we examined the waveform data to recognize fault-zone trapped waves (FZTWs) characterized by large amplitudes and long wavetrains with dispersive feature following *S*-arrivals in seismograms recorded at near-fault stations for aftershocks occurring within the rupture zone. These FZTWs illustrate the coherent interference phenomenon of wave propagation in the highly fractured low-velocity fault zone bounded by high-velocity surrounding rocks, which forms a waveguide to trap seismic energy within it.

Then, we tested the trapping efficiency of a thrusting fault with various geometry and velocities using a 3-D finite-difference computer code. Eventually, we used the testing result as a guideline to simulate FZTWs recorded at the across-fault array and network stations near the south Longmen-Shan fault for earthquakes occurring within the fault zone at different depths and epicentral distances. We obtained the best-fit model parameters (width, velocities, Q values and shape) applicable to the subsurface rupture zone along the south LSF in the source region of 2008 Wenchuan earthquake.

Finally, we examined the data recorded at the same stations for local earthquakes occurring in the Wenchuan earthquake source region before the *M*8 mainshock on May 12, 2008. We selected the similar events located at the same location within the south Longmen-Shan fault zone but occurring before and after the 2008 Wenchuan earthquake. We simulated FZTWs generated by the similar events and compared the model parameters that best fit observations for similar events before and after the 2008 Wenchuan earthquake. We found the seismic velocities within the south LSF were reduced by ~10%-15% immediately after the Wenchuan earthquake in 2008, lower than those in 2006 and 2007, likely due to the co-seismic damage of fault rocks caused by the *M*8 mainshock. The moving-window cross-correlations of waveforms for repeated aftershocks occurring at the same place show that seismic velocities near the LSF increased by ~5% or more in the first year after the Wenchuan earthquake, indicating that the post-mainshock fault heals with rigidity recovery of damaged rocks with time. The measurements of fault healing are not conclusive because the cross-correlation coefficient of waveforms for the repeated aftershocks is not great enough, probably due to location errors of aftershocks. Alternatively, the medium has been disturbed by the large aftershocks occurring between the repeated events used in the measurements.

The significance of this investigation is to illuminate the subsurface rock damage along a dip fault at seismogenic depths associated with the 2008 *M*8 Wenchuan earthquake. The data from the 2008 Wenchuan earthquake add the information into previous results of the spatio-temporal variations of fault zone properties associated with large earthquakes (Li et al., 1994, 1998, 2002, 2003, 2004, 2006; Vidale and Li, 2003). The spatial extent of fault-zone damage and the loss and recouping of strength across the earthquake cycle are critical ingredients in understanding of fault mechanics and physics. A comparison of the results of a reverse-thrusting dip faults, like the Longmen-Shan fault, with those of the strike-slipping faults, like the San Andreas fault at Parkfield in California (refer to Chapter 3 in this book), helps us to examine if the magnitude of fault rock damage is a function of earthquake size and to evaluate potential earthquake risk in seismogenic regions globally. These results also provide the useful information of subsurface rock damage caused by the 2008 Wenchuan earthquake in site selection for reconstruction in the earthquake hazardous areas.

4.2 Data and Results

4.2.1 Data Collection

Immediately after the *M*8 Wenchuan earthquake on May 12, 2008, the leading author of this article together with researchers of China Seismological Administration (CSA) surveyed the rupture zone along the ~200-km-long Longmen-Shan fault (LSF) segment between Yingxiu and Nanba to select the sites for deployment of portable seismic arrays to record fault-zone trapped waves (FZTWs) (Li, 2008a, b) (See Appendix). One selected seismic array site is located at the south LSF, ~35 km northeast of the mainshock epicenter near Yingxiu (Fig. 4.2). The main rupture shows ~4-m vertical slip exposed at the surface (see photos in this chapter). The fault scarp with sleek surface caused by one rock mass sliding over another is seen clearly at this site. Damaged rocks spread over the ground surface on both sides of the fault scarp. This zone with severely damaged rocks along the fault trace at the ground surface is about a couple hundred meters wide, extending more on the hanging wall block.

Sixteen seismographs provided by CSA were deployed in a line across the main rupture at this site. The station spacing is 25 m for stations in the central part of the array and 50 m for two stations at ends of the array. Three components of the 4-Hz sensor were oriented vertically, parallel and perpendicularly to the fault strike. The recorders worked in continuous mode with the sample rate of 100 samples per second. The internal clocks of recorders were synchronized by an external GPS clock when the seismographs were deployed. The recorders were of power supplied by internal batters. This seismic array worked for about 2 weeks in August of 2008 to record fault-zone trapped waves generated by aftershocks occurring within the rupture zone. After the 2008 *M*8 Wenchuan earthquake, a 1.2-km-deep borehole was drilled at the hanging-wall of the south LSF nearby this array site by Chinese Academy of Geological Sciences (CAGS). The borehole met the dipping fault at depths of ~650-800 m, showing a ~150-m-wide damage zone with severely fractured rocks and a thin clay layer (Xu et al., 2008a, b).

Left photo: The main rupture along the south Longmen-Shan fault shows the near-vertical slip of ~4 m on the fault scarp caused by the 2008 *M*8 Wenchuan earthquake. The leading author of this chapter surveyed this site with researchers from CSA to deploy a dense linear array of 16 seismographs across the rupture at surface to record fault-zone trapped waves generated by aftershocks. **Right photo:** The sleek surface on the nearly-vertical fault scarp sliding between the hanging-wall and foot-wall along the south LSF at this site.

We then collected the waveform data recorded at 8 stations of Sichuan Seismic Network (SSN) around Zipingpu Reservoir in the source region. This data set includes waveform data recorded for ~6,000 Wenchuan aftershocks in the selected time windows during the first year after the mainshock, and for ~500 local earthquakes occurring in 2006, 2007 and February of 2008. Fig. 4.2 shows locations of these aftershocks and stations. These 8 network stations are installed with FBS-3B sensors with 0.05-40 Hz frequency band. The 24-bit EDAS seismographs record the data of 100 samples per second. All these stations are located on the outcrops of hard rock at ground surface (Zhang et al., 2010). Stations

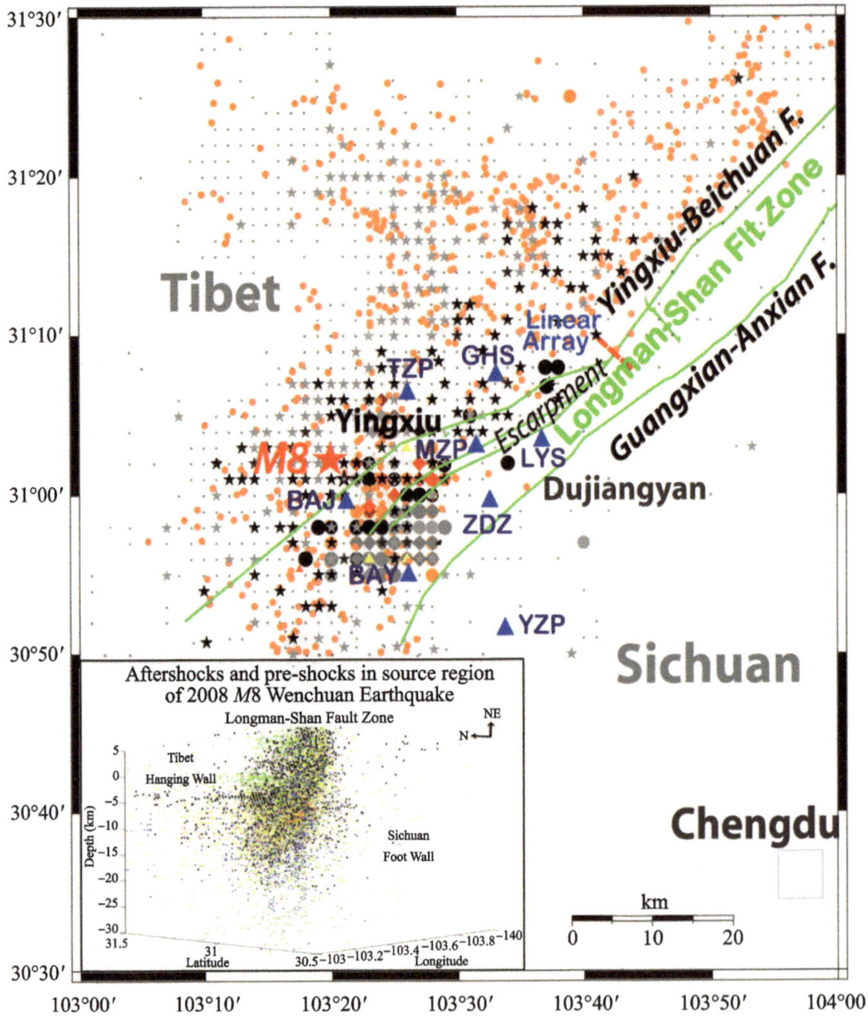

Fig. 4.2 Map view shows locations of 8 stations (blue triangles) among Sichuan Seismic Network (SSN), the linear dense seismic array (red line) deployed across the main rupture, aftershocks in the first year after the mainshock and local earthquakes in 2006-2008 recorded at 8 SSN stations in the Wenchuan earthquake source region(color symbols). Waveform data from these quakes are used in this study. Black symbols denote aftershocks generating prominent fault-zone-trapped waves (FZTWs) at stations close to the ruptures along the south Longmen-Shan fault zone. Brawn dots are relocated aftershocks using D-D method. Green lines denote faults and surface ruptures in the May 12, 2008 *M*8 Wenchuan earthquake, adapted from Xu et al. (2009) and Zhang et al. (2010). **Inset:** A 3-D volume shows these pre-shocks and aftershocks, indicating the aftershock zone dipping to northwest at seismogenic depth.

MZP and LYS are located close to the escarpments along the south Longmen-Shan fault in Shenxigou Valley, which were formed by the 2008 *M*8 mainshock (Xu et al., 2009, 2010; Zhang et al., 2010). Station BAJ is located near the southernmost segment of the Yingxiu-Beichuan fault (YBF) where no obvious slips were observed at the ground surface in the 2008 Wenchuan earthquake. Stations GHS and TZP are located on the hanging wall of the YBF while station YZP is located at the foot-wall of the Guangxian-Anxian fault (GAF) which dips at low angle of ~20°. These three stations are far away from the main rupture zone of the 2008 Wenchuan earthquake. Stations BAY and ZDZ are located on the hanging wall of GAF. The waveform data collected from these SSN network stations are continuous recordings. We sorted the data for interesting events used in this study. The locations of these events are listed in the SSN Catalog. Some of them have been relocated using D-D method. We examined the waveforms with high signal-to-noise ratio for aftershocks located in the SSN Catalog in our investigation.

4.2.2 Examples of Waveform Data

In this section, we first show seismograms recorded at the dense linear array deployed across the south Yingxiu-Beichuan fault (YBF) which was ruptured with nearly vertical slips of ~4 m accompanied by severely damaged rocks on both sides of the fault scarp at ground surface in the 2008 *M*8 Wenchuan earthquake. For example, Fig. 4.3 shows

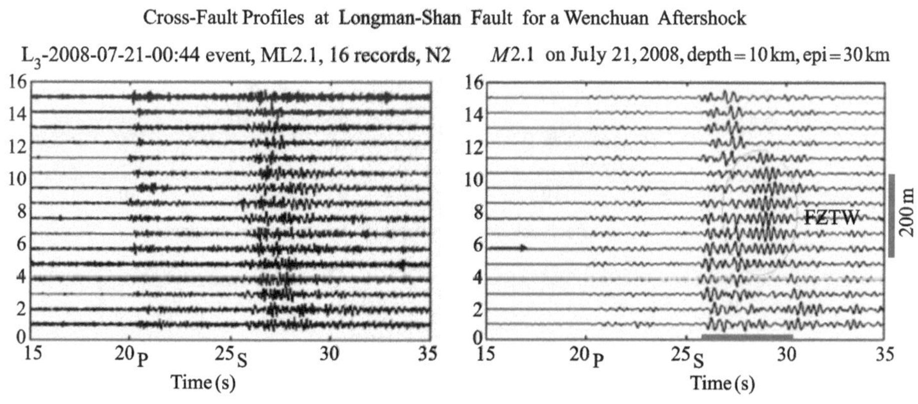

Fig. 4.3 Vertical-component seismograms recorded at the dense linear array of 16 stations across the Yingxiu-Beichuan fault (YBF) ruptured in the 2008 *M*8 Wenchuan earthquake for an *M*2.1 aftershock occurring at ~10 km depth within the rupture zone on July 21, 2008. The epicentral distance of this aftershock is ~30 km NNE of the array. The array is ~400-m long with the station spacing of 25 m in the central part of the array and 50 m for the end-members. The red arrow denotes the main slip in the Wenchuan earthquake (corresponding to the escarpment in Shenxigou Valley formed by the *M*8 mainshock), which exposed on the surface at the array site (Fig. 4.2). Seismograms in the left panel are raw data and < 3 Hz filtered in the right panel. Prominent fault-zone trapped waves (FZTWs) with large amplitudes and long wavetrains (denoted by a horizontal grey bar) after *S*-arrivals are recorded at stations in the ~200-m-wide rupture zone (marked by a vertical bar) along the YBF, within which rocks were severely damaged by the 2008 Wenchuan earthquake. Blue bars denote the FZTWs durations, in which amplitude envelopes of FZTWs are more than twice the level of the background noise coda.

seismograms recorded at the cross-fault array for an *M*2.1 aftershock occurring at ~10 km depth within the rupture zone of the 2008 *M*8 Wenchuan earthquake and ~30 km from the array. Prominent FZTWs with large amplitudes and long wavetrains following *S*-waves are dominant in the low-pass filtered seismograms. These waves appear clearly at stations located within the ~200-m-wide rupture zone exposed at the surface. The amplitudes of FZTWs decrease rapidly at stations away from the rupture zone. The duration of FZTW wavetrains after *S*-arrivals is approximately 5 s, indicating a distinct low-velocity waveguide formed by severely damaged rocks along the south LSF at depth

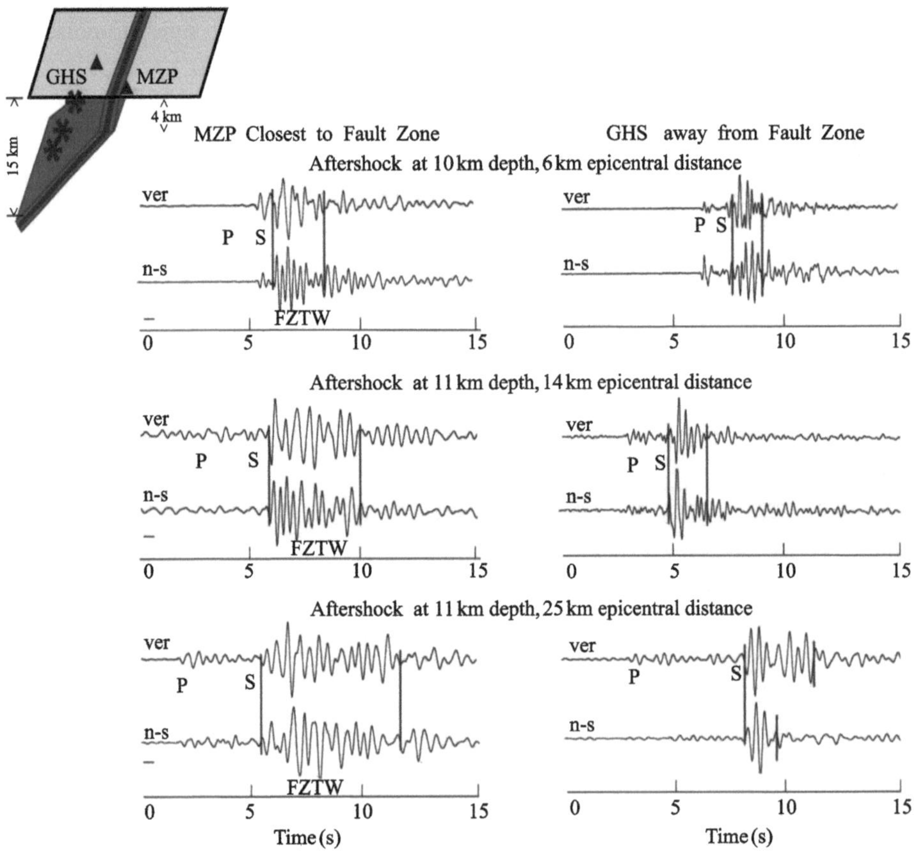

Fig. 4.4 Vertical and north-south components of seismograms recorded at Sichuan Seismic Network stations MZP and GHS for 3 on-fault aftershocks occurring at similar depths of ~10 km and epicentral distances of 6 km, 14 km and 25 km show prominent fault-zone trapped waves (FZTWs) with large amplitudes and long wavetrains after *S*-arrivals at station MZP close to the main rupture along the south LSF, but brief *S*-waves with much shorter wavetrains at station GHS located far away from ruptures for these aftershocks. Seismograms have been < 3Hz filtered. Time durations (marked by two vertical lines) of FZTWS recorded at station MZP increase with epicentral distances of these events, showing the existence of a distinct low-velocity waveguide formed by severely damaged rocks along the south LSF. Amplitudes of FZTWs between two vertical red lines are more than twice the level of background noise coda.

to trap seismic waves generated by the aftershock occurring within it. Observations of these FZTWs suggest that the width of the core damage zone is ~200-m wide at this site, within which seismic velocities are reduced by in average, ~40% from surrounding rocks at seismogenic depths above 10 km. The FZTWs recorded at stations across the main fault show that the damage zone is not asymmetry at the surface, but wider as it extends northwestward in the hanging wall of the south LSF with reverse-thrusting.

Secondly, we exhibit the data recorded at stations of Sichuan Seismic Network located in the Wenchuan earthquake source region. For example, Fig. 4.4 shows seismograms recorded at stations MZP and GHS for 3 Wenchuan aftershocks occurring within the rupture zone at similar depths of ~10-11 km and epicentral distances of 6 km, 14 km and 25 km from the stations. We observed prominent fault-zone trapped waves characterized by large amplitudes and long wavetrains after *S*-waves at station MZP located close to the main rupture with escarpments along the south Longmen-Shan fault. By contrast, only brief wavetrains after *S*-arrivals with flat changes are registered at station GHS located on

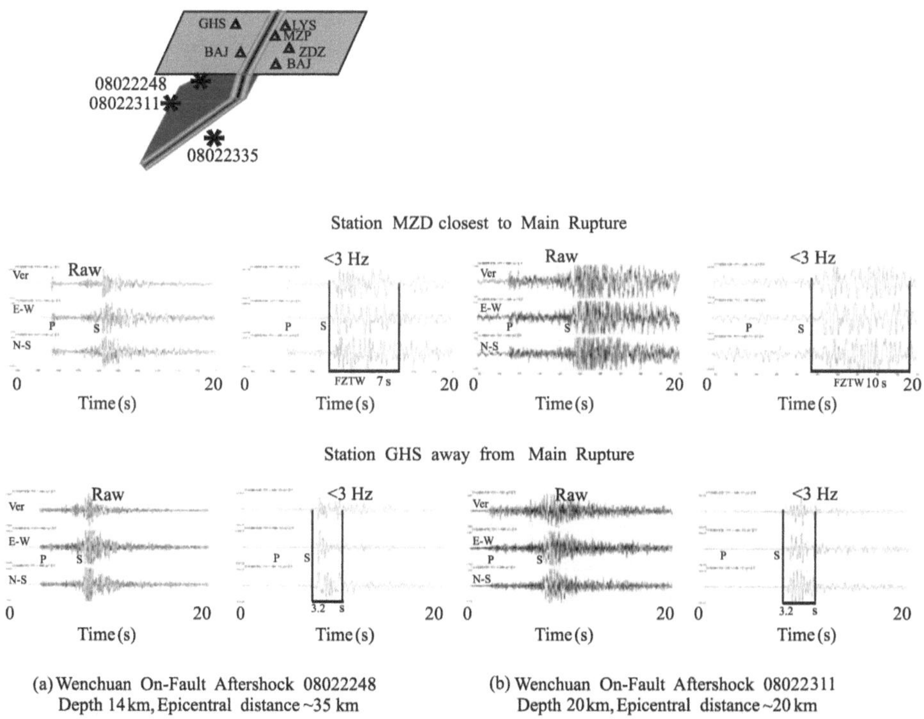

Fig. 4.5 (a) Raw and low-pass (<3 Hz) filtered three-component seismograms recorded at stations MZP and GHS for an *M*1.9 on-fault aftershock (ID 08022248) occurring at ~14 km depth and ~35 km northeast of the stations show prominent fault-zone trapped waves (FZTWs) with large amplitudes and long wavetrains (denoted by two vertical bars) after *S*-arrivals at station MZP located close to the main rupture along the south LSF, but much shorter wavetrains after *S*-arrivals at station GHS farther away from the rupture zone. (b) Same as (a) but for an *M*2.8 on-fault aftershock (ID 08022311) occurring at ~20 km depth and ~20 km north of the stations, which show prominent FZTWs at station MZP, not at station GHS.

the hanging-wall of the LSF and away from the main rupture for these aftershocks. We note that the time duration of FZTW wavetrains appearing at station MZP increases from ~2.5 s to ~6.5 s as the epicentral distance of aftershocks increases from 6 km to 25 km, showing a move-out in traveltime. These observations indicate a distinct low-velocity waveguide along the south LSF, which likely extends to ~10 km focal depth of these aftershocks.

We then exhibit seismograms recorded at 6 SSN stations for 3 Wenchuan aftershocks occurring within and away from the rupture zone at different depths and epicentral distances to show the sensitivity of trapped waves to source location with respect to the low-velocity fault-zone. Fig. 4.5a shows prominent FZTWs with large amplitudes with respect to body waves and long wavetrains (~10 s in time) after *S*-arrivals in the low-pass (< 3 Hz) filtered seismograms at close-fault station MZP for an aftershock occurring within the main rupture zone at ~14-km depth and ~35 km northeast of the station. By contrast, much shorter wavetrains (~3.2 s) after *S*-waves are registered at station GHS located away from the main rupture zone for the same aftershock. Fig. 4.5b exhibits raw and low-pass filtered seismograms at stations MZP and GHS for another on-fault aftershock at ~20-km depth and ~20-km northeast of the stations. Fault-zone trapped waves with ~7-s wavetrains after *S*-arrivals appear at station MZP, which are shorter than the wavetrains of FZTWs shown in Fig. 4.5a, indicating that the FZTWs travelling over a longer distance within the low-velocity fault-zone produce the longer wavetrains. By contrast, much shorter wavetrains (~3.2 s) after *S*-arrivals are registered at the away-fault station GHS for both aftershocks at different distances.

Fig. 4.5c and Fig.4.5d shows seismograms recorded at SSN stations LYS and ZDZ for these two on-fault aftershocks. Prominent fault-zone trapped waves have larger ampli-

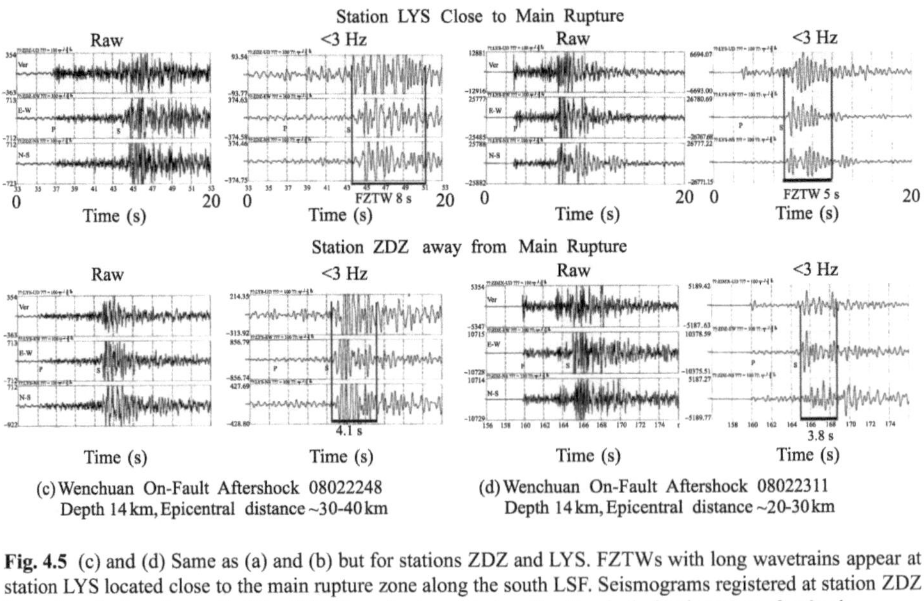

Fig. 4.5 (c) and (d) Same as (a) and (b) but for stations ZDZ and LYS. FZTWs with long wavetrains appear at station LYS located close to the main rupture zone along the south LSF. Seismograms registered at station ZDZ located on the hanging wall of the GAF show short wavetrains after *S*-arrivals for these two aftershocks.

tudes with respect to body waves and longer wavetrains at station LYS located close to the main rupture than those registered at station ZDZ located away from the LSF in the foot-wall. The wavetrains of FZTWs travelling over a greater distance from aftershock 08022248 to station LYS are ~8 s, longer than ~5 s wavetrains of FZTWs from aftershock 08022331. However, ~4 s wavetrains after *S*-arrivals are registered at station ZDZ for these two aftershocks. These observations show again the existence of a continuous low-velocity damage zone along the south LSF at seismogenic depth.

By contrast, Fig. 4.5e and Fig.4.5f exhibits short wavetrains (~1-2.5 s) after *S*-arrivals in seismograms recorded at stations MZD, GHS, LYS and ZDZ for an off-fault aftershock (ID 08022335) occurring at ~15 km depth. The epicenter of this event is ~8 km away from the main rupture and ~15-25 km from these stations. The wavetrains after *S*-arrivals at stations MZP and LYS close to the main rupture zone for this off-fault aftershock are much shorter than those at the same stations for on-fault aftershocks shown in Fig. 4.5a to Fig.4.5d, while seismograms at stations GHS and ZSZ away from the main rupture show short wavetrains after *S*-arrivals for either on-fault or off-fault event. However, we note that the wavetrains in *S*-coda at stations MZD and LYS located close to the main rupture are slightly longer than those at station GHS and ZDZ located farther away from the main rupture for the off-fault event in the foot-wall of the LSF, probably due to some seismic energy trapped in the nearly-vertical upmost part of the low-velocity rupture zone on the south LSF.

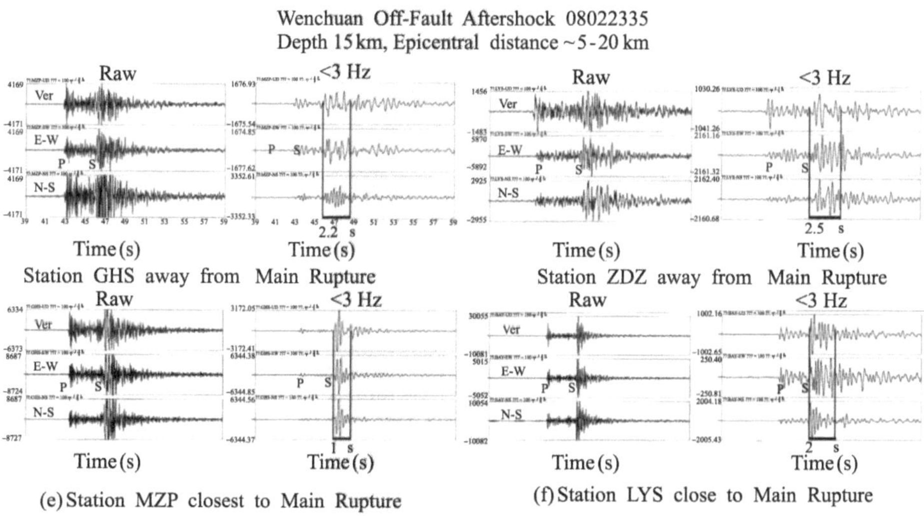

Fig. 4.5 (e) and (f) Raw and low-pass (<3 Hz) filtered three-component seismograms recorded at 4 SSN stations MZD, GHS, LYS and ZDZ for an *M*2.3 off-fault aftershock (ID 08022335) occurring at ~15 km depth and ~5-25 km southeast from these stations show shorter wavetrains after *S*-arrivals at all these 4 stations because the source is ~8 km away from the low-velocity zone along the main rupture. Other notations are the same as in Fig. 4.5a and Fig.4.5b.

Finally, Fig. 4.5g and Fig. 4.5h shows seismograms recorded at network stations BAJ and BAY for these aftershocks. Station BAJ is located close to the southernmost

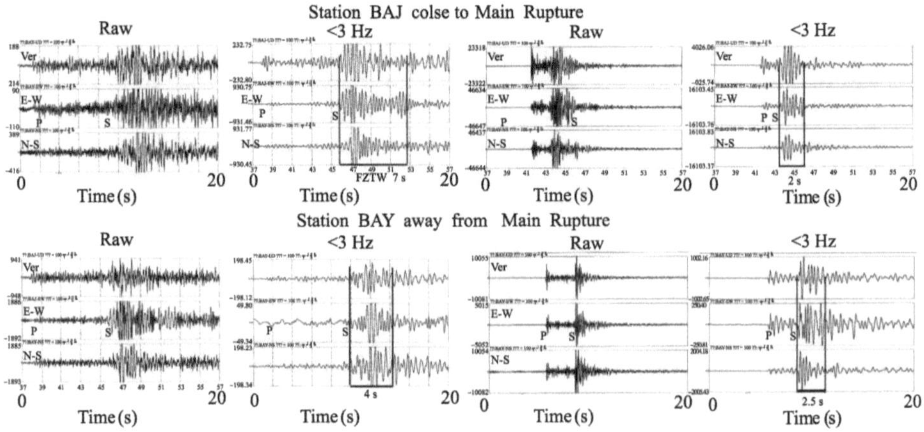

(g) Wenchuan On-Fault Aftershock 08022248
Depth 20 km, Epicentral distance ~ 40 km

(h) Wenchuan off-Fault Aftershock 08022335
Depth 20 km, Epicentral distance ~ 15 - 20 km

Fig. 4.5 (g) and (h) Raw and low-pass (<3 Hz) filtered three-component seismograms recorded at stations BAJ and BAY for the on-fault aftershock 08022248 and the off-fault aftershock 08022335. FZTWs with long wavetrains appear at station BAJ close to the southernmost YBF for the on-fault aftershock. Other notations are the same as in Fig. 4.5a and Fig. 4.5b.

Yingxiu-Beichuan Fault (YBF) on which there were no surface breaks seen in the 2008 *M*8 Wenchuan earthquake. Station BAY is located on the hanging wall of the low-angle thrusting Guangxian-Anxian fault (GAF) which dips at low angle of ~20° at the surface. FZTWs appearing at station BAJ for the on-fault aftershock show ~7-s wavetrains which are shorter than those appearing at stations MZD and LYS for the same event although the epicentral distance from this event to station BAJ is longer than distances to stations MZD and LYZ, indicating the weaker trapping efficiency of the low-velocity waveguide along the southernmost YBF with lower damage magnitude of rocks than the main rupture along the south LSF.

Seismograms shown in Fig. 4.5 illuminate that the amplitude and wavetrain length of FZTWs are sensitive to locations of source and receiver with respect to the low-velocity waveguide formed by highly damaged rocks along the Wenchuan rupture zone. These observations also imply the extension of the damage zone along the fault and with depth. We observed prominent fault-zone trapped waves at stations MZD and LYS located close to the main rupture along the south Longmen-Shan fault (LSF) and station BAJ close to the southernmost Yingxiu-Beichuan fault (YBF). By contrast, shorter wavetrains after *S*-arrivals appear at station GHS located on the hanging wall of the south LSF and at stations ZDZ and BAY located on the hanging wall of the Guangxian-Anxian fault (GAF), which are far away from the main rupture zone. Seismograms at other two SSN stations TZP and YZP farther away from the main rupture showing short wavetrains similar to those registered at station GHS are not exhibited in this chapter.

We have examined waveforms for ~350 Wenchuan aftershocks recorded at 8 network stations in the Wenchuan earthquake source region, and divided these aftershocks into two groups. The first group includes aftershocks showing prominent FZTWs with large amplitudes and long wavetrains in three-component seismograms recorded at stations MZP, LYS and BAJ close to the main rupture along the south LSF and Yingxiu-Beichuan fault. The second group includes aftershocks for which seismograms recorded at 8 network stations show brief *S*-waves but are lack of FZTWs. In summary, Fig. 4.6 shows locations of these two groups of aftershocks. We interpret the zone in which aftershocks were located and generated prominent FZTWs as the damage zone formed by severely fractured rocks caused by dynamic rupture of the 2008 Wenchuan earthquake. This zone might be co-incident with the principal slip plane of the *M*8 mainshock at seismogenic depth. This zone with highly damaged rocks likely extends to the depth of at least 10 km. An et al. (2009, 2010) have presented deep ruptures around the hypocenter of the 2008 Wenchuan earthquake deduced from aftershock observations. Recently, Fu et al. (2011) have computed the possible rupture depth of Longmen-Shan fault in the 2008 *M*8 earthquake at ~23 km from zero-strain points of co-seismic surface deformation.

Locations of those aftershocks generating FZTWs also show that the damage zone

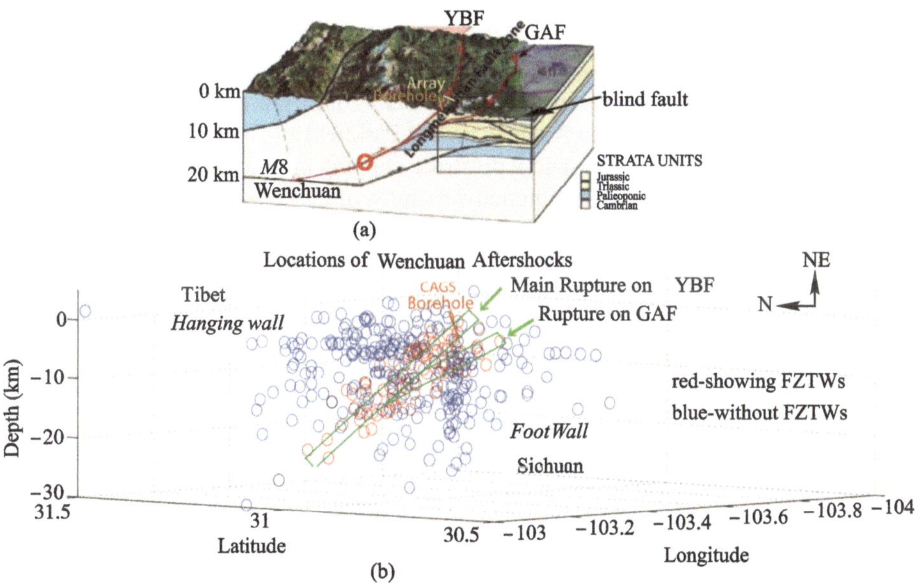

Fig. 4.6 (a) Geological structure model of the southern Longmen-Shan fault (LSF) zone shows thrust-faulting and bifurcation of ruptures along the Yingxiu-Beichuan fault (YBF) and Guangxian-Anxian fault (GAF). The main rupture along the YBF dips at a high angle at shallow depth, but becomes low angle at lower levels. The branch rupture along the GAF dips at a low angle of ~20°. The yellow-color bar denotes the cross-fault seismic array deployed at the YBF near Yingxiu. The brawn bar denotes the borehole drilled by CAGS near the array site after the 2008 *M*8 Wenchuan earthquake. (b) Locations of ~350 Wenchuan aftershocks (circles) whose waveforms have been examined in this study. Red circles denote aftershocks for which we observed fault-zone trapped waves (FZTWs) with large amplitudes and long wavetrains at stations located close to surface ruptures, but not at stations away from ruptures. Blue stars denote aftershocks for which we did not observe clear FZTWs at all these stations. Green boxes are our interpreted damage zones based on locations of those aftershocks generating prominent FZTWs, co-incident with the principal slip plane of Wenchuan earthquake on the YBF. The branch rupture segment is along the GAF dips at low angle.

dips northwestward at a low angle at seismogenic depth, agreeable with the geometry of reverse-thrusting Longmen-Shan fault in the geological model. Core samples taken from the 1.2-km-deep CAGS borehole show a severely damaged zone about ~150-m wide with highly fractured rocks at the depths of ~650-800 m including a ~20-m thin clay layer (Xu et al., 2008a, b). Locations of those aftershocks with FZTWs also likely show a bifurcation at shallow depth, co-incident with rupture branches along the YBF and GAF mapped after Wenchuan earthquake.

4.3 3-D Finite-Difference Investigations of Trapping Efficiency at the Dipping Fault

Highly damaged rocks along the Longmen-Shan fault (LSF) caused by the 2008 *M*8 Wenchuan earthquake form a distinct low-velocity waveguide which is able to trap seismic energy generated by aftershocks as the sources occurred within or close to the waveguide. Since fault-zone trapped waves (FZTWs) are produced by constructive interference of reflected waves from the boundaries between the low-velocity fault zone and high-velocity surrounding rocks, the amplitude and dispersion feature of FZTWs are sensitive to the geometry and physical properties of the fault zone. Observations and numerical simulations of FZTWs at the San Andreas fault at Parkfield and other active faults in California, which are nearly vertical and with strike-slipping mechanism, have been carried out for characterization of the fine internal structures of these fault zones at seismogenic depth (e.g., Li et al., 1990, 1994, 1997, 2001, 2002, 2004; see Chapter 3 in this book). Li and Vidale (1996) have used finite difference codes (Vidale et al., 1985; Graves, 1996) for simulation of FZTWs to numerically investigate trapping efficiency of the low-velocity fault zones with various geometry (varying width and depth, kink, branch and layering) and physical properties (velocity reduction and *Q* values).

The south Longmen-Shan fault shows reverse-thrusting and dips with varying angles at seismogenic depths. This geometry of the low-velocity fault zone may reduce trapped wave excitation and trapping efficiency, and affect observation of FZTWs at surface stations adversely. In order to simulate the FZTWs recorded at the south LSF with varying dip angles at depth, we have investigated the effects of various possible fault dipping structures that might obscure the signature of faults and disrupt the FZTWs, using a 3-D finite difference code. This numerical study helps us to design the field experiments to record significant FZTWs at the thrusting south LSF and identify FZTWs in the seismograms recorded at surface stations around the Wenchuan earthquake source region.

Based on the geological mapping of surface ruptures along the Longman-Shan fault in the 2008 *M*8 Wenchuan earthquake (The Emergency Science and Research Team of Chinese Earthquake Administration, 2008) and the information from fault-zone drilling at the south LSF after Wenchuan earthquake (Xu et al., 2008a, b), we constructed plausible models for the dipping LSF (Fig. 4.7). The shape and extension of the low-velocity rupture zone along the south LSF at depth are also constrained by the coseismic slip model of the 2008 Wenchuan earthquake derived from joint inversion of interferometric synthetic aperture radar, GPS, and field data (Tong et al., 2010; Xu et al., 2010). They

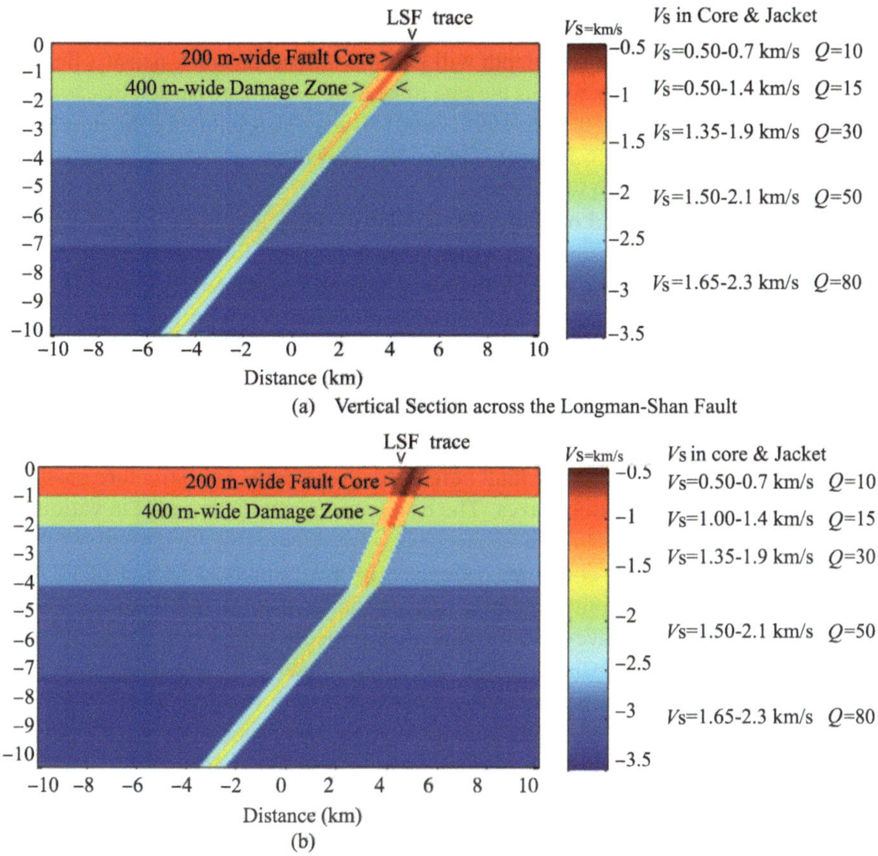

Fig. 4.7 Vertical sections of velocity and Q models across the fault used in numerical test of trapping efficiency of a dip fault zone with various geometry and material properties. A 200-m-wide fault core zone is sandwiched in the 400-m-wide damage zone (jacket). The velocities within the fault core are reduced by 25%-50% from wall-rock velocities with the maximum reduction within the fault core at shallow depth. (a) The fault dips at 45°. (b) The fault dips at 75° above 4 km and 45° below. The shear velocities and Q values within the fault core and damaged jacket at certain depths are shown at right to the model.

used four sub-faults with variable geometry and dips to capture the simultaneous rupture of both the Yingxiu-Beichuan fault (YBF) and Guangxian-Anxian fault (GAF). Most of the slip asperities along the YBF are above 10-km depth, except for the southwest part near the hypocenter where the rupture may exceed 20 km. The best fit model has fault dipping at 35° in average along the deep portion of the south YBF, on which the thrust-slip is mostly ~13 m and principally in the upper 10 km of the crust, while their rupture model is complex with variations in both depth and rake along the YBF and GAF. The rupture progressively transformed to right-lateral strike slip as it propagated northeast. Velocities of surrounding rocks in these models are constrained by Sichuan Province velocity model (Sun et al., 2004; An et al., 2009, 2010) and recent results from low-frequency 3-D wave propagation modeling of the 2008 *M*8 Wenchuan earthquake (Chavez et al., 2010).

The structural models used in the test procedure are depth-dependent because the increasing pressure with increasing depth will strongly affect the crack density, fluid pressure, and amount of fluids, as well as the rate of healing of damage caused by earthquakes (Byerlee, 1990; Rice, 1992; Sibson, 1996, 2000). It may also influence the development of fault gouge (Scholz, 1990; Marone, 1998a, b). For all these reasons, a realistic fault zone is probably not uniform with depth.

Our simulations are based on geometry similar to those inferred for the 3-D geological structure model of the south Longmen-Shan fault with thrust-faulting in the 2008 *M*8 Wenchuan earthquake and the subsurface damage zone inferred by locations of aftershocks showing prominent fault-zone trapped waves (Fig. 4.6). In the simulations, our reference model has a 10-km source depth, a 200-m-wide fault core zone sandwiched in the 400-m-wide damage zone (jacket), and a double-couple source is placed against one edge of the fault core zone. The receiver array is centered on the surface fault trace. The velocities within the fault core are reduced by 25%-50% from wall-rock velocities with the maximum reduction within the fault core at shallow depth. This structure most effectively traps 1- to 10-Hz seismic waves. The basic model parameters used in various fault structures are shown in Fig. 4.7.

The 3D finite-difference computer code used in this numerical testing of FZTWs is second order in time and fourth order in space (Vidale et al., 1985; Graves, 1996). It propagates the complete wave-field through elastic media with a free surface boundary and spatially variable anelastic damping (an approximate Q). The calculation used a 400-by-800-by-200 element grid in x–y–z coordinates with the grid spacing of 50 m to simulate a volume of 20 km in width, 40 km in length, and 10 km in depth. We decreased or increased the grid volume for a shorter or longer distance between the source and receiver. The low-velocity waveguide composed by a fault core zone with maximum velocity reduction sandwiched by a wider damage jacket with milder velocity reduction is embedded in the high-velocity half space with a free surface where the receiver array is placed across the waveguide at the middle of the volume. The low-velocity fault zone dips at varying angles. A double-couple source with radiation patterns was included in the volume.

4.3.1 Effect of Fault-Zone Dip Angle

First, we tested the effect of a dip angle of the fault on generation and propagation of fault-zone trapped waves. Fig. 4.8 shows 3-D finite-difference synthetic seismograms at the cross-fault array for a fault dipping at angles of 15°, 30°, 45°, 60°, 75° and 90° with respect to the horizontal. Prominent FZTWs with relatively large amplitudes and long-duration wavetrains after *S*-arrivals appear at stations located within the 200-m-wide fault core zone which has the largest velocity reductions. Some trapping energy appears in the jacket having slight velocity reductions. The length and amplitude of FZTW wavetrains decrease dramatically for the fault zone dipping at an angle lower than 30°, indicating the weak creation capacity and trapping efficiency for a fault dipping at a very low angle. We note the asymmetry of amplitudes of FZTWs across the fault with the larger amplitudes of body waves in the hanging wall of the dip fault, probably due to the wedge effect at the edge of the hanging wall near surface. These simulations suggest that it is not favorable to observe FZTWs at a reverse-thrust fault dipping at an angle lower than 30°, such as the Guangxian-Anxian fault. We also note the different wavetrain lengths of FZTWs in three-component seismograms, probably due to anisotropic trapping efficiencies to waves with different polarizations.

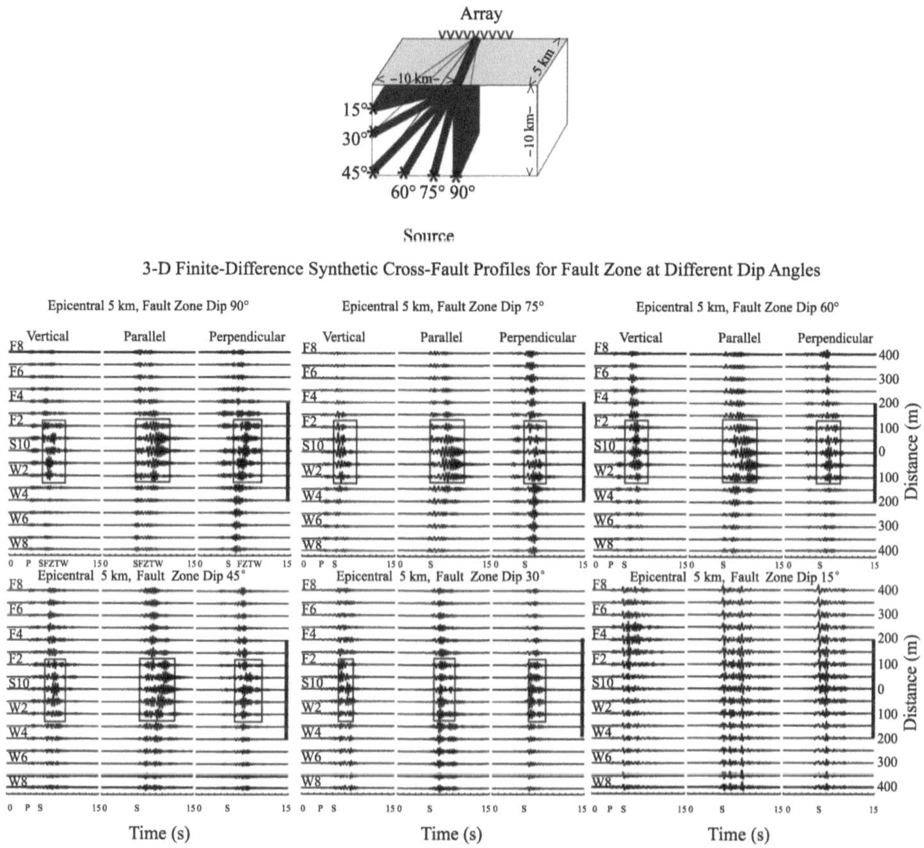

Fig. 4.8 Top: The geometry of fault zones dipping at different angles. The ray-path (grey line) between the source within the fault zone and the receiver at the center of seismic array is plotted. **Bottom:** Three-component finite-difference synthetic seismograms at the cross-fault array for fault-zone structural model A in Fig. 4.7 with a low-velocity fault zone dipping at angles of 15°, 30°, 45°, 60°, 75°, 90° with respect to the ground surface. The 200-m wide fault core (marked by a dark grey bar right to seismograms) with maximum velocity reduction is sandwiched by a 400-m-wide pocket (marked by a light grey bar) with slight velocity reduction. Shear velocities and Q values within the fault core and damaged jacket shown at right to the model. A double-couple source is placed within the fault zone at 5 km epicentral distance from the cross-fault array. The array is 800-m long with 50 m station spacing. Upper traces are for stations in the hanging wall of the dip fault. Seismograms have been $<$ 5 Hz filtered and are plotted using a fix amplitude scale for each panel. Prominent FZTWs with large amplitudes and long wavetrains (in rectangular boxes) appear at stations located within 200-m-wide fault core zone. The width of box denotes the post-S wave duration, in which amplitude envelopes of FZTWs are in average more than twice the level of the background noise coda at 5 stations within the 200-m-wide fault zone. The time duration and amplitude of FZTWs decrease dramatically as the dip angle of the fault zone is less than 30°.

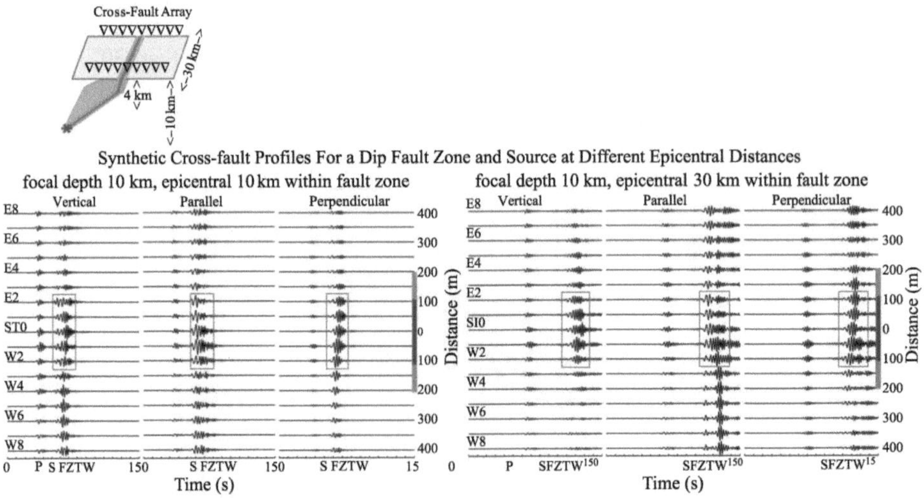

Fig. 4.9 (a) Three-component synthetic seismograms at the cross-fault array for a fault zone dipping at 75° above 4-km depth and at 45° below as shown in model B of Fig. 4.7 for a double-couple source located within the fault zone at focal depth of 10 km. The cross-fault array is 10 km and 30 km from the source epicenter at the surface. Prominent fault-zone trapped waves (FZTW) with large amplitudes and long wavetrains (in rectangular boxes) appear at stations located within 200-m-wide fault core zone (marked by a dark grey bar). The FZTWs from the source at 30-km epicentral distance show longer wavetrains than those for the source at 10-km epicentral distance. Other notations are the same as in Fig. 4.8.

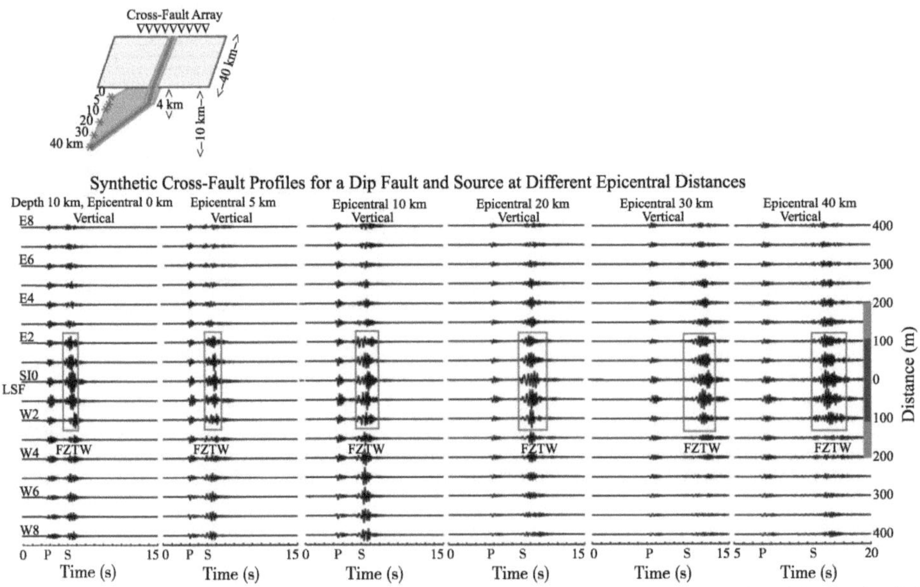

Fig. 4.9 (b) Vertical-component synthetic seismograms at the cross-fault array in terms of model B in Fig. 4.7 for a double-couple source located within the fault zone at focal depth 10 km and different epicentral distances of 0, 5, 10, 20, 30, and 40 km from the array. Prominent fault-zone trapped waves (FZTW) with large amplitudes and long wavetrains (in rectangular boxes) appear at stations located within 200-m-wide fault core zone. The time durations of FZTW wavetrains after *S*-arrival increase from ~1 s to ~4 s as epicentral distances increase from 0 to 40 km. Other notations are the same as in Fig. 4.8.

4.3.2 Effect of Epicentral Distance

Secondly, we tested the effect of source epicentral distance on propagation of fault-zone trapped waves (FZTWs) for a dipping fault zone. Fig. 4.9a exhibits 3-D finite-difference synthetic profiles at the cross-fault array for a double-couple source located at 10-km depth and epicentral distances of 10 km and 30 km, respectively, within the fault zone using model B in Fig. 4.7. Prominent FZTWs with large amplitudes and long wave-trains with respective to those of body waves are registered at stations located within 200-m-wide fault core zone. The FZTWs show wavetrains of ~3.5 s in time duration after *S*-arrival for the source at 30-km epicentral distance and ~2 s for the source at 10-km epicentral distance. The longer duration of FZTW wavetrains is produced as these waves travel over a greater distance within the low-velocity fault-zone waveguide. This is shown clearly in Fig. 4.9b for the source located within the fault zone at epicentral distances of 0, 5, 10, 20, 30, and 40 km from the receiver array. The time durations of FZTW wave-trains after *S*-arrival increase from ~1 s to ~4 s as epicentral distances increase from 0 to 40 km.

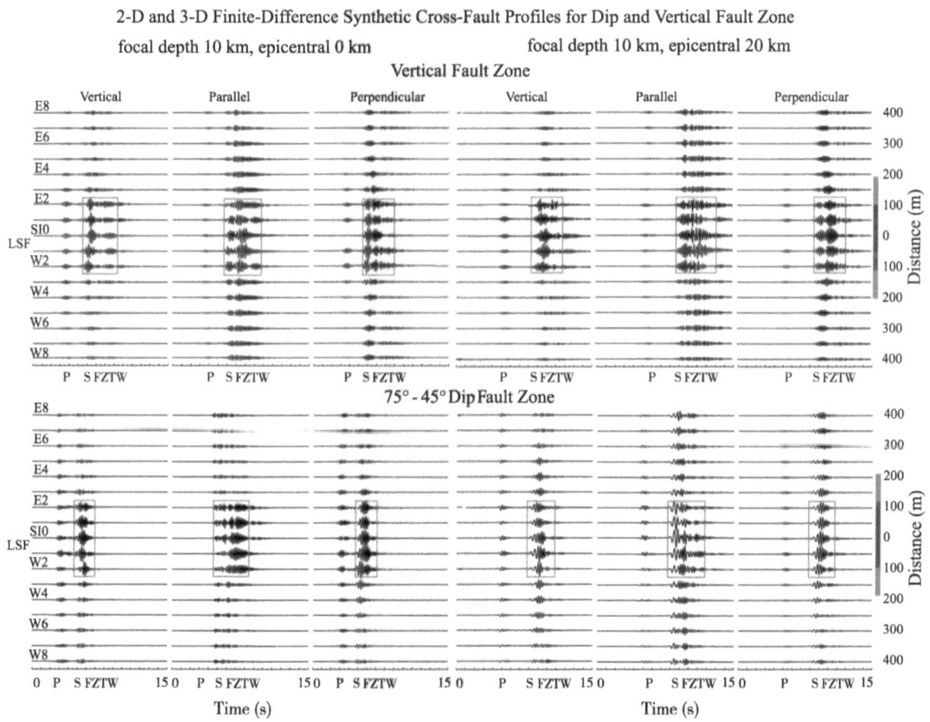

Fig. 4.9 (c) Three-component finite-difference synthetic seismograms at the cross-fault array (top row) for a vertical low-velocity fault zone, and (bottom row) for a fault zone dipping at 75° and at 45° above and below 4-km depth as shown in model B of Fig. 4.7. A double-couple source is located within the fault core zone at 10-km depth with 0-km and 20-km epicentral distances from the array, respectively. The wavetrains of FZTWS (marked by rectangular boxes) elongate as the epicentral distance increases. Other notations are the same as in Fig. 4.8.

We compared a dip fault zone with a vertical fault zone for their trapping efficiencies in this test. Fig. 4.9c shows 3-D finite-difference simulations with the source at 10 km depth within the low-velocity fault core zone with 0-km and 20-km epicentral distances from the receiver array. It is more clearly seen that the wavetrain length of FZTWs increases with the increasing epicentral distance for a vertical fault zone other than the dip fault zone. We also note that amplitudes of FZTWs with respect to body wave amplitudes for the vertical fault zone are greater than those for a dip fault zone. These observations show that the vertical fault zone has a stronger excitation of FZTWs and trapping efficiency than the dip fault zone does.

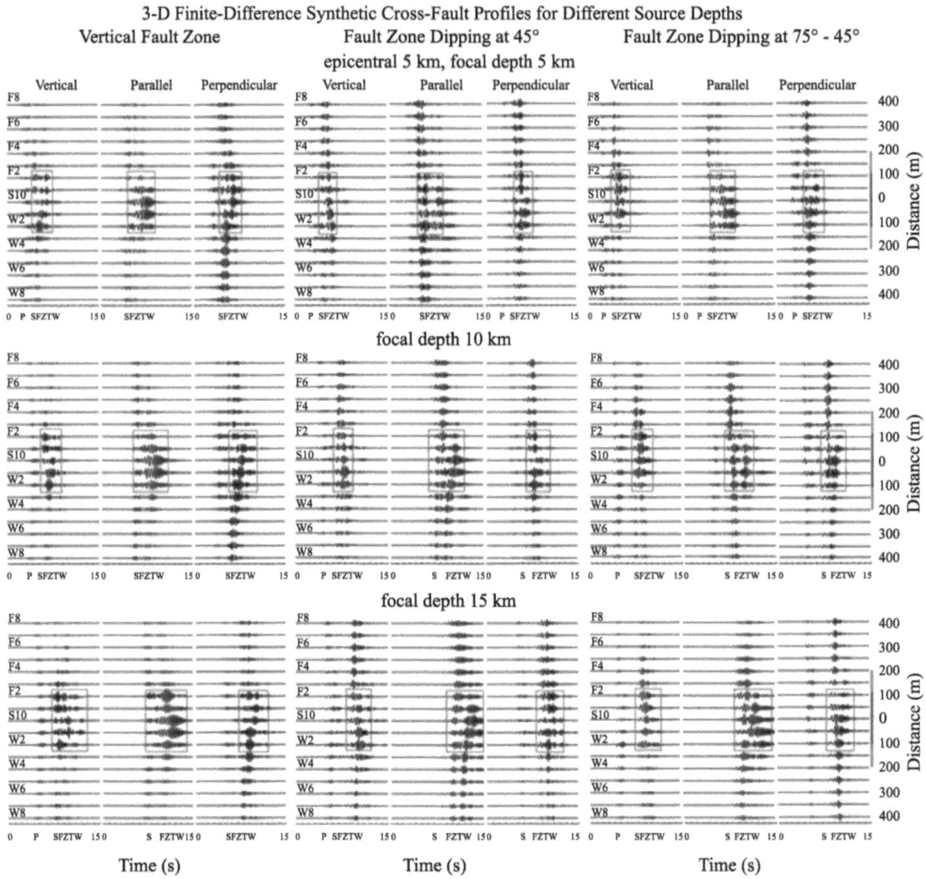

Fig. 4.10 Three-component finite-difference synthetic seismograms at the cross-fault array (left column) for a vertical low-velocity fault zone (middle column) for a 45° dipping fault zone in model A of Fig. 4.7, and (right column) for a fault zone dipping at 75° and at 45° above and below 4-km depth as shown in model B of Fig. 4.7. A double-couple source is located within the fault core zone with the epicentral distance of 5 km from the array and at focal depths of 5 km, 10 km and 15 km, respectively. The wavetrains of FZTWS (marked by rectangular boxes) elongated as the source depth increases although the elongation of FZTW wavetrain is not linear with the increase of source depth because the velocity reduction within the waveguide decreases with the increasing depth. Other notations are the same as in Fig. 4.8.

4.3.3 Effect of Source Depth

Thirdly, we tested the effect of the source depth on the excitation and propagation of FZTWs using depth-dependent fault-zone models with different dipping angles. Fig. 4.10 shows 3-D finite-difference simulations with the source depths of 5, 10, and 15 km. The

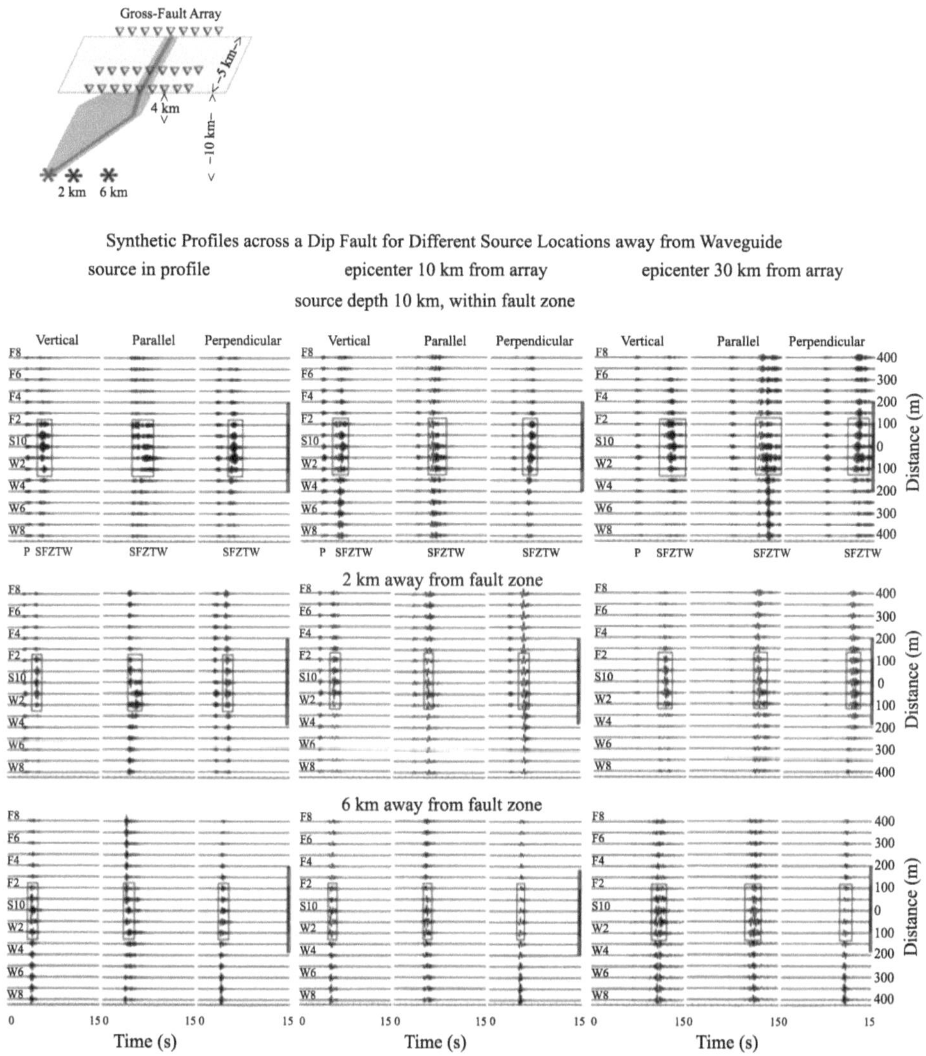

Fig. 4.11 (a) Three-component finite-difference synthetic seismograms at the cross-fault array for a low-velocity fault zone dipping at 75° and at 45° above and below 4-km depth as shown in model B of Fig. 4.7. A double-couple source is located at 10-km depth within the fault core zone, 2-km and 6-km away from the fault zone. The cross-fault array is 0 km, 10 km and 30 km from the source epicenter at the surface, respectively (from the left to right column). The wavetrains after *S*-arrivals are marked by rectangular boxes. Other notations are the same as in Fig. 4.8.

source is located within the low-velocity fault core zone at 5-km epicentral distance from the receiver array. We observe the increase in wavetrain length of FZTWs (the move-out in traveltime) as the source depth increases. We note that amplitudes of FZTWs with respect to body wave amplitudes for the vertical fault zone are larger than those for dip fault zones, showing again that the vertical fault zone has stronger trapping efficiency than a dip fault zone. We also notice that FZTWs show longer wavetrains in parallel-fault component seismograms than those in vertical and perpendicular-fault components, probably due to the different coherent effects for waves with different particle motions.

4.3.4 Effect of Source away from Vertical and Dip Fault Zones

Fourthly, we tested the effect of source location away from the low-velocity fault-zone waveguide. Trapped waves are most efficiently excited by a seismic source that is located within the low-velocity waveguide. This is clear in the following examples. Fig. 4.11a shows the FZTWs generated by a source at 10-km depth. The source is located within the core zone of a dip fault, and then horizontally moves 2 km and 6 km away from the low-velocity fault zone in the foot-wall. The cross-fault array is located at 0-km, 10-km and 30-km from the source epicentral distance at surface. We observe the most prominent FZTWs with largest amplitudes with respect to body wave amplitudes and longest wave-

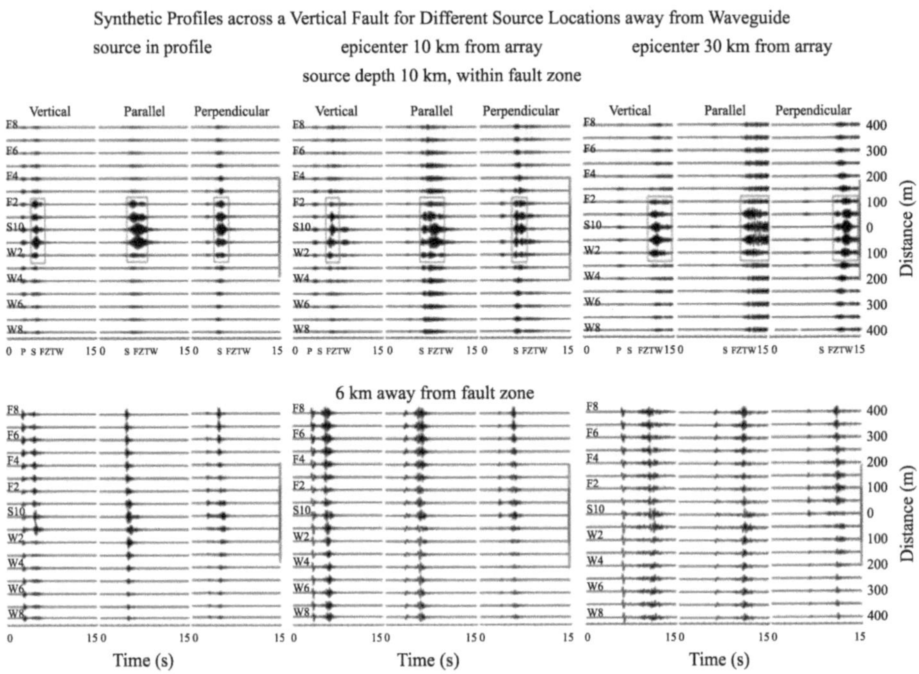

Fig. 4.11 (b) Same as in Fig. 4.11a, but for a vertical fault zone. Fault-zone trapped waves (FZTW) diminish as the source moves 6 km far away from the low-velocity waveguide of the vertical fault zone.

trains for the source located within the fault core zone having the lowest velocities. We notice that FZTW wavetrains generated by the source within the fault zone elongate as its epicentral distance increases because the FZTWs travel over a longer distance within the low-velocity waveguide for the farther source. By contrast, as the sources are located at 2 km and 6 km away from the fault zone 1, the lengths of wavetrains after *S*-arrivals and their amplitudes with respect to body wave amplitudes decrease rapidly, but do not change obviously with the source epicentral distance.

We also tested for a vertical fault zone in this case. Fig. 4.11b shows that wavetrain

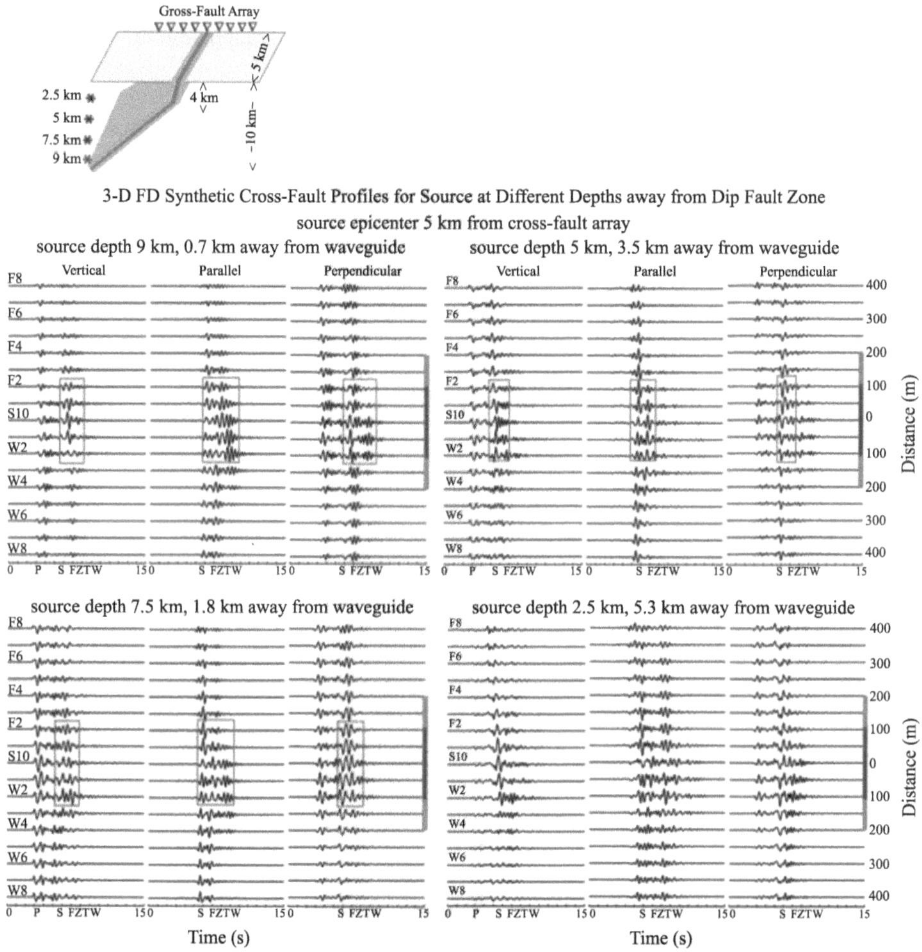

Fig. 4.11 (c) Three-component finite-difference synthetic seismograms at the cross-fault array for a fault zone dipping at 75° and at 45° above and below 4-km depth as shown in model B of Fig. 4.7. A double-couple source is located with its epicenter being 5 km from the cross-fault array and 10 km from the fault strike, but at different depths of 9 km, 7.5 km, 5 km and 2.5 km, respectively. The wavetrain length of FZTWs (marked by rectangular boxes) decreases as the source moves away from the low-velocity fault zone. Other notations are the same as in Fig. 4.8.

lengths of FZTWs and their amplitudes with respect to body wave amplitudes decrease more dramatically as the source moves away from the vertical fault zone other than the dip fault zone. We note that seismograms for the source located at 6 km away from the vertical fault zone show brief *S*-waves with shorter wavetrains other than those for the dip fault, suggesting some seismic waves from the off-fault source might be trapped by the shallow part of the dip fault zone.

We then tested the trapping efficiency of a dip fault zone for a source moving vertically away from the fault zone. Fig. 4.11c shows synthetic seismograms at the cross-fault array for a source moving vertically from 9-km to 7.5-km, 5-km and 2.5-km depths in the hanging wall of the dip fault zone. The source epicenter is 5 km from the cross-fault array. FZTWs are not visible as the off-fault source is located at depth above ~5 km. This phenomenon has been discussed in (Igel et al., 2002).

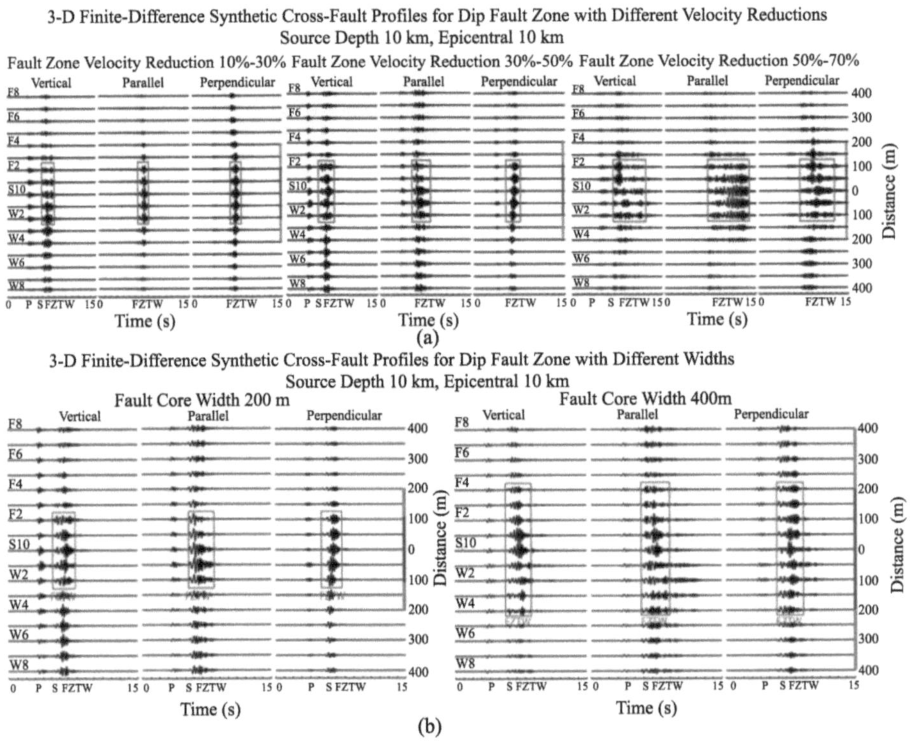

Fig. 4.12 (a) Three-component finite-difference synthetic seismograms at the cross-fault array for a fault zone dipping at 75° and at 45° above and below 4-km depth as shown in model B of Fig. 4.7. A double-couple source is located with 10-km epicentral distance and at 10-km depth within the fault zone. The velocities within the fault zone core zone and jacket are reduced (left) by 30% and 10%, (middle) by 50% and 30%, (right) by 70% and 50% from wall-rock velocities, respectively. The longest wavetrains are produced by a lowest-velocity fault zone. (b) Three-component finite-difference synthetic seismograms at the cross-fault array for a fault-zone (left) with 200-m-wide core zone and 400-m-wide jacket, and (right) with 400-m-wide core zone and 800-m-wide jacket. Velocities are shown in Fig. 4.7. A double-couple source is located with 10-km epicentral distance and at 10-km depth within the fault zone. Fault-zone trapped waves for a wider fault zone show longer wavetrains at lower frequencies than those for the narrower fault zone.

4.3.5 Effect of Fault-Zone Width and Velocity Reduction

We finally tested the effect of the fault zone with different widths and velocity reductions on trapping efficiency. The amount of reduction of velocity within the fault zone determines the amount of dispersion and wavetrain length of the fault-zone trapped waves. A 10%-30% velocity reduction results in a much more compact wavetrain of FZTWs than a 50%-70% velocity reduction (Fig. 4.12a). The frequency content of trapped waves also depends on the velocity within the fault zone, since lower-velocity material within a fixed width fault zone causes longer period resonance, but this effect is weak. The width of the low-velocity fault zone, however, controls the frequency of the trapped waves. A 400-m-wide fault core zone shows a clear shift toward lower-frequency in longer wavetrains of trapped waves compared with the reference model of a 200-m wide fault core zone (Fig. 4.12b).

Through this numerical test procedure of trapping efficiency for dip faults with various geometries and physical properties, we obtained necessary experience to simulate fault-zone trapped waves (FZTWs) recorded at the Longmen-Shan fault in our study area.

4.4 3-D Finite-Difference Simulations of FZTWs at the South Longmen-Shan Fault

Taking the results from geological mapping of the surface ruptures conducted by researchers of China Seismological Administration (The Emergency Science and Research Team of China Seismological Administration, 2008) and fault-zone drilling at the southern LSF by Chinese Academy of Geological Sciences (Xu et al., 2009) as additional constraints, we construct a depth-varying structural model for the LSF at seismogenic depth (based on model B in Fig. 4.7) to simulate FZTWs observed at the south Longmen-Shan fault in Wenchuan earthquake source region.

To find the model parameters best fit fault-zone trapped waves generated by Wenchuan aftershocks and recorded at the cross-fault array and SSN network stations, we tested various values for the fault zone width, velocity, and Q, the wall-rock velocity and Q, the layer depths, and the source location. Previous articles on modeling of trapped waves observed at the Landers rupture zone in California (Li et al., 2000, 2002) have shown the sensitivity of these model parameters to synthetic trapped waveforms. A wider fault zone produces trapped waves with lower frequencies, and a slower fault zone produces longer dispersive wavetrains of trapped waves. A lower-Q fault zone produces trapped waves with smaller amplitudes and shorter wave-trains at lower frequencies. The variation of wall-rock velocities and layer depths affects the arrival times of P and S waves, but variation of wall-rock Q produces minimal variation in modeling results.

However, these model parameters trade off with each other in modeling, so they are not uniquely determined. This problem has been discussed in previous studies for a delineation of fault zone structure using trapped waves (e.g., Li and Leary, 1990; Li and Vidale, 1996; Ben-Zion, 1998). When we have independent estimates of some parameters to be used as constraints in modeling, such as group velocities and Q-values estimated from the dispersion and attenuation of trapped waves, and results of other technologies, such as seismic tomography and fault-zone drilling in the region of investigation (e.g. at the San Andreas Fault, Parkfield of California; see Chapter 3), the trade-offs among the model parameters can be reduced.

Through a trial-and-error forward modeling procedure to simulate fault-zone trapped waves recorded at the linear seismic array deployed across the south LSF and Sichuan Seismic Network station located near the main rupture zone for Wenchuan aftershocks occurring at different depths and epicentral distances, we obtained the best-fit model parameters shown in Table 4.1. In this model, the fault-zone dips at 75° above 4 km and then dips at 45° at lower depths. Velocities within the 200-m-wide fault core are reduced by 30%-60% from wall-rock velocities, with the maximum reduction at depths above 2 km. The fault core is sandwiched by a 400-m-wide zone with slighter velocity reductions. The Q values within the fault zone are 10-60, with the lowest value within the fault core at shallow depth. This low-velocity waveguide is formed by severely damaged rocks along the LSF during the 2008 $M8$ Wenchuan earthquake.

For example, Fig. 4.13 show 3-D finite-difference synthetic seismograms to best fit the seismic profile recorded at the cross-fault dense array for an $M2.1$ near-fault aftershock occurring at depths of 10 km and ~30 km from the array, using the model parameters in Table 4.1. Prominent FZTWs with large amplitudes and long wavetrains appear at stations within the 200-m-wide fault core-zone that has the lowest velocities. Seismic waves trapped in the 400-m-wide damage zone with moderate velocity reduction show shorter wavetrains at lower frequencies than those trapped within the fault core zone. Synthetic seismograms fit observations quite well using the model parameters in Table 4.1.

In the modeling procedure, we simulated seismograms recorded at the Sichuan Seismic Network (SSN) stations in the source region for Wenchuan aftershocks occurring within the rupture zone. For example, Fig. 4.14 shows 3-D finite-difference synthetic waveforms using model parameters in Table 4.1 to fit observed seismograms at stations BAJ, MZP, LYS and ZDZ for an $M2.3$ on-fault aftershock occurring beneath these stations at 7-km depth with epicentral distances of ~25 km, ~20 km, ~18 km and ~11 km, respectively. Fault-zone trapped waves with large amplitudes and long wavetrains after S-waves appear at stations MZD, LYS and BAJ located within a range of ~2 km to the rupture zones of Wenchuan earthquake. Stations MZD and LYS located close to the main rupture register

Table 4.1 Model Parameters for the South Longmen-Shan Rupture Zone

			Best Fit		
Model parameters	Layer 1	Layer 2	Layer 3	Layer 4	Layer 5
Depth of the layer bottom, km	1.0	2.0	4.0	7.0	15.0
Waveguide width, m (Damage zone/core)	400/200	400/200	400/200	400/200	400/200
Waveguide S velocity, km/s	0.75/0.5	1.1/0.8	2.25/1.5	2.5/1.65	2.6/1.75
Waveguide P velocity, km/s	1.5/1.0	3.0/2.0	4.1/2.75	4.6/3.1	4.9/3.25
Waveguide Q-value	10	15	30	50	80
Wall-rock S velocity, km/sec	1.0	2.2	3.0	3.3	3.5
Wall-rock P velocity, km/sec	2.0	4.0	5.5	6.2	6.5
Wall-rock Q-value	40	80	100	150	200

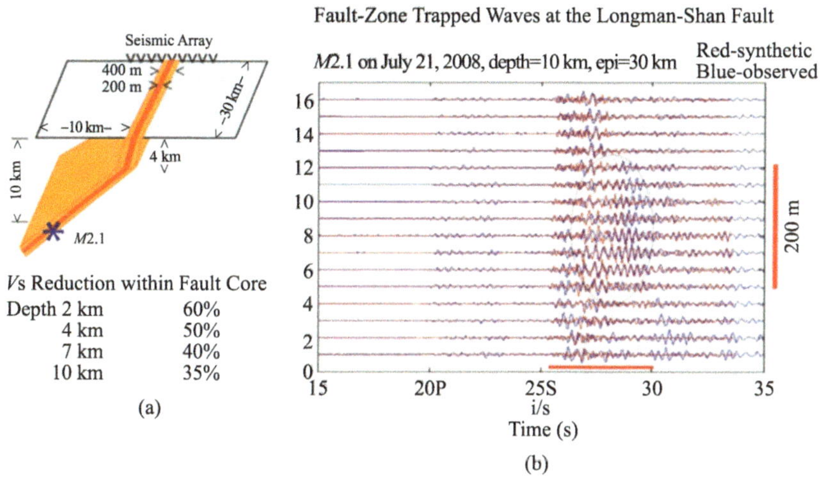

Fig. 4.13 (a) The schematic model of fault zone used for simulations of FZTWs recorded at the dense linear array deployed across the rupture zone along the south Longmen-Shan fault. The depth-dependent model parameters are shown below the model given in Table 4.1. (b) 3-D finite-difference synthetic seismograms (red lines) using the best-fit model parameters to match observations (blue lines) at the cross-fault dense array for an *M*2.1 near-fault aftershock occurring at depths of ~10 km and ~30 km from the array. Seismograms have been low-pass (< 3 Hz) filtered. Prominent fault-zone trapped waves (FZTWs) appear at stations within the 200-m-wide fault core (marked by vertical red bar at right). The horizontal red bar denotes the duration of FZTWs, in which amplitude envelopes of FZTWs are more than twice the level of the background noise coda. A double-couple source is located within the fault core zone.

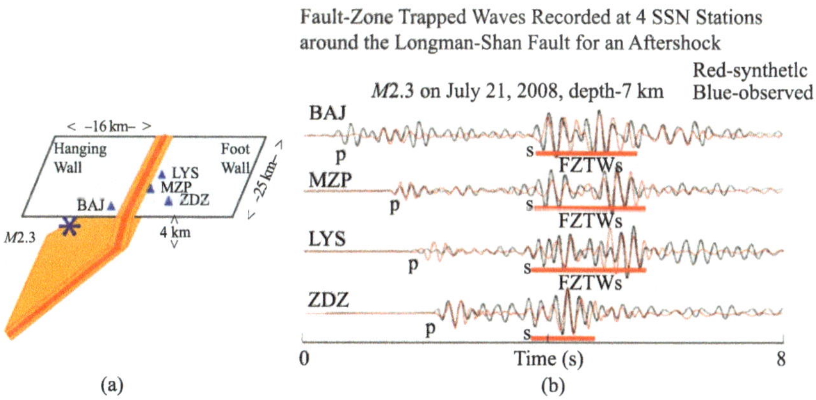

Fig. 4.14 (a) The schematic model shows the geometry and locations of 4 network stations LYS, MZP, ZDZ and BAJ and an *M*2.3 aftershock occurring at 7-km depth within the fault-zone. Model parameters are shown in Table 4.1. Other notations are the same as in Fig. 4.13a. (b) Synthetic (red lines) and observed (grey lines) vertical-component seismograms recorded at 4 network stations for this on-fault aftershock, showing FZTWS with large amplitudes and long wavetrains after *S*-arrivals and long wavetrains (denoted by red bars) at stations MZP, LYZ and BAJ located close to the south LSF. Seismograms registered at station ZDZ farther away from the main rupture zone show shorter wavetrain after *S*-waves.*S*-arrivals at 4 stations are aligned at the same time in plot. Seismograms have been < 3 Hz filtered and plotted in trace-normalization. A double-couple source is located within the fault core zone at depth of 7 km in computation.

most prominent FZTWs with longest wavetrains with respect to the travel distance among these network stations. FZTWs recorded at station BAJ near the geological fault trace of the southernmost Yingxiu-Beichuan fault segment which did not rupture to the ground surface in the 2008 Wenchuan earthquake show moderate wavetrain length of FZTWs. By contrast, much shorter wavetrains after *S*-arrivals are registered at station ZDZ located at ~3-4 km away from the main rupture.

Fig. 4.15a exhibits 3-D finite-difference synthetic seismograms to fit observations network stations GHS and MDZ for other two Wenchuan aftershocks occurring within the Wenchuan rupture zone at depths of 14 km and 25 km. The epicentral distances of these two events to station GHS and MZD are approximately ~32 km and ~20 km, respectively. The hypocentral distances between the two aftershocks and stations are slightly similar. Fault-zone trapped waves with long wavetrain and large amplitude appear at station MZD located close to the main rupture zone of Wenchuan earthquake for comparison with much shorter wavetrains after *S*-arrivals at station GHS located far away from the rupture zone in the hanging wall of the LSF for the same events. We note that the FZTWs recoded at station MZD for the deeper aftershock at 25 km depth are shorter than those at the same station for the aftershock at 14-km depth while the travel distances of seismic waves in the two examples are similar, because the velocity structure is depth-dependent, with the larger velocity reduction in the shallower part of the fault zone. The velocity reduction within the fault zone becomes slighter as the depth increases, corresponding to the closure of cracks due to the higher confining stress in the deeper crust.

Fig. 4.15b exhibits synthetic and observed seismograms at other 4 SSN network stations for the aftershock at 14-km depth, showing the longer wavetrains after *S*-arrivals at station LYS closer to the low-velocity waveguide along the main rupture zone than

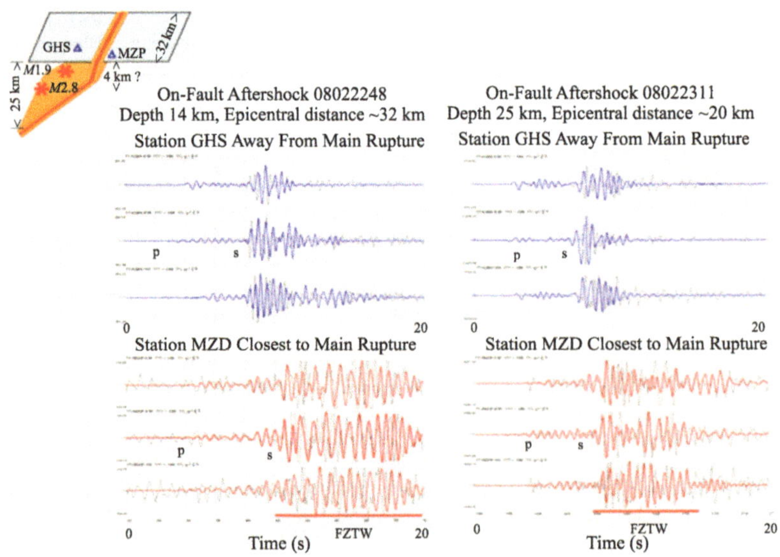

(a) Observed and 3-D Finite-Difference Simulations of Fault-Zone Trapped Waves for Wenchuan Aftershocks

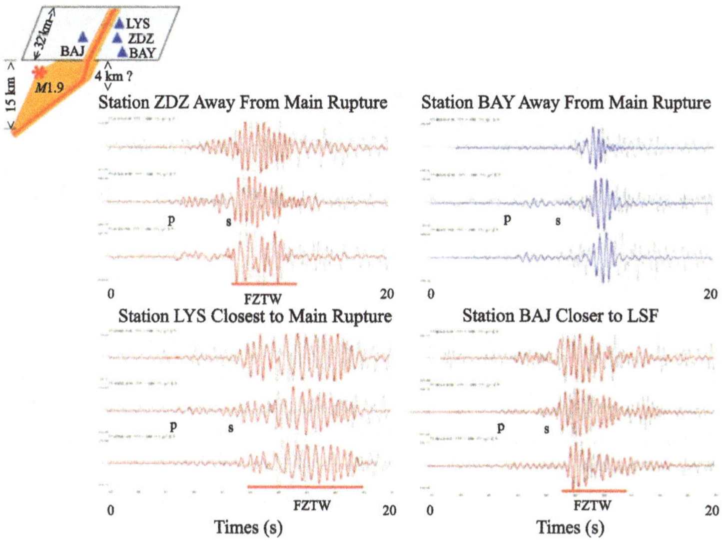

(b) Observed and 3-D Finite-Difference Simulations of Fault-Zone Trapped Waves for Wenchuan On-Fault Aftershock 08022248

Fig. 4.15 (a) 3-D finite-difference synthetic seismograms fit observations at stations MZD and GHS for 2 Wenchuan aftershocks of *M*2.8 and *M*1.9 occurring within the rupture zone at depths of 14 km and 25 km with epicentral distances of 32 km and 20 km, respectively. Synthetic seismograms (red lines) at station MZD closest to the main rupture show much longer wavetrains than those (blue lines) at station GHS far away from the rupture. Synthetics fit recorded seismograms (grey lines). Model parameters are given in Table 4.1. (b) Synthetic and observed seismograms at other 4 SSN stations for the on-fault aftershock at 14 km depth and 32 km epicentral distance. FZTWs recorded at station LYS closer to the main rupture along the south LSF show longer wavetrains than those at stations BAJ, ZDZ and BAY. Seismograms at station BAY show shortest wavetrains after *S*-arrivals because it is located farthest away from the main rupture among these 4 stations. Other notations are the same as in Fig. 4.14.

those at stations BAJ, ZDZ and BAY. We obtain a good fit of synthetic seismograms to observations in these examples, using the depth-dependent model parameters in Table 4.1.

The results of 3-D finite-difference simulations of observed FZTWs at the rupture zone in the Wenchuan earthquake source region illuminate a ~200-m-wide low-velocity fault core zone within which seismic velocities are reduced remarkably by 30%-60% from surrounding rock velocities at seismogenic depths. We interpreted this distinct low-velocity zone formed by damaged rocks in dynamic rupture of the 2008 *M*8 mainshock although it accumulated damage from historical earthquakes. Although the structural model shown in Fig. 4.7b with model parameters in Table 4.1 accounts for the observations of FZTWs at the main rupture zone along the south Longmen-Shan fault, it represents a gross average of the actual rupture zone structure of the 2008 *M*8 Wenchuan earthquake at seismogenic depths. The true structure along the entire length of the LSF will certainly be more complicated, and the damage magnitude and extent will vary along the fault strike and with depth due to rupture distributions and stress variations over multiple length and time scales. A systematic numerical simulation procedure with inversion technique (e.g.

the full-3D waveform tomography described in Chapter 1) for FZTWs using more aftershocks at various depths and epicentral distances will allow us to document more detailed subsurface structure, including branching and segmentation of the LSF zone in true 3-D.

4.5 Fault Rock Co-Seismic Damage and Post-Mainshock Heal

In order to relate present-day crustal stresses and fault motions of the geological structures formed by previous ruptures, we must understand the evolution of fault systems on many spatial and temporal scales. The spatial extent of fault weakness, and the loss and recouping of strength across the earthquake cycle are critical ingredients in understanding of fault mechanics. Extensive research in the field, in laboratories, and with numerical simulations has illuminated that the fault zone undergoes high, fluctuating stress and pervasive cracking during an earthquake (e.g., Aki, 1984; Mooney and Ginzburg, 1986; Scholz, 1990; Rice, 1992; Kanamori, 1994). Rupture models involving variations in fault-zone fluid pressure over the earthquake cycle have been proposed (e.g., Dieterich, 1979; Blanpied et al., 1992; Olsen et al., 1998). Structural fault variations and rheological fault variations (e.g., Sibson et al., 1975; Angevine et al., 1982; Taira et al., 2009) as well as variations in strength and stress may affect the earthquake rupture (e.g., Vidale et al., 1994; Marone et al., 1995; Massonnet et al., 1996; Karageorgi et al. 1997; Korneev et al., 2000; Schaff and Beroza, 2004; Yasuhara et al., 2004; Rubinstein and Beroza, 2005; Zhang et al., 2007; Liu et al., 2008).

Previous results of observations and 3-D finite-difference simulations of fault-zone trapped waves (FZTWs) recorded at ruptures of the 1992 *M*7.4 Landers, 1999 *M*7.1 Hector Mine, and the 2004 *M*6 Parkfield earthquakes in California show distinct low velocity zones existing on these active faults, which are a couple hundred meters wide with velocities reduced by up to 50% from wall-rock velocities. These low-velocity fault zones have undergone strong dynamic stresses and pervasive cracking during historical major ruptures partly healed after the mainshock (Li et al. 1998, 2003, 2006, 2007; Li and Vidale, 2001; Vidale and Li, 2003).

In order to evaluate the co-seismic damage magnitude of fault-zone rocks along the south Longmen-Shan fault zone caused by the dynamic rupture of the *M*8 Wenchuan earthquake, we examined the data recorded at the Sichuan Seismic Network stations in the source region for the local earthquakes occurring at the similar locations and with similar magnitudes before and after the 2008 May 12 *M*8 mainshock. We examined the waveform data recorded at these stations for similar earthquakes in the SSN Catalog. However, we find that the similarity of waveforms recorded at the same station for repeated events occurring before and after Wenchuan earthquake is not strong enough, but with low (<0.7) waveform cross-correlation coefficient, probably due to the location errors of these repeated events because the same velocity model was used for locating earthquakes in SSN Catalog before and after the 2008 Wenchuan earthquake. However, the prior velocity structure in Wenchuan region might have been changed by the 2008 *M*8 mainshock since the rock matrix in the region has been disturbed strongly by this big earthquake.

Alternatively, we examined the variations in dispersion features of fault-zone trapped waves rather than travetimes of body waves for these similar events to evaluate the temporal variations in seismic velocity due to rock damage caused by the *M*8 Wenchuan mainshock. Fig. 4.16a shows that seismograms recorded at 8 Sichuan Seismic Network stations in the Wenchuan earthquake source region for an *M*2.4 local earthquake occurring at 9 km depth and within the south Longmen-Shan fault zone southwest of the stations

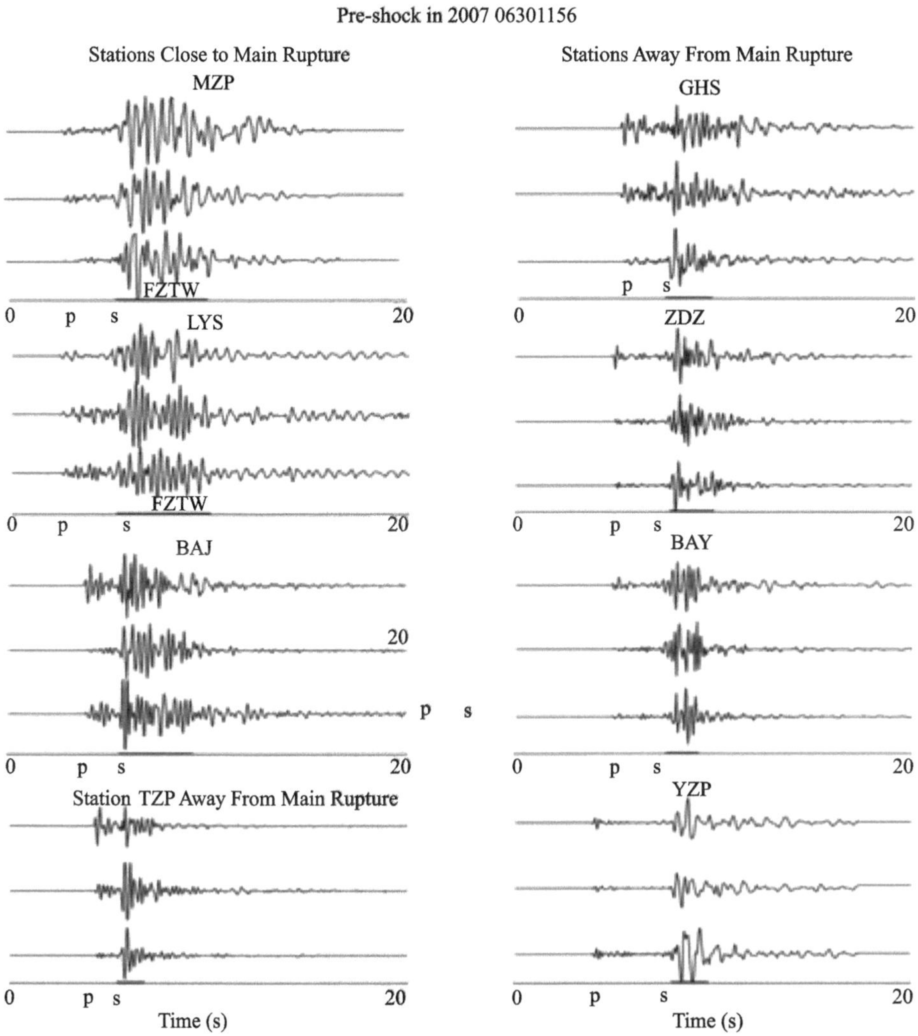

Fig. 4.16 (a) Three-component seismograms recorded at 8 Sichuan Seismic Network stations (their locations shown in Fig. 4.2) for an *M*2.4 local earthquake occurring at 9 km depth and within the south Longmen-Shan fault zone southwest of the stations on June 30, 2007 show large amplitude (with respect to *P* wave amplitudes) and long wavetrain following *S*-arrivals at stations MZP and LYS closer to the south Longmen-Shan fault, but short S-wavetrains at other stations away from the fault zone. Other notations are the same as in Fig. 4.5.

on June 30, 2007 show larger amplitude (with respect to *P* wave amplitudes) and longer wavetrain following *S*-arrivals at stations MZP and LYS closer to the south Longmen-Shan fault other than those at other stations located farther away from the fault zone. We interpret the long-duration wavetrains following *S*-arrivals registered at stations MZP, LYS and BAJ being fault-zone trapped waves (FZTWs) although they are not as long as those generated by Wenchuan aftershocks occurring at the same place (Fig. 4.16b). These observations suggest that there existed a prior waveguide with moderate velocity reduction on the south LSF which has accumulated rock damage in historical large earthquakes before the 2008 *M*8 Wenchuan earthquake.

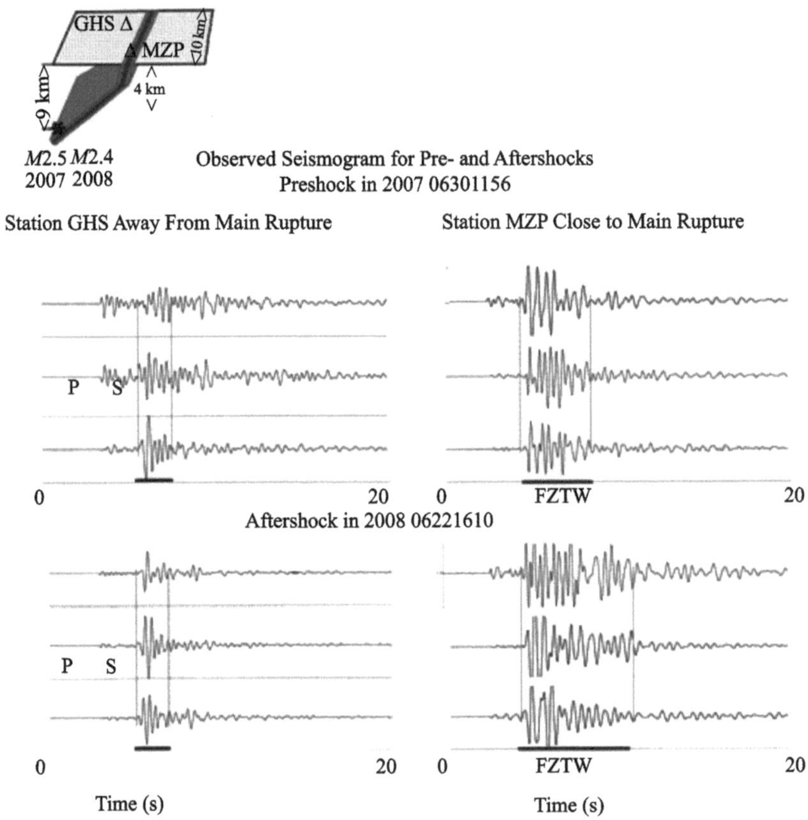

Fig. 4.16 (b) Seismograms recorded at Sichuan Seismic Network stations GHS and MZP for an *M*2.4 earthquake occurring on June 20, 2007 (shown in Fig. 4.16a) and an *M*2.5 Wenchuan aftershock on June 22, 2008, occurring at the similar place show large amplitudes (with respect to *P* wave amplitudes) and long wavetrains of FZTWs following S-arrivals at station MZP close to the main rupture of 2008 *M*8 Wenchuan earthquake on the south LSF. By contrast, short wavetrains are shown in seismograms recorded at station GHS away from the rupture zone. We note that the FZTWs recorded at station MZP for the Wenchuan aftershock show longer wavetrains than those for the pre-shock occurring at the same place in 2007. However, brief S-waves with the similar short lengths are registered at station GHS away from the main rupture zone for both preshocks and aftershocks. These observations indicate additional rock damage on the south LSF caused by the 2008 *M*8 Wenchuan earthquake.

Fig. 4.16b exhibits the difference in waveforms recorded at stations MZD and GHS for this local earthquake in 2007 from those for an *M*2.5 Wenchuan aftershock occurring at the same location on June 22, 2008. Prominent FZTWs with large amplitudes and long wavetrains appear in seismograms recorded at station MZD close to the main rupture along the south LSF for both of them. However, the FZTWs generated by the Wenchuan aftershock show longer wavetrains (~6.2 s duration in time) after *S*-arrivals than those (~4.5 s duration) generated by the similar event in 2007, indicating that seismic velocities within the south LSF zone decreased remarkably due to co-seismic damage of fault rocks in dynamic rupture during the 2008 *M*8 Wenchuan earthquake. By contrast, seismograms

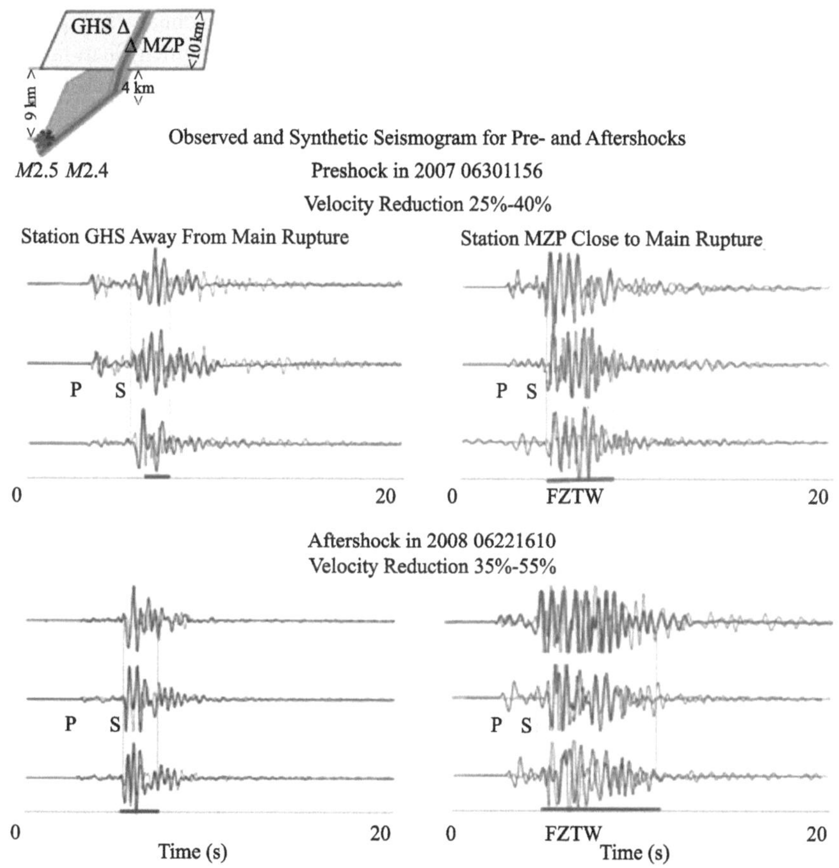

Fig. 4.16 (c) Observed (grey lines) and 3-D finite-difference synthetic seismograms at stations MZD and GHS located close and away from the surface main rupture for 2 earthquakes occurring at the same place on June 30, 2007 (blue lines) and June 22, 2008 (red lines), respectively, show longer wavetrains of FZTWs at station MZD for the aftershock than those for preshocks, indicating that fault rocks along the south LSF were severely damaged in dynamic rupture of the *M*8 Wenchuan earthquake, causing the remarkable decrease of seismic velocities within the fault zone. Synthetic seismograms are computed using the structural model shown in Fig. 4.7b and Table 4.1, but with fault-zone velocity reductions of 25%-40% for pre-shock but 35%-55% for aftershock.

recorded at station GHS located away from the main rupture zone do not show significant changes before and after Wenchuan earthquake.

We simulated the seismograms recorded at stations MZD and GHS for these two similar events occurring before and after the 2008 Wenchuan earthquake to evaluate the co-seismic velocity reduction caused by the *M*8 mainshock (Fig. 4.16c). We computed synthetic seismograms using the structural model given in Fig. 4.7b and Table 4.1 with velocities reduced by ~30%-55% within the fault core zone for the Wenchuan aftershock but with velocity reductions of ~20%-40% within the fault zone for the similar earthquake before Wenchuan earthquake, respectively. We obtain an acceptable goodness of synthetic seismograms to fit observations at these two stations for the preshock and aftershock. The larger velocity reductions within the fault zone used for the aftershocks other than those for the preshocks illuminate that the 2008 *M*8 Wenchuan earthquake had caused additional damage of fault-zone rocks along the south Longmen-Shan fault. We estimate that seismic velocities within the main rupture zone along the south LSF fault in Wenchuan earthquake source region were reduced by approximate 10%-15% owing to the *M*8 mainshock. This value is a rough estimation; it might be affected by other factors. A systematic examination of all available data recorded before and after the 2008 Wenchuan earthquake is needed for further constraints. However, Liu and Huang (2010) also found remarkable temporal changes of seismic velocity near the epicenter of the Wenchuan earthquake from ambient noise correlation.

We also examined the post-mainshock heal of rocks along the south Longmen-Shan fault, which were severely damaged during the 2008 *M*8 Wenchuan earthquake, using waveform cross-correlation for the similar aftershocks to measure the difference in traveltime of seismic waves from these repeated events. In our preliminary examination of the data, we selected 10 pairs of repeated aftershocks occurring at different depths and epicentral distances in the Wenchuan earthquake source region. For example, Fig. 4.17a exhibits seismograms recorded at station LYS close to the main rupture along the south LSF for 2 repeated *M*3.2 on-fault aftershocks occurring at ~10-km depth and ~30-km northeast of station LYS at the same place on May 20 and June 24, 2008. Seismograms show similar waveforms for them, but waves travelled faster from the June 24 aftershock than those from the May 20 aftershock. Waveform cross-correlation for these two repeated events shows ~250 ms decrease in traveltime for the June 24 event. Assuming velocity changes were uniform in the crust sampled by these waves, the traveltime decreases are straightforward to interpret. The fault-zone trapped waves (FZTWs) following *S*-waves arrived at station LYS with ~250 ms advance for the later aftershock, with traveltime of ~12-14 s from the aftershock origin time, so the shear velocity within the fault zone might increase by ~2% in about a month between May 20 and June 24 after the *M*8 Wenchuan mainshock on May 12, 2008. It indicates that fault-zone rocks co-seismically damaged by the *M*8 Wenchuan earthquake have been subsequently healed with time after the mainshock.

Fig. 4.17b exhibits seismograms recorded at station LYS for 2 repeated on-fault aftershocks occurring at 8-km depth (near the place of the events in Fig. 4.17a on July 10 and September 7. We observe the similar waveforms for them, and ~200 ms advance in traveltime for the later event. We estimated that *S*-wave velocities within the fault zone increased by ~1.5% in the following 2 months between July 10 and September 7, 2008, indicating the post-mainshock fault healing (recovery of rock rigidity) with time. We note

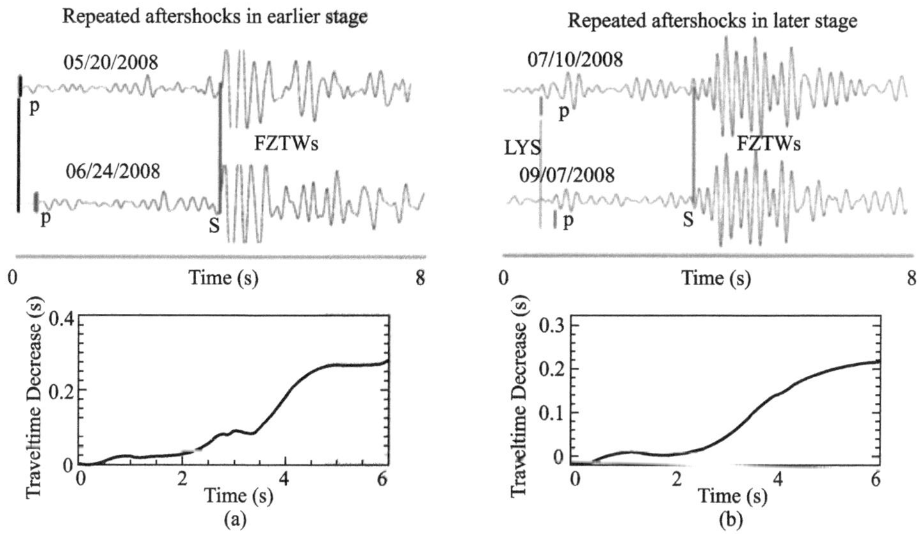

Fig. 4.17 (a) Seismograms recorded at station LYS close to the main rupture of 2008 *M*8 Wenchuan earthquake for 2 repeated *M*3.2 on-fault aftershocks occurring at 10-km depth and 25-km epicentral distance on May 20 and June 24, 2008 show that seismic waves from the later event travelled faster than the earlier event. *S*-arrivals of 2 repeated events are aligned at the same time in plot. We measured the maximum traveltime advance of ∼250 ms *S* coda for the later aftershock using the moving-window waveform cross-correlation method with *P*-arrivals of the 2 repeated events aligned at the same time. (b) Same as in (a) but for 2 on-fault repeated *M*2.2 aftershocks occurring at 8-km depth on July 10 and September 7 showing ∼200 ms traveltime advance for the later event.

that the healing rate is not constant but with the larger healing rate in the earlier stage after the mainshock.

In summary, Fig. 4.18 shows fault-zone seismic velocity changes with date associated with the May 12, 2008 *M*8 Wenchuan earthquake. Seismic velocities within the south Longmen-Shan fault (LSF) decreased by ∼10%-15% likely due to the co-seismic damage of fault rocks caused by the *M*8 Wenchuan earthquake on May 12, 2008. Seismic velocities were consequently increased by ∼5% or more in the first year after the Wenchuan earthquake, indicating the post-mainshock fault heal with rigidity recovery of damaged rocks with time. The measurements of fault damaged and healing in the current study are not conclusive because the cross-correlation coefficients of waveforms for the repeated aftershocks are not high enough although the catalog shows the similar locations of these repeated events, probably due to location errors of aftershocks as mentioned above and/or variations in rock matrix caused by large aftershocks between the repeated events used in measurements. Our preliminary results will be further refined through a systematic analysis of the data from more well-located earthquakes. We compared the magnitude of rock damage and heal at the Longmen-Shan fault zone caused by the 2008 *M*8 Wenchuan earthquake with those in the previous study at the rupture zones of the 2004 *M*6 Parkfield earthquake, the 1992 *M*7.4 Landers and 1999 *M*7.1 Hector Mine earthquakes (Li et al., 1998; 2003; Vidale and Li, 2003), showing the greater damage of fault-zone rocks likely caused by the larger earthquake.

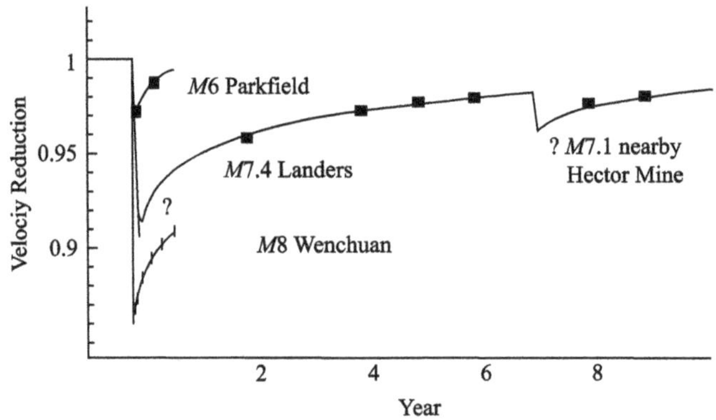

Fig. 4.18 Seismic velocity changes with date before and after the *M*8 Wenchuan earthquake on May 12, 2008. The co-seismic decrease in seismic velocity of fault-zone rocks is inferred by simulations of FZTWs generated by preshocks and aftershocks occurring at the same location. The post-mainshock increases in seismic velocity are measured from traveltime advances for repeated aftershocks. Each data point in the plot is the average of measurements of waveforms recorded at 3 network stations close to the main rupture for pairs repeated over a 2-month period in the first year after the mainshock. For a comparison, we also show our previous observations of rock damage and heal at the ruptures of the *M*6 Parkfield earthquake (Li et al., 2006; 2007), the *M*7.4 Landers and *M*7.1 Hector Mine earthquakes (Li et al., 1998; Vidale and Li, 2003), showing fault heal likely as a logarithmic function of time, and with the greater damage caused by the larger magnitude of the earthquake.

4.6 Conclusion and Discussion

We document the rock damage and heal along the southern Longmen-Shan fault (LSF) zone ruptured in the 2008 *M*8 Wenchuan earthquake using the data acquired from Sichuan Seismic Network (SSN) and portable stations in the earthquake source region where the south LSF is characterized by reverse thrusting (Fig. 4.1 and Fig. 4.2). Fault-zone trapped waves (FZTWs) have been recognized in seismograms recorded at a linear dense seismic array deployed across the south LSF and SSN stations located close to the main rupture for aftershocks occurring at different depths and epicentral distances within the rupture zone (Fig. 4.3 to Fig. 4.5). These FZTWs illuminate the coherent interference phenomenon of wave propagation in a low-velocity zone formed by severely damaged rocks along the south LSF. Results of observations and 3-D finite-difference simulations of these FZTWs show a distinct low-velocity wave-guide (LVWG), being several hundred meters wide along the south LSF, within which the maximum velocity reduction is up to ~50%-60% at shallow depth. This LVGW along the south LSF with varying dip angles likely extends across seismogenic depths. The deep ruptures along the LSF have been derived from aftershock observations around the hypocenter of the 2008 Wenchuan earthquake (An et al., 2010). The computation from zero-strain points of co-seismic surface deformation indicates that the possible rupture depth of Longmen-Shan fault in the 2008 *M*8 earthquake is ~23 km (Fu et al., 2011).

Because of the sensitivity of fault-zone trapped wave excitation to the source location from the LVGW, it allows us to depict the principal slip of the *M*8 Wenchuan earthquake at seismogenic depth inferred by locations of aftershocks which generate prominent FZTWs

(Fig. 4.6). The width, velocity and geometry of the south LSF at shallow depth delineated by FZTWS in our study are generally consistent with the results of geological mapping (Burchfield et al., 2008; Xu et al., 2009, 2010; Lin et al., 2009) and fault-zone drilling at the southern LSF (Xu et al., 2008a, b). We interpreted this remarkable LVGW along the south LSF as a damage zone in dynamic rupture that accumulated damage from historical earthquakes, mainly from the 2008 *M*8 Wenchuan earthquake.

The south Longmen-Shan fault dips with varying angles at seismogenic depths. This geometry may affect the trapping efficiency of FZTWs. In order to obtain better simulations of the FZTW recorded at the reverse-thrusting south LSF, we tested the possible dipping fault structures (Fig. 4.7) that might obscure the signature of faults and disrupt the FZTWs using a 3-D finite difference code in the present chapter. In the testing procedure, our reference model has a 10-km source depth, a 200-m wide fault core zone sandwiched in the 400-m wide damage zone (jacket), and a double-couple source is placed against one edge of the fault core zone. The receiver array is centered on the surface fault trace. The velocities within the fault core are reduced by 25%-50% from wall-rock velocities with the maximum reduction within the fault core at shallow depth. This structure most effectively traps 1 to 10-Hz seismic waves. We have tested the effects of fault-zone dip angle, width and velocity reduction, source epicentral distance, focal depth and location away from the fault zone, and *Q* values on excitation and propagation of FZTWs (Fig. 4.8 to Fig. 4.12). This numerical study of trapping efficiency for dip faults with various geometry and physical properties has helped us to design the field experiment to record significant FZTWs at the surface stations near the south LSF and identify FZTWs in the recorded seismograms.

Based on the results of surface rupture mapping along the Longman-Shan fault after the 2008 *M*8 Wenchuan earthquake (The Emergency Science and Research Team of China Seismological Administration, 2008), the fault-zone drilling at the south LSF immediately after the *M*8 2008 Wenchuan earthquake (Xu et al., 2008a, b), and the coseismic slip model of the 2008 Wenchuan earthquake (Tong et al., 2010; Xu et al., 2010), we constructed plausible models for the reverse-thrusting LSF dipping at various angles at depth (Fig. 4.7). Velocities of surrounding rocks in these models are constrained by Sichuan Province velocity model (Sun et al., 2004; An et al., 2009) and recent results of low-frequency 3D wave propagation modeling of the 2008 *M*8 Wenchuan earthquake (Chavez et al., 2010). We demonstrate examples of 3-D finite-difference simulations of FZTWs recorded at the cross-fault array and network stations close to the south LSF using the best-fit model parameters shown in Table 4.1 (Fig. 4.13 to Fig. 4. 15). In modeling, we permit ~2 km variation in hypocentral distance of the events, allowance for location errors in the SSN Catalog and the lateral heterogeneity along the fault zone. Our model includes a 200-m-wide fault core sandwiched in the 400-m-wide damage zone. The velocities within the fault core are reduced by 30%-60% from wall-rock velocities with the maximum reduction at shallow depth. The low-velocity fault-zone extends to the depth of at least 10-km. While the model parameters shown in Table 4.1 are not uniquely constrained by this forward modeling because there exist trade-offs among them, but these values are applicable to a preliminary documentation of the subsurface rock damage structure along the south Longmen-Shan fault.

The depth to which a fault zone penetrates the crust is usually not well known. Some discrepancies exist in the fault-zone depth estimates in early studies of the San Andreas fault using seismological and gravity methods (Mooney and Ginzburg, 1986). Recently, Ben-Zion et al. (2003) and Lewis and Ben-Zion (2010) argue that the low-velocity damage zone on faults is a near-surface feature that reaches only down to the top of seismogenic zone at ~2-3 km. Others demonstrate that the low-velocity zone formed by damaged rocks on the San Andreas fault at Parkfield and rupture zones of the 1992 Landers and 1999 Hector Mine earthquakes active in California extends across the seismogenic depths from 3 km to 10 km (e.g., Korneev et al. 1993; Li et al., 1994, 2001; Li and Malin 2008; Wu et al., 2010). The different conclusions illustrate the non-uniqueness of FZTWs interpretations based on surface measurements. However, results of core sampling and well logs in the San Andreas Fault Observatory (SAFOD) borehole at Parkfield show a ~200 m-wide zone of high porosity material, with multiple slip planes and average velocity reductions of ~30%-35% at 3 km depth (Hickman et al., 2007). Recently, Ellsworth and Malin (2010) document a profound zone of rock damage on the SAF downwards to at least half way (> 5-6 km) through the seismogenic crust at Parkfield using both Rayleigh-type and Love-type of FZTWs recorded at the SAFOD MH seismograph. Wu et al. (2010) also show that the low-velocity waveguide on the SAF at Parkfield extends to the depth > 10 km using SAFOD borehole data. A comprehensive analysis of FZTWs recorded at Parkfield surface and borehole stations (Li et al., Chapter 3 in this book) shows that the low-velocity zone on the SAF at Parkfield extends to the depth of at least ~7 km or deeper although the velocity reduction within the damage zone decreases with decreasing depth due to the increasing confining pressures.

The fault zone co-seismically weakening during the major earthquake and subsequently healing (partially) on the LSF has been studied using similar earthquakes occurring before and after the 2008 Wenchuan earthquake (Fig. 4.16 to Fig. 4.17). We examined the changes in amplitude and wavetrain length of FZTWs recorded at the same seismic station for earthquakes occurring at the same places within the LSF zone between 2007 and 2008. Results suggest that seismic velocities within the south LSF zone could be co-seismically reduced by ~10%-15% due to the rock damage caused by the *M*8 mainshock on May 12, 2008. Using repeated aftershocks, we then measured ~5% increase of seismic velocities within the LSF in the first year after the 2008 Wenchuan earthquake, with the larger healing rate in the earlier stage, indicating the post-mainshock healing with rigidity recovery of fault-zone rocks damaged in the *M*8 earthquake. Consistently, remarkable temporal changes of seismic velocity have been found near the epicenter of the Wenchuan earthquake from ambient noise correlation (Liu and Huang, 2010).

We attribute the measured co-seismic velocity reduction within the fault zone to rocks opened during the dynamic rupture of the *M*8 Wenchuan earthquake. Crack opening is likely favored at shallow depths with soft rock and lower confining crustal stress. Our stations are located close to the main rupture zone Compared with the network stations. The data presented here from the shots and aftershocks located on or close to the fault mainly document the changes within the fault zone rather than the changes in surrounding rocks. We think the observed co-seismic velocity decrease within the LSF zone is mainly due to crack opening caused by the dynamic rupture of the 2008 *M*8 Wenchuan earthquake while shaking-induced weakening may also effect the rupture propagation because the pre-existing weak, low impedance fault zone is susceptible to damage (Fialko et al., 2002;

Vidale and Li, 2003). Calculation of a ~10%-15% decrease in velocity using formula for cracked media (O'Connell and Budiansky, 1974) revealed that the apparent crack density within the main rupture zone along the south LSF increased by ~0.14-0.21, which caused ~20%-30% decrease in shear rigidity of the fault-zone rock during dynamic rupture of the 2008 *M*8 Wenchuan earthquake. The subsequent ~5% increase in *S* velocity suggests the apparent crack density within the rupture zone decreased by 0.07 in the following ~1 year.

During the fault healing, the reduction of crack density may be controlled by a combination of mechanical and chemical processes on the active fault. Fault healing may be affected by time-dependent frictional strengthening (Vidale et al., 1994; Marone, 1998), rheological fluid variations or changes in the state of stress (Blanpied et al., 1992), cementation, re-crystallization, pressure solution, crack sealing and grain contact welding (Hickman and Evans, 1992; Olsen et al., 1998) and the fault-normal compaction of the rupture zone (Massonnet et al., 1996; Boettcher and Marone, 2004) as well as chemical healing from mineralogical lithification of gouge materials over longer time period at seismogenic depth (Angevine et al., 1982). In addition, the 'crack dilatancy' mechanism (Nur, 1972) associated with the earthquake is likely to operate for co-seismic fault damage and post-mainshock healing even if other processes are active too. The stress-related temporal changes in seismic velocity caused by the 1989 Loma Prieta, California earthquake have been reported (Ellsworth et al., 1992; Dodge and Beroza, 1997; Schaff and Beroza, 2004). Baisch and Bokelmann (2002) suggest that coseismic deformation caused by this earthquake might lead to crack opening either by localizations of shear stress or by elevated pore fluid pressure. Concentrated deformation at low-strength fault zones may help to cause damage. After the earthquake, relaxation processes, such as crack healing, fluid diffusion, and post-seismic deformation cause the cracks to close again with an approximately logarithmic recovery rate (Dieterich, 1972; Richardson and Marone, 1999). As rocks heal, a contribution can be from either continued right-lateral deformation due to the regional stress field that dominated the coseismic displacements or fault-normal compression owing to a reduction in crack volume. The variation in apparent crack density inferred by seismic velocity measurements reflects changes in either crack volume or rearrangement of aspect ratio caused by the earthquake. We tentatively conclude that the cracks that opened during the mainshock closed soon thereafter. This is consistent with our interpretation of the soft low-velocity fault-zone waveguide on the LSF as being at least partially weakened in the 2008 *M*8 Wenchuan mainshock, but with possible significant cumulative effects as well.

The magnitudes of rock damage and heal observed at the Longmen-Shan fault (Fig. 4.18) are larger than those observed at the San Andreas fault ruptured in the 2004 *M*6 Parkfield earthquake (Li et al., 2006, 2007) and on ruptures of the 1992 *M*7.4 Landers and 1999 *M*7.1 Hector Mine earthquakes (Li et al., 1998, 2000; Vidale and Li, 2003). This difference is probably related to the different earthquake magnitude, faulting mechanism, stress drop, and other factors. The spatial extent of fault-zone damage and the loss and recouping of strength across the earthquake cycle are critical ingredients in understanding of fault mechanics and physics. With a comparison of major earthquakes at active faults in California (Chapter 3 in this book) and the Longmen-Shan fault in Sichuan (Chapter 4), we obtain the most basic information on the *in-situ* states of fault-zone physical properties with large earthquakes, and thus further understand earthquake processes and hazards globally.

Acknowledgements

This investigation is awarded by NSF Grant EAR-0910911 and Southern California Earthquake Center. The lead author of this article was invited by Foreign Expert Bureau of State Counsil and China Earthquake Administration (CSA) to join the investigation of Wenchuan Earthquake in 2008. Special thanks to P. Z. Zhang, and other researchers in CSA for their co-operation in site survey and collection of the data used for this study. He is also grateful for the assistance provided by Overseas Chinese Affairs Office of Sichuan Province. The lead author carried out this research project and prepared this article partially when he was an invited professor at Fudan University and Ningbo Nottingham University in China.

References

Aki, K., Asperities, barriers, characteristic earthquakes, and strong motion prediction. J. Geophys. Res., 89, 5867-5872, 1984.

An, M. J., M. Feng, and C. X. Long Deep ruptures around the hypocenter of the 12 May 2008 Wenchuan earthquake deduced from aftershock observations. Tectonophysics, 491, 96-104, 2010.

An, M. J., M. Feng, S. W. Dong, C. X. Long, Y. Zhao, N. Yang, W. J. Zhao, J. Z. Zhang, Seismogenic structure around the epicenter of the May 12, 2008 Wenchuan earthquake from micro-seismic tomography. Acta Geologica Sinica (English Edition), 83(4): 724-732, 2009.

Angevine, C. L., D. L., Turcotte, and M. D. Furnish, Pressure solution lithification as a mechanism for the stick-slip behavior of faults. Tectonics, 1, 151-160, 1982.

Baisch. S, and G. H. R. Bokelmann, Seismic waveform attributes before and after the Loma Prieta earthquake: scattering change near the earthquake and temporal recovery. J. Geophys. Res. 106, 16,323-16,337, 2001.

Ben-Zion, Y., Properties of seismic fault zone waves and their utility for imaging low velocity structures. J. Geophys. Res., 103, 12567-12585, 1998.

Ben-Zion, Y., Z. Peng, D. Okaya, L. Seeber, J. G. Armbruster, N. Ozer, A. J. Michael, S. Barris, and M. Aktar, A shallow fault zone structure illuminated by trapped waves in the Karadere-Dusce branch of the North Anatolian Fault, Western Turkey. Geophys. J. Int., 152, 699-717, 2003.

Blanpied, M. L., D. A. Lockner, and J. D. Byerlee, An earthquake mechanism based on rapid sealing of faults. Nature, 359, 574-576, 1992.

Boettcher, M, S., and C. Marone, Effects of normal stress variation on the strength and stability of creeping faults. J. Geophys. Res. **109**, B03406, doi:10.1029/2003JB002824,

2004.
Burchfield, B. C., L. H. Royden, van der Hilst, B. H. Hager, Z. Chen, R. W. King, C. Li, J. Lu, H. Yao, and E. Kirby, A geological and geophysical context for the Wenchuan earthquake of 12 May 2008, Sichuan, PRC. GSA Today, 18(7), doi: 10.1130/GSATG18A.1, 2008.
Burchifild, B. C., Z. Chen, Y. Liu and L. H. Royden, Tectonics of the Longmen-Shan and Ajacent regions, central China. International Geological Review 37 (8). 661-735, 1995.
Byerlee, J., Friction, overpressure and fault-normal compression, Geophys. Res. Lett., 17, 2109-2112, 1990.
Chavez, M, E. Cabrera, R. Madariaga, H. Chen, N. Perea, D. Emerson, A. Salazar, M. Ashworth, Ch. Moulinec, X. Li, M. Wu, and G. Zhao, Low-frequency 3D wave propagation modeling of the 12 May 2008 *M*w 7.9 Wenchuan earthquake. Bull. Seism. Soc. Am., 100, No. 5B, 2561-2573, doi: 10.1785/0120090240, 2010.
Chen et al., Analysis of the Source of *M*8 Wenchuan Earthquake on May 12. CEA, Beijing, 2007 (in Chinese).
Densmore, A.L., M. A. Ellis, Y. Li, R. Zhou, G. S. Hancock, and N. Richardson, Active tectonics of the Beichuan and Pengguan faults at the eastern margin of the Tibetan Plateau. Tectonics, V 26, TC4005, doi:10.1029/2006TC001987, 1-17, 2007.
Dieterich, J. H., Modeling of rock friction: 1. Experimental results and constitutive equations, J. Geophys. Res., 84, 2,161-2,168, 1997.
Dodge, D., and G. C. Beroza, Source array analysis of coda waves near the 1989 Loma Prieta, California mainshock: Implications for the mechanism of coseismic velocity changes. J. Geophys. Res. 102, 24,437-24,458, 1997.
Ellsworth W. L. and P. E. Malin, Deep rock damage in the San Andreas fault revealed by *P*- and *S*-type fault zone guided waves. Sibson's volume, New Zealand, in press, 2011.
Ellsworth, W. L, A. T. Cole, G. C. Beroza, and M. C. Verwoerd, Changes in crustal wave velocity association with the 1989 Loma Prieta, California earthquake. EOS 73, 360, 1992.
Emergency Science and Research Team of China Seismological Administration, Investigation Report of Surface Ruptures in the Wenchuan Earthquake, China Seismological Administration Special Report of Wenchuan Earthquake, Beijing, 2008 (in Chinese).
Fialko, Y., D. Sandwell, D. Agnew, M. Simons, P. Shearer, B. Minster, Deformations on nearby faults induced by the 1999 Hector Mine earthquake. Science, 297, 1858-1862, 2002.
Fu, Zhen, C. B. Hu, H. M. Zhang, Y. E. Cai, and Y. J. Zhou, The possibility of inferring rupture depths of fault earthquakes from zero-strain points of co-seismic surface deformation. Seism. Res. Lett., v82, No. 1, 89-96, doi: 10.1785/gssrl.82.1.89, 2011.
Graves, R. W., Simulating seismic wave propagation in 3D elastic media using staggered-grid finite differences. Bull. Seismol. Soc. Am., 86, 1091-1106, 1996.
Hickman, S. H., and B. Evans, Growth of grain contacts in halite by solution-transfer: Implications for diagenesis, lithification, and strength recovery, in Fault Mechanics and Transport Properties of Rocks, pp. 253-280, Academic, San Diego, Calif., 1992.
Hickman, S. H., M. D. Zoback, W. L. Ellsworth, N. Boness, P. Malin, S. Roecker and C. Thurber, Structure and Properties of the San Andreas Fault in Central California: Recent results from the SAFOD experiment. Scientific Drilling, Special Issue, No.1,

doi:10.2204 /iodp.sd.s01.39.2007, 29-32, 2007.

Igel, H., G. Jahnke, and Y. Ben-Zion, Numerical simulation of fault zone guided waves: accuracy and 3-D effects. Pure Appl. Geophys., 159, 2067-2083, 2002.

Ji, C., Preliminary result of the May 12, 2008 *M*w 7.97 Shi Chuan earthquake, personal website, UCSB, 2008.

Kanamori, H., Mechanics of earthquakes. Ann. Rev. Earth Planet., Sci., 22, 207-237, 1994.

Karageorgi, E. D., T. V. McEvilly, and R. W. Clymer, Seismological studies at Parkfield: IV: Variations in controlled-source waveform parameters and their correlation with seismicity, 1987 to 1995. Bull. Seism. Soc. Am. 87, 39-49, 1997.

Korneev, V. A., R. M. Nadeau, and T. V. McEvilly, Seismological studies at Parkfield IX: Fault-zone imaging using guided wave attenuation. Bull. Seism. Soc. Am. 80, 1245-1271, 2003.

Korneev, V. A., T. V. McEvilly, and E. D. Karageorgi, Seismological studies at Parkfield Ⅷ: Modeling the observed travel-time changes. Bull. Seism. Soc. Am. 90, 702-708, 2000.

Lewis, M. A. and Y. Ben-Zion, Diversity of fault zone damage and trapping structures in the Parkfield section of the San Andreas Fault from comprehensive analysis of near fault seismograms. Geophys. J. Int., doi: 10.1111/j.1365-246X.2010.04816.x, 2010.

Li, Y. G. and P. E. Malin, San Andreas Fault damage at SAFOD viewed withfault-guided waves. Geophys. Res. Lett., 35, L08304, doi:10.1029/2007GL032924, 2008a.

Li, Y. G. J. Y. Sue, T. C. Chen, E. S., Cochran, P. Chen, Seismic Documentation of Rock Damage and Heal on the Longmen-Shan Fault of the *M*8 Wenchuan Earthquake Compared with the San Andreas Fault at Parkfield, Western Pacific Geophysical Meeting, Taipei, June 22-25, 2010.

Li, Y. G., and J. E. Vidale, Low-velocity fault-zone guided waves: numerical investigations of trapping efficiency. Bull. Seism. Soc. Am. 86, 371-378, 1996.

Li, Y. G., and P. C. Leary (1990), Fault-zone trapped seismic waves. Bull. Seism. Soc. Am., 80, 1245-1271.

Li, Y. G., Fault Damage in the 2008 *M*8 Wenchuan Earthquake Epicentral Region. Academic Prospective, ISSN 1936-1246, Volume 6, 2-16, 2010.

Li, Y. G., Fault-Zone Trapped Waves at the Thrusting Longmen-Shan Fault of the Wenchuan Earthquake: 3-D Finite-Difference Investigations of Trapping Efficiency, Western Pacific Geophysical Meeting, Taipei, June 22-25, 2010.

Li, Y. G., J. E. Vidale, and K. Aki, Depth-dependent structure of the Landers fault zone using fault zone trapped waves generated by aftershocks. J. Geophys. Res., 105, 6237-6254, 2000.

Li, Y. G., J. E. Vidale, E. S. Cochran, P. Chen, J. Y. Sue, and T. C. Chen, Seismic documentation of rock damage and heal on the Parkfield San Andreas Fault and the Longmen-Shan Fault ruptured in the 2008 *M*8 Wenchuan earthquake. International Symposium on Earthquake Seismology and Predictability, Abstracts Volume, p18, Beijing, China, July 6-9, 2009.

Li, Y. G., J. E. Vidale, K. Aki, F. Xu, T. Burdette, Evidence of shallow fault zone strengthening after the 1992 M7.5 Landers, California, earthquake. Science, 279, 217-219, 1998.

Li, Y. G., J. E. Vidale, S. M. Day and D. Oglesby, Study of the M7.1 Hector Mine, California, earthquake fault plan by fault-zone trapped waves, Hector Mine Earthquake Special Issue. Bull. Seism. Soc. Am., 92, 1318-1332, 2002.

Li, Y. G., J. E. Vidale, S. M. Day, D. D. Oglesby, and E. Cochran, Post-seismic fault healing on the 1999 *M*7.1 Hector Mine, California earthquake. Bull. Seism. Soc. Am., 93, 854-869, 2003.

Li, Y. G., J. E., Vidale, and S. E. Cochran, Low-velocity damaged structure of the San Andreas fault at Parkfield from fault-zone trapped waves. Geophy. Res. Lett., 31, L12S06, 2004.

Li, Y. G., P. C. Leary, K. Aki, and P. E. Malin, Seismic trapped modes in Oroville and San Andreas fault zones. Science, 249, 763-766, 1990.

Li, Y. G., P. Chen, E. Cochran and J. Vidale, Seismic velocity variations on the San Andreas Fault caused by the 2004 *M*6 Parkfield earthquake and their implications. Earth Planets Space, 59, 21-31, 2007.

Li, Y. G., P. Chen, E. S. Cochran, J. E., Vidale, and T. Burdette, Seismic evidence for rock damage and healing on the San Andreas fault associated with the 2004 *M*6 Parkfield earthquake, Special issue for Parkfield *M*6 earthquake. Bull. Seism. Soc. Am., 96, No.4, S1-15, doi:10.1785/0120050803, 2006.

Li, Y. G., Seismic Study of the San Andreas Fault in California and the Longmen-Shan Fault Ruptured in the 2008 *M*8 Wenchuan Earthquake in China. Academic Prospective, V4, 4-25, 2008b.

Li, Y.-G., J. E. Vidale, K. Aki, C. Marone, and W. H. K. Lee, Fine structure of the Landers fault zone; segmentation and the rupture process. Science, 256, 367-370, 1994.

Lin, A. M., Z. K. Ren, D. Jia and X. J. Wu, Co-seismic thrusting rupture and slip distribution produced by the 2008 Mw7.9 Wenchuan earthquake, China. Tectonophysics, 471(3-4), 203-215, 2009.

Liu, Q., J. Chen, S. Li, Y. Li, J. Wang, B. Guo, and S. Qi, Wenchuan Ms 8.0 earthquake: The crustal velocity structure and stress field from the western-Sichuan seismic array observations. Eos Trans. AGU, 89(53), Fall Meet. Suppl., abs. U22B-01, 2008.

Liu, Z. and J. Huang, Temporal changes of seismic velocity near the epicenter of the Wenchuan earthquake from ambient noise correlation. *EOS Trans. AGU Fall Meet.*, S13B2003L, 2010.

Marone, C., J. E. Vidale and W. L. Ellsworth, Fault healing inferred from time dependent variations in source properties of repeating earthquakes. Geophys. Res. Lett., 22, 3095-3098, 1995.

Marone, C., Laboratory-derived friction laws and their application to seismic faulting. Ann. Revs. Earth & Plan. Sci., 26, 643-696, 1998b.

Marone, C., The effect of loading rate on static friction and the rate of fault healing during the earthquake cycle. Nature, 391, 69-72, 1998a.

Massonnet, D, W. Thatcher, and H. Vadon, Detection of postseismic fault-zone collapse following the Landers earthquake. Nature, 382, 612-616, 1996.

Mooney, W. D., and A. Ginzburg, Seismic measurements of the internal properties of fault zones. Pure Appl. Geophys., 124, 141-157, 1986.

Nur. A., Dilatancy, Pore fluid, and premonitory variations of ts/tp travel times. Bull. Seismol. Soc. Am., 62, 1217-1222, 1972.

O'Connell, R. J., and B. Budiansky, Seismic velocities in dry and saturated cracked solids. J. Geophys. Res., 79, 5412-5426, 1974.

Olsen, M., C. H. Scholz, and A. Leger, Healing and sealing of a simulated fault gouge under hydrothermal conditions for fault healing. J. Geophys. Res., 103, 7421-7430, 1998.

Rice, J. R., Fault stress states, pore pressure distributions, and the weakness of the San Andreas fault, in Fault Mechanics and Transport Properties of Rocks, edited by B. Evans and T.-F. Wong, 475-503, Academic, San Diego, Calif, 1992.

Rubinstein, Justin, and G. C. Beroza, Depth constrain on nonlinear strong ground motion from the 2004 Parkfield earthquake. Seism. Res. Lett., 32, L14313, doi: 10.1029/2005GL023189, 2005.

Schaff, D. P., and G. C., Beroza, Coseismic and postseismic velocity changes measured by repeating earthquakes. J. Geophys. Res., 109, B10302, doi: 10.1029/2004JB003011, 2004.

Scholz, C. H. (1990), The Mechanics of Earthquakes and Faulting. Cambridge Univ. Press, New York.

Sibson, R. H., J. M. Moore and A. H. Rankin, Seismic pumping – A hydrothermal fluid transport mechanism. Geological Society of London Journal, 131, 653-659, 1975.

Sibson, R. H., Structural permeability of fluid-driven fault-fracture meshes. Journal of Structural Geology, 18: 1031-1042, 1996.

Sibson, R. H., Tectonic controls on maximum sustainable overpressure: fluid redistribution from stress transitions. Journal of Geochemical Exploration, 69-70, 471-475, 2000.

Sun, Y. S., S. Kuleli, F. D. Morgan, w. Rodi, N. Toksoz, W. B. Han and Z Y. Lu, Location robustness of earthquakes in Sichuan province, China. Seism. Res. Lett., 75(1), 54-62, doi: 10.1785/gssrl.75.1.54, 2004.

Taira T., Paul G. Silver, F. Niu, and R. M. Nadeau, Remote triggering of fault-strength changes on the Parkfield. Nature, 461, 636-639, dio:10.1038/nge:10.1038/nature 08395, 2009.

Tong, X. P., D. T. Sandwell, and Y. Fialko, Coseismic slip model of the 2008 Wenchuan earthquake derived from joint inversion of interferometric synthetic aperture, radar, GPS, and field data. J. Geophys. Res., V 115, B04314, doi:10.1029/2009JB006625, 2010.

Vidale, J. E. and Y. G. Li, Damage to the shallow Landers fault from the nearby Hector Mine earthquake. Nature, 421, 524-526, 2003.

Vidale, J. E., D. V. Helmberger, and R. W. Clayton, Finite-difference seismograms for SH waves. Bull Seism. Soc. Am. 75, 1765-1782, 1985.

Vidale, J. E., W. L. Ellsworth, A. Cole, and C. Marone, Rupture variation with recurrence interval in eighteen cycles of a small earthquake. Nature, 368,624-626, 1994.

Wu, J., J. A. Hole, and J. A. Snoke, Fault-zone structure at depth from differential dispersion of seismic guided waves: Evidence for a deep waveguide on the San Andreas fault. Geophys. J. Int., 182, 343-354, doi: 10.111.j.1365-246X.2010.04612.x, 2010.

Xu Z. et al., Uplift of the Longmen Shan range and the Wunchuan earthquake. Episodes. 31(3), 291-301, 2008a.

Xu Z. et al., Wenchuan Earthquake and Scientific Drilling. Acta Geologica Sinica (Chi-

nese Edition), 82(12), 1613-1622, 2008b.
Xu, C. J., Y. Liu, and Y. M. Wen, Coseismic slip distribution of the 2008 M_w 7.9 Wenchuan earthquake from joint inversion of GPS and InSAR data. Bull. Seism. Soc. Am., 100, 2736-2749, DOI: 10.1785/0120090253, 2010.
Xu, X., X. Wen, G. Yu, G. Chen, Y. Klinger, J. Hubbard, and J. Shaw, Coseismic reverse- and oblique-slip, surface faulting generated by the 2008 Mw 7.9 Wenchuan earthquake, China. Geology, 37 (6), 515-518, 2009.
Yasuhara, H., C. Marone, and D., Ellsworth, Fault zone restrengthening and frictional healing: the role of pressure solution. J. Geophys. Res. 110, B06310, doi:10.1029/2004JB003327, 2005.
Zhang, P. Z., et al., Continuous deformation of the Tibetan Plateau from global positioning system data. Geology, 32, 809-812, 2004.
Zhang, Y. J., Y. Gao, Y. T. Shi and L. X. Tai, The shear-wave splitting study of Sichuan Zipingpu reservoir region. Chinese Journal of Geophysics, 53(5), 750-861, 2010.

Author Information

Yong-Gang Li

Department of Earth Sciences, University of Southern California, Los Angeles, California 90089, USA
E-mail: ygli@usc.edu

Jin-Rong Su and Tian-Chang Chen

Seismological Bureau of Sichuan Province, Chengdu, China

Appendix

Immediately after the *M*8 Wenchuan earthquake on May 12, 2008, the lead author of this article with researchers of China Earthquake Administration (CEA) surveyed the rupture zones along the Longmen-Shan fault (LSF) between Yingxiu and Nanba to select the sites for deployment of portable seismic arrays to record fault-zone trapped waves (FZTWs). These pictures show main deformation characteristics of co-seismic surface ruptures and landslides as well as destroyed buildings and road structures in the earthquake epicentral areas. This information is helpful with us to understand the severe damage caused by a big earthquake and the importance of seismic hazard mitigation, ranging from improved building code to upgrading structures, earthquake education and other lifelines, for reducing the losses of lives and properties during devastating earthquakes in Sichuan Province and elsewhere in the world.

Fig. 4A-1 (a) The vertical fault wall with 4-m coseismic surface slip on the main rupture along the Yingxiu-Beichuan fault near Bajiaomiao village in the 2008 *M*8 Wenchuan earthquake. (b) The low-angle ($<20°$) co-seismic slip on the branch rupture along the Guangxian-Anxian fault at Bailu Middle School. (c) The co-seismic slip with 6-m vertical displacement along the reverse-thrusting southern segment of Yingxiu-Beichuan fault about 6 km north of Yingxiu. (d) The co-seismic slips with ~4-m vertical displacement and ~3-m horizontal dextral displacement along the northern segment of Yingxiu-Beichuan fault near Leigu town. (e) The co-seismic slips with ~3-m vertical displacement and ~3-m horizontal dextral displacement along the northern segment of Yingxiu-Beichuan fault near Pingtong town. (f) ~3-m dextral slip on the northernmost strike-sliping segment of Longmen-Shan fault near Nanba town.

(a)

Fig. 4A-2 Pictures show damage magnitudes of destroyed buildings increasing with the distance from the Longmen-Shan fault ruptured in the 2008 *M*8 Wenchuan earthquake along the roads (a) from Jiangyu to Nanba, (b) from Jiangyu to Pingtong, (c) from Anxian to Leigu, and (d) from Pengzhou to Bailu, and (e) from Yingxiu to Hongkou.

(a)

(b)

Fig. 4A-3 (a) The concrete high-ways and bridges along the road from Beichuan to Hongkou destroyed by the 2008 *M*8 Wenchuan earthquake. (b) Land-sliding caused by the 2008 *M*8 Wenchuan earthquake in the southern Longmen-Shan areas.

Fig. 4A-4 (a) The lead author of this article (right) with Prof. Pei-Zhen Zhang (left), the chief scientist in Wenchuan Earthquake Scientific Investigation Headquarter of China Earthquake Administration settled in Chengdu, Sichuan Province after the 2008 Wenchuan earthquake. (b) Director Zhou Ming-Qian (left) of Overseas Chinese Affairs Office of Sichuan Province and his office colleague Cao Hui (right) offered assistances when the lead author of this article stayed in Sichuan for his research after the 2008 Wenchuan earthquake. (c) Prof. Pei-Zhen Zhang guided to the main rupture along the southern Yingxiu-Beichuan fault with the nearly vertical co-seismic slip of $\sim$4 m near Bajiaomiao. (d)-(h) The lead author of this article visited Beichuan, Bailu, and Juyuan high-schools which educational buildings were destroyed in the 2008 *M*8 Wenchuan earthquake. (i) The road to Beichuan town was blocked by sliding rocks. (j) and (k) Rescue teams formed by soldiers provided administering first aids in severely destroyed towns. (l) Temporary houses built near Leigu town that had been completely damaged in the 2008 M8 Wenchuan earthquake.

Chapter 5
Ground-Motion Simulations with Dynamic Source Characterization and Parallel Computing

Benchun Duan

This chapter illuminates an emerging trend in earthquake ground-motion simulations. In this trend, earthquake sources are characterized by spontaneous rupture models and high performance computing systems are used to calculate time histories of ground motion with frequency contents up to several Hz in seismically active regions such as Southern California. We first review the basics of spontaneous rupture models. Then, a finite element method (FEM) algorithm EQdyna is introduced and two examples of ground-motion related applications are discussed. Finally, recent development of the FEM algorithm in parallelization using a hybrid MPI/OpenMP approach and its application to the convergence test of a benchmark problem are presented.

Keywords: Ground motion prediction, Dynamic rupture models, Parallel Computing

5.1 Introduction

Synthetic seismograms for possible future moderate and large earthquakes are one of the key products that seismologists can provide to engineers. These seismograms are used by engineers in the design of earthquake-resistant structures. Thus, ground-motion simulations are a vital bridge between earthquake seismology and earthquake engineering. Particularly, for close-in distances near active faults, the ground-motion recordings from real earthquakes are sparse. Synthetic seismograms can fill this gap. In addition, with increasing usage of nonlinear analyses in the seismic design of structure, time histories of ground motion become more important for completely determining structure response and damage estimation from future significant earthquakes. Synthetic seismograms also allow engineers to examine the variability in the structure response to different earthquake scenarios with different rupture directivity and slip distributions, which would not be available in recorded data.

Ground-motion simulations are routinely performed using kinematic source characterization. In this type of source characterization, fault kinematics (i.e., slip as function of space and time on the fault) is defined from simple, empirically-guided rules, or derived from kinematic source inversions of recent earthquakes based on recorded data. These kinematic-source-based ground-motion simulations have enjoyed substantial success, such as recognition of potential importance of seismic energy channeling and focusing by sedimentary basins (e.g., Olsen et al., 2006; Day et al., 2008). However, near-field ground motions are sensitive to complex source processes that are difficult to be characterized by kinematic source models. For example, TeraShake2 (Olsen et al., 2008) using dynamic source models with small-scale stress-drop heterogeneity predicts smaller PGV extremes by factors of 2-3 relative to TeraShake1 (Olsen et al., 2006) that uses kinematic source models. Both TeraShake1 and TeraShake2 simulate large earthquakes of Mw 7.7 on the southern segment of the San Andreas fault. Olsen et al (2008) attribute this reduction of ground-motion extremes largely to the less coherent wave field radiated by the dynamic sources. In these dynamic sources, fluctuations in the speed, shape, and propagation direction of the slip pulse spontaneously arise from stress-drop heterogeneities. Rupture complexity of this type is very difficult to be parameterized in a purely kinematic source model, and even in a pseudo-dynamic source model (e.g., Guatteri et al., 2004).

With rapid development of modern high performance computing systems, particularly cluster systems with CMPs (Chip MultiProcessors), parallel computing has been becoming increasingly important and popular in numerical simulations of rupture dynamics and ground motion. Parallel computing allows seismologists to explore small-scale rupture complexities observed in large earthquakes and to augment high frequency limits of deterministically simulated ground motions.

In this chapter, we first review the basics of the spontaneous, dynamic rupture model (hereafter called "the spontaneous rupture model"). Then we introduce an explicit finite element method (FEM) algorithm EQdyna for simulating spontaneous rupture on realistically complex faults and wave propagation in complex geologic structure. Two examples of its applications are presented. Finally, recent advancement in parallelization of the algorithm using a hybrid MPI/OpenMP approach and its application to a benchmark problem are discussed.

5.2 The Spontaneous Rupture Model

In contrast to a kinematic source model, a dynamic source model starts with a given prestress field on the fault (and the medium in some cases) and solves equations of motion for slip evolution and distribution on the fault. Particularly, with a failure criterion and a friction law, rupture propagation itself can be a part of solution of the dynamic source model, rather than a priori (e.g., a fixed rupture velocity). This type of dynamic source models in which rupture propagation spontaneously evolves is referred to the spontaneous rupture model. The defining characteristic of a spontaneous rupture model is that it obeys the conservation laws of continuum mechanics and incorporates (to varying degrees) some of our understanding of rock friction, rock strength and rheology, and state of stress in the crust.

In any material, there must be a limit to the stress state that the material can support. Thus, stress singularity at a rupture front in earlier studies of dynamic rupture (e.g., Burridge, 1973) is not physical. This theoretical consideration, evidence in some rock friction experiments, and the interest of simplicity have promoted applications of slip-weakening friction laws, particularly a linear slip-weakening law (e.g., Ida, 1972; Andrews, 1976b; Day, 1982). In such a friction law, the frictional coefficient on the fault plane drops linearly from a static value μ_s to a dynamic (sliding) value μ_d over a critical slip-weakening distance D_0 (Fig. 5.1). Another type of friction laws used in spontaneous rupture models is the rate-and state-dependent friction law derived from laboratory experiments of rock friction (Dieterich, 1979; Ruina, 1983), in which frictional coefficient is a function of slip velocity and state variables.

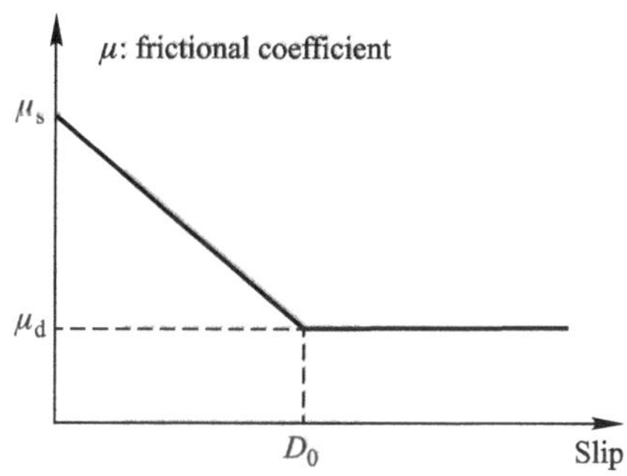

Fig. 5.1 A slip-weakening friction law widely used in rupture dynamics and ground-motion simulations. In this law, frictional coefficient μ drops from the static value μ_s to the dynamic value μ_d over a critical slip-weakening distance D_0.

Inputs that a spontaneous rupture model needs mainly include the geologic structure (including fault geometry and material properties), the initial pre-stress field in the model (on fault only if elastic off-fault response is assumed), the failure criterion (e.g., the Coulomb criterion), and the friction law that governs the evolution of fault friction. Because of involvement of fault friction, there are no analytical solutions for spontaneous rupture problems and numerical simulations are required. Various numerical methods, including finite difference methods (FDM), finite element methods (FEM), boundary element methods (BEM) and spectral element methods (SEM), have been used in simulating spontaneous rupture propagation (e.g., Harris et al., 2009). One challenging issue these numerical methods face is verification and validation of computer codes implemented by different researchers based on these methods. Verification refers to comparison of results from different codes on an identical problem, while validation generally means comparison of simulation results with ground-motion recordings from natural events and involves validation of not only the source process, but also the path effect (including the velocity structure and the local site condition). A broad, rigorous community-wide exercise on verification of dynamic rupture codes has been underway in the Southern California Earthquake Center/U.S. Geological Survey (SCEC/USGS) community (Harris et al., 2009). This exercise is to compare the computer codes used by SCEC and USGS researchers to verify that these codes are functioning as expected for studying earthquake source physics. More than 15 computer codes have been involved in the exercise and results of some benchmark problems from some of these codes are publically accessible on the Web site, http://scecdata.usc.edu/cvws. Figure 5.2 gives an example of the publically available comparisons from the Web site, showing the rupture-front contours and the time-histories of down-dip slip rate at a point on the fault from three codes, including two FEM codes and one SEM code, which treated the benchmark problem TPV210 at an

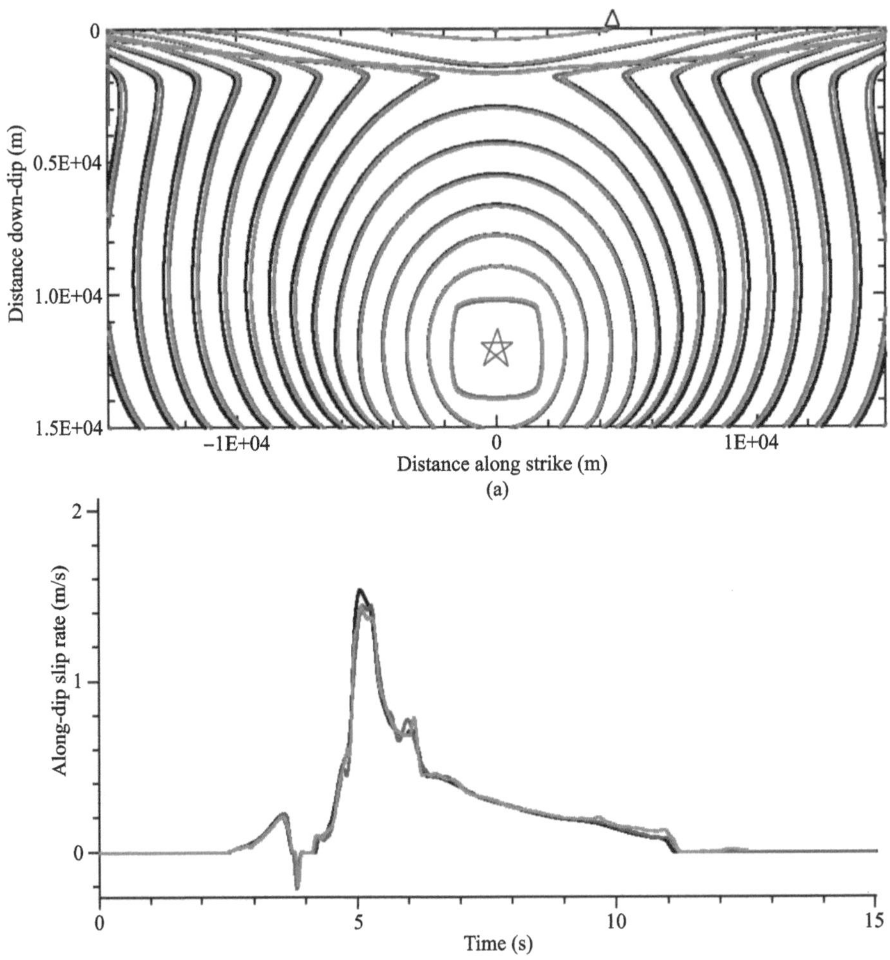

Fig. 5.2 (a) The rupture-front contours on the fault plane, and (b) the down-dip slip rate at an on-fault station from three spontaneous rupture codes of the SCEC/USGS community that ran the TPV210 benchmark, with an on-fault element size of 50 m. The star and triangular in (a) represent the hypocenter and the station location, respectively. Black and red are results from two FEM codes (red from EQdyna), and green is from an SEM code. These plots are publically accessible from the Web site: http:/scecdata.usc.edu/cvws. Both the rupture contours and the time histories of slip rate match well among the three codes.

on-fault element size of 50 m. The red curves in these plots are the results from EQdyna that will be discussed in detail in the next section. Both the contours and time histories match well among the three codes, indicating that they work as expected.

5.3 EQdyna: An Explicit Finite Element Method for Simulating Spontaneous Rupture on Geometrically Complex Faults and Wave Propagation in Complex Geologic Structure

EQdyna is one of major codes in the SCEC/USGS dynamic code verification exercise and has been verified by more than 10 benchmark problems in the exercise. The code has been under development since 2005. Its various versions have been used to investigate dynamics of geometrically complex faults over multiple earthquake cycles (Duan and Oglesby, 2006; 2007), off-fault damage distributions and effects on rupture dynamics and ground motion (Duan and Day, 2008; Duan, 2008a, b), spontaneous rupture propagation in recent large earthquakes (Duan, 2010a), pre-existing fault zone response to nearby earthquakes (Duan, 2010b; Duan et al., 2011), and physical limits to ground motion with critical facilities (Duan and Day, 2010). We will review two examples of these applications in the next section.

EQdyna solves the dynamic problem in an isotropic medium containing surfaces (faults) across which the displacement vector may have a discontinuity. The equations of motion for the medium are

$$\rho \boldsymbol{u} = \nabla \cdot \boldsymbol{\sigma} + \rho \mathbf{b}, \tag{5.1}$$

in which $\boldsymbol{\sigma}$ is the stress tensor, $\boldsymbol{u}$ is the displacement vector, $\mathbf{b}$ is the body force vector, ρ is density, and double dots on $\boldsymbol{u}$ represent the second derivative in time (thus the acceleration). The first term on the right-hand side of Equation (5.1) with the dot product of the operator ∇ and the stress tensor gives the divergence of the stress field, resulting in a vector.

Following Day et al. [2005], we can formulate the jump conditions at the fault surfaces as

$$\tau_c - \tau \geqslant 0, \tag{5.2a}$$

$$\tau_c \dot{\mathbf{s}} - \boldsymbol{\tau} \dot{s} = 0. \tag{5.2b}$$

Equation (5.2a) stipulates that magnitude of the shear traction $\boldsymbol{\tau}$ on the fault surfaces should be bounded by the nonnegative frictional strength τ_c. Equation (5.2b) stipulates that any nonzero slip velocity vector $\dot{\mathbf{s}}$ (with magnitude $\dot{s}$) must be opposed by an anti-parallel traction vector $\boldsymbol{\tau}$ with magnitude equals to the frictional strength τ_c. Equation (5.2b) remains valid even if $\dot{s}$ is zero. When equality does not hold in (5.2a), (5.2b) can be satisfied only with $\dot{s}$ equal to zero. In the framework of a slip weakening friction law, the frictional strength τ_c is the product of compressive normal stress $-\sigma_n$ (positive in tension) and the coefficient of friction $\mu(l)$,

$$\tau_c = -\sigma_n \mu(l), \tag{5.3}$$

where l is the path-integrated distance the fault node has slipped over the time period t, i.e.,

$$l(t') = \int_0^t \dot{s}(t')dt'$$

For a linear slip-weakening friction law illustrated in Figure 5.1, $\mu(l)$ is given by

$$\mu(l) = \mu_s - (\mu_s - \mu_d)\min\{l, D_0\}/D_0. \quad (5.4)$$

Equations (5.2)-(5.4), combined with given initial stress conditions on the fault surfaces, govern fault behavior at all times, including prerupture, initial rupture, arrest of sliding, and possible subsequence episodes of reactivation and arrest. When the initial transition from inequality to equality occurs in (5.2a) at a point on the fault surfaces, rupture starts at the point. When (5.2a) undergoes a transition from equality back to inequality at the point, sliding arrests. If (5.2a) switches back again from inequality to equality, reactivation of slip occurs at the point.

There are different schemes used to implement the fault-jump conditions in numerical algorithms for spontaneous rupture simulations, including the traction-at-split-node (TSN) method (e.g., Andrews, 1999; Day et al., 2005), the thick fault (TF) method (e.g., Madariaga et al., 1998), and the stress glut (SG) method (e.g., Andrews, 1976a; Andrews, 1999). Comparison of these different representation methods shows that TSN attains power-law convergence, while SG achieves qualitatively meaningful solutions but convergence is uncertain and TF does not achieve qualitatively meaningful solutions to a 3D test problem (Dalguer and Day, 2006).

EQdyna uses TSN to represent fault surfaces in the finite element mesh. Early versions of EQdyna (Duan and Oglesby, 2006, 2007) followed the implementation of TSN summarized by Andrews (1999), which requires separate treatments for initiation of sliding, continuous sliding, and arrest of sliding. As discussed above, a new formulation of TSN proposed by Day et al. (2005) and characterized by Equation (5.2) treats fault behavior at all times in a concise and robust way. New versions of EQdyna (Duan and Day, 2008; Duan, 2008a, 2008b, 2010a, 2010b; Duan and Day, 2010; Duan et al., 2011) follow the new formulation of TSN.

The TSN method represents the faults by split nodes: a fault node is split into plus-side and minus-side parts. All three components of displacement and velocity may have discontinuities between the two halves of the split node, which give slip and slip velocity (i.e., relative displacement and velocity between the two halves, respectively) at the fault node. The two halves of the split node interact through a traction acting on the fault surface between them, which is limited by the frictional strength τ_c. As an FEM method, EQdyna can deal with fault surfaces in arbitrary orientations. Using a local coordinate system (**n, s, d**) defined by the unit normal, strike, and dip vectors of the fault surface at a split node, which may be different from the global coordinate system (**x, y, z**) of the model region, EQdyna first solves for the trial traction between the two halves of a split node (Duan, 2010a) as

$$\tilde{T}_v = \frac{\Delta t^{-1} M^+ M^- [(\dot{u}_v^+ - \dot{u}_v^-) + (u_v^+ - u_v^-)\delta_{vn}] + M^- R_v^+ - M^+ R_v^-}{a(M^+ + M^-)} + T_v^0, v = s, d, n \quad (5.5)$$

where δ_{vn} is the Kronecker delta (i.e., it equals 1 only when $v = n$, otherwise 0), $M, R, \dot{u}_v, u_v$ with superscripts plus and minus signs are mass, elastic restoring force, particle velocity, particle displacement of plus-side and minus-side split nodes, respectively, T_v^0 is the initial stress (in equilibrium), a is the area of the fault surface associated with the split node under consideration, and Δt is the dynamic simulation time step. These fault surface nodal traction components would enforce continuity of tangential velocity (i.e., $\dot{u}_v^+ = \dot{u}_v^-$ for $v = s$ or $v = d$) and continuity of normal displacement (i.e., $u_n^+ = u_n^-$). Then, the true fault tangential (shear) traction components at the fault node that enforce jump conditions of Equation (5.2) are calculated by

$$T_v = \begin{cases} \tilde{T}_v & v = s,d, \quad T \leqslant \tau_c, \\ \dfrac{\tau_c \tilde{T}_v}{T} & v = s,d, \quad T > \tau_c, \end{cases} \tag{5.6}$$

where $T = [(\tilde{T}_s)^2 + (\tilde{T}_d)^2]^{1/2}$ is the magnitude of shear traction on the fault surface. Equation (5.6) states that if the magnitude of the trial shear traction is not greater than the frictional strength, the trial shear traction components are the true shear traction components. Otherwise, the true shear traction components are limited by the frictional strength and must be adjusted from the trial components. In general, the normal traction on the fault surface may fluctuate due to reflected waves in complex fault systems and/or geologic structure. Under some circumstances (e.g., at shallow depth), portions of the fault surface may undergo separation if there is a transient reduction of the compressive normal traction to zero. In this case, the trial normal traction ($v = n$) in Equation (5.5) may become tensile. Following Day et al. (2005), we calculate the true normal traction in EQdyna as

$$T_n = \begin{cases} \tilde{T}_n & \tilde{T}_n \leqslant 0 \\ 0 & \tilde{T}_n > 0 \end{cases}. \tag{5.7}$$

Equation (5.7) states that the normal traction on the fault surface cannot be tensile.

EQdyna follows the standard FEM procedure (e.g., Hughes, 2000) to solve the equations of motion (5.1). Duan and Oglesby (2006) gave a brief description of the procedure. After discretizing in space, a matrix equation can be obtained from (5.1)

$$\mathbf{M}\ddot{\mathbf{u}} + \mathbf{K}(\mathbf{u} + q\dot{\mathbf{u}}) = \mathbf{F}, \tag{5.8}$$

where **M** and **K** are the mass matrix and the stiffness matrix, respectively, **F** is the vector of applied forces (e.g., body forces), q is a stiffness-damping parameter used to selectively damp high-frequency noises near the resolution limit of the FEM mesh. q can be specified through a non-dimensional parameter β (Duan and Day, 2008), so that $q = \beta\Delta t$, or equivalently,

$$q = \beta\alpha\Delta x / v_p, \tag{5.9}$$

where Δx is the minimum element size, α is the Courant-Friedrich-Lewy (CFL) number and v_p is the P wave velocity. Equation (5.8) can be solved by the central difference time integration method, which is explicit (e.g., no need for solving a coupled set of equations) when **M** is diagonal (e.g., Hughes, 2000). The central difference method is conditionally stable, and the simulation time step is determined by the minimum element size and the wave speed in the model to ensure numerical stability.

The one-point quadrature element is widely adopted in dynamic problems. Its rate of convergence is comparable to that of fully integrated elements (Belytschko et al., 1984), while it is much more computationally efficient. EQdyna uses the one-point quadrature element. However, one needs to treat possible hourglass modes of deformation (which is not physical) with this type of elements. EQdyna adopts the method proposed by Kosloff and Frazier (1978) to determine the element restoring force (referred to as the hourglass force hereafter) to resist hourglass modes. This introduces a new vector of the hourglass forces $\mathbf{H}$ into Equation (5.8). For split nodes along the fault surfaces, a surface force vector from the coupling traction in Equations (5.6) and (5.7) is also incorporated into Equation (5.8) to solve for particle acceleration $\ddot{\mathbf{u}}$, and then particle velocity $\dot{\mathbf{u}}$ and displacement $\mathbf{u}$.

5.4 Two Examples of Ground-Motion Related Applications of EQ-dyna

EQdyna has been used in rupture dynamics studies, physical limits to ground motion at critical facilities, and pre-existing compliant fault zones' response to nearby earthquakes. It has been under parallelization with OpenMP and MPI techniques since 2008 (see the next section). MPI parallelization of EQdyna is necessary for cross-computer-node parallel computing, which is needed for large-scale ground-motion simulations, similar to TeraShake simulations (Olsen et al., 2006, 2008). TeraShake2 (Olsen et al., 2008) used dynamic source characterization, but with a significantly simplified fault representation. Because of the limitation of the finite difference method, the non-coplanar southern San Andreas fault is projected onto a single planar fault surface in TeraShake2 (Olsen et al., 2008). EQdyna has potential to simulate realistically complex fault systems and to be a major player in TeraShake-type ground-motion simulations. We are moving toward this direction. Here, we review two application examples of EQdyna that are relevant to ground-motion simulations. One is to examine physical limits to extreme ground motion at a critical facility (Duan and Day, 2010). The other is to examine ground-motion features associated with pre-stress rotations along strike of a dipping fault (Duan, 2010a).

5.4.1 Sensitivity of Physical Limits on Ground Motion at Yucca Mountain

Current practice in probabilistic seismic hazard analysis (PSHA) assumes untruncated lognormal distributions for the ground-motion parameters. When PSHA is applied at very low probability of exceedance levels required for critical facilities, ground-motion estimates are controlled by the upper tails of the distribution functions (e.g., Stepp et al., 2001) and extremely high. For example, at exceedance levels of 10^{-7}/yr and 10^{-8}/yr, the peak ground velocity (PGV) is inferred to be 7.0 m/s and 13.0 m/s, respectively, at a potential high-level radioactive waste storage site on Yucca Mountain (Stepp et al., 2001). These large magnitudes of PGVs are widely regarded as physically unrealizable and will significantly increase costs of construction of the facility. We need to place physical limits on PGVs that can never be exceeded. These physical limits may be determined from physical principles operating at earthquake sources (e.g., the maximum possible stress drop) and along the path seismic waves propagate (e.g., the finite strength of the material). Dynamic source characterization is necessary in this type of ground-motion calculations. With the physical constraint on the nearly maximum stress drop on the Solitario Canyon fault near the repository site, which results in surface slip of about 15 m, Andrews et al. (2007) obtained vertical PGV (V-PGV) of 5.78 m/s and horizontal PGV (H-PGV) of 4.33 m/s at the repository site (about 0.3 km depth and 1 km away from the fault plane on the footwall side) in a two-dimensional (2D) normal faulting framework, with an assumption of off-fault material strength being infinite (i.e., elastic response to seismic waves). Adding the physical constraint on a finite strength of the off-fault material (i.e., the material yields when stress reaches the strength), these two PGVs are reduced to

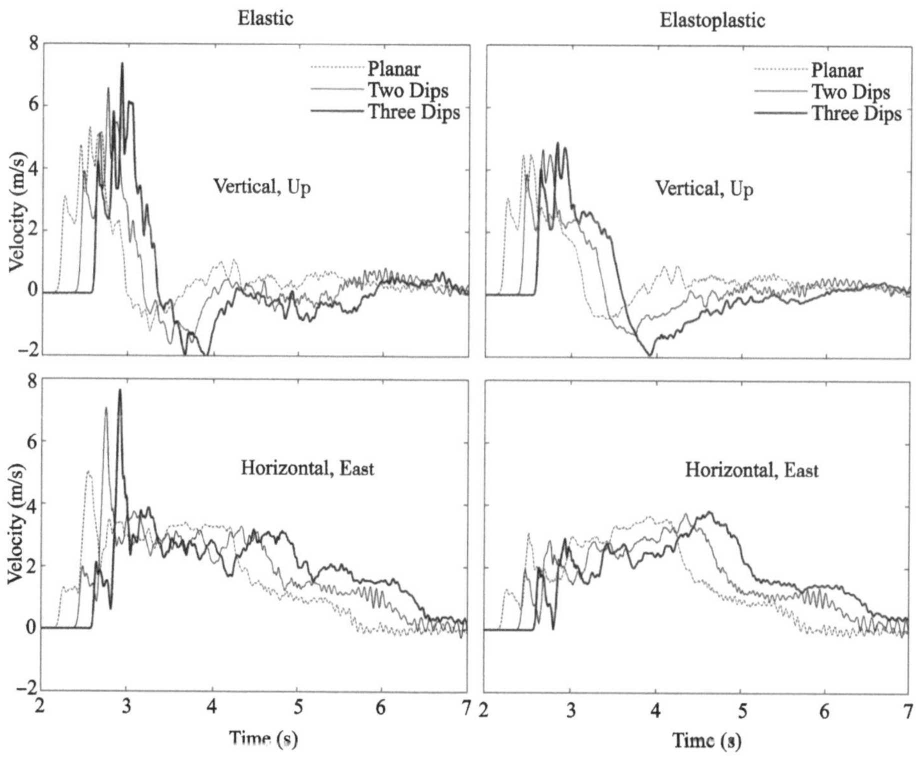

Fig. 5.3 Sensitivity of ground motion at the Yucca Mountain repository site to fault geometry at depth. Off-fault rocks are assumed to be infinitely strong on left panels and to have finite strength on right panels. See text for details of different models (After Duan and Day, 2010).

4.29 m/s and 3.59 m/s, respectively. With a factor of 1.33 for the maximum stress drop on the fault, physical limits of PGVs at the site are placed based on these dynamic, 2D calculations by Andrews et al. (2007).

Using a 2D version of EQdyna, Duan and Day (2010) explored the sensitivity of ground motion at the site to uncertainties in fault geometry at depth, material strength, poroelastic response of the fluid pressure, and fault zone structure. For example, Figure 5.3 shows the sensitivity of time histories of ground velocity at the site to fault geometry at depth, with the same depth profiles of other model parameters, including pre-stresses, frictional properties, and velocity structure. Planar represents a planar fault geometry with a dip of 60°, which was also used in Andrews et al. (2007). The fault dip decreases from 60° to 50° below 1 km depth in Two Dips, while an addition change of the dip from 50° to 40° below 6 km depth is present in Three Dips. In this nearly maximum stress drop case, PGVs increase with shallower dipping fault geometry if off-fault material is strong and responds to seismic waves elastically (left panels). This may be primarily explained by larger fault slip with shallower dip. However, if the material has a finite strength (with

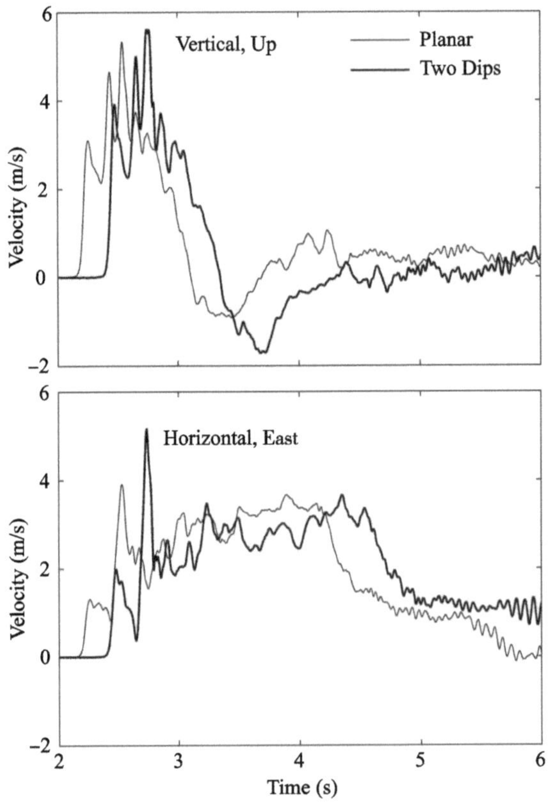

Fig. 5.4 Sensitivity of ground motion at the Yucca Mountain repository site to fault geometry at depth when cohesion values of shallow geologic units are doubled relative to those used in models shown on the right panels of Figure 5.3 (Revised from Duan and Day, 2010).

values of internal friction and cohesion used by Andrews [2007]) and it yields when the stress state reaches the strength, PGVs are relatively insensitive to the fault geometry at depth (right panels). With the finite strength of the off-fault material, enhanced plastic yielding essentially cancels the effects of larger slip associated with shallower dipping fault geometry. However, when cohesion values of rock units at shallow depth are doubled (i.e., the material at shallow depth is stronger), PGVs increase with shallower dipping fault geometry as illustrated in Figure 5.4, the H-PGV in particular. Therefore, physical limits on PGVs at the repository site are sensitive to the strength parameters of the shallow geologic units, and to the deep fault geometry of the Solitario Canyon fault if the shallow units are stronger than those assumed in Andrews et al. (2007).

5.4.2 Effects of Faulting Style Changes on Ground Motion

The 2008 Mw 7.9 Wenchuan (China) Earthquake exhibits obvious changes in faulting style along the Beichuan fault. From the southwest to the northeast along the fault, thrust faulting changes to oblique faulting with comparable amounts of thrust and right-lateral motions, and ends with predominantly right-lateral faulting. A shear-wave splitting study (Liu et al., 2008) suggests that the prestress field of the earthquake rotates along the Longmen Shan fault zone, with the maximum horizontal compressive stress (σ_1) making a very high angle to the fault strike along the south portion of the fault zone and smaller angles to the northeast. These rotations in the prestress field alone can explain the changes in faulting style in the Wenchuan earthquake. Motivated by these observations, Duan (2010a)

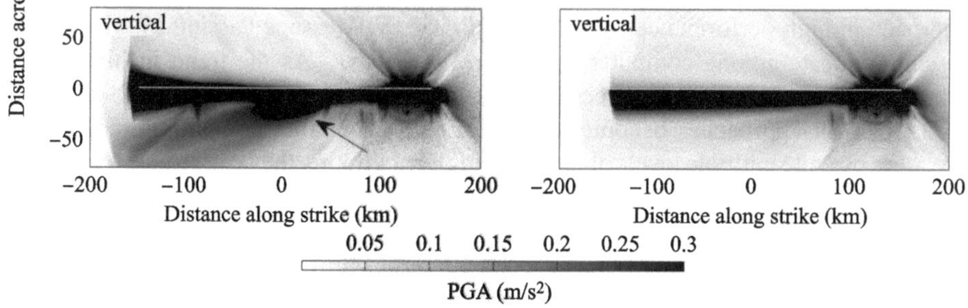

Fig. 5.5 Signals in peak ground acceleration (indicated by arrows on the left panels) caused by a prestress rotation along strike of a shallow dipping fault. The top panels are the strike-parallel horizontal component and the bottom panels are the vertical component. Stress rotations are present in the model OBLIQUE, while being absent in the model THRUST. See text for details (Revised from Duan, 2010a).

examined the effects of prestress rotations on rupture dynamics and ground motion, using a 3D version of EQdyna. Figure 5.5 shows two components (horizontal along strike, top panels, and vertical, bottom panels) of peak ground acceleration (PGA) from two models OBLIQUE and THRUST. In these models, a planar fault with a dip of 33° (the white line denoting the fault trace) is embedded in a homogeneous half-space, with negative distance across the fault being on the hanging wall. Prestresses on the fault in these models are depth-dependent. In THUST, there is no change in the principal prestress orientation along strike (with σ_1 making an angle of 90° with respect to the fault trace). In OBLIQUE, σ_1 makes an angle of 90° with the fault trace along the portion between 50 km and 150 km along strike, an angle of 60° between −100 km and 50 km, and an angle of 30° between −150 and −100 km. By comparing the two components of PGA from the two models, one can clearly see that the rotation of the prestress orientation at the along-strike distance of 50 km in OBLIQUE results in a relatively large PGA lobe on the hanging wall (indicated by black arrows) in the two components, which is absent in THRUST. A similar feature in PGA was observed in the Wenchuan strong ground motion recordings, which may be caused by rotations in the prestress field (thus changes in faulting style).

5.5 Hybrid MPI/OpenMP Parallelization of EQdyna and Its Application to a Benchmark Problem

In the simulations shown in Figure 5.5, the node spacing along the fault surface is 500 m. With degenerated wedge elements around the fault surface, the node spacing in the vertical and strike-normal directions is somewhat smaller (Duan, 2010a). With typical crustal rock properties, the highest frequency in simulated seismic waves is about 1 Hz. Each of the above models has about 47 million elements. With a dynamic simulation time step of 0.02 second and a termination time of 100 seconds, each model takes about 25 hours and 42 GB RAM memory to run on a 8-core SUN server with OpenMP parallelization of EQdyna.

To simulate higher frequency phenomena in rupture and wave propagations, we need to go beyond one-computing-node (e.g., using OpenMP) parallelization. A widely used approach in high performance computing is to use the Message Passing Interface (MPI) to communicate among computer nodes (and/or CPUs). As the trend in high performance computing systems has been shifting towards cluster systems with CMP, hybrid MPI/OpenMP approaches become more desirable in parallel computing. Hybrid approaches provide multiple levels of parallelism. Furthermore, they can reduce the communication overhead of MPI within a CMP node, by taking advantage of the shared address space and on-chip high inter-core bandwidth and low inter-core latency.

We have been parallelizing EQdyna using a hybrid MPI/OpenMP approach (Wu et al., 2011). In this approach, we use an element-based partitioning scheme for the MPI implementation because most time-consuming computations in EQdyna are element-based. For example, nodal forces (including the internal and hourglass forces) of a finite element node are calculated by assembling contributions from its adjacent elements, which are computed on the element basis. At each time step, nodal forces along boundaries between two adjacent MPI processes are transferred before calculating the trial fault tractions (i.e., Equation (5.5)). Finding nodes along boundaries of adjacent MPI processes can be very time-consuming. With structured mesh schemes we used, we make these boundary node numbers continuous and record the first and last nodes of a boundary between two adjacent MPI processes during the mesh generation phase. Within each MPI process, we use OpenMP directives in the code to parallel large loops to make use of shared memory and low cost in communication within a computing node. This hybrid MPI/OpenMP version of EQdyna has been verified by the SCEC/USGS benchmark problem TPV 11 (Wu et al., 2011). The results from this version overlap those from a sequential version of EQdyna, suggesting the implementation of the hybrid approach is correct. The performance analysis of two relatively small CMP systems tested by Wu et al. (2011), using several fixed model sizes (e.g., strong scaling), indicates that the hybrid version of EQdyna has good scalability.

Here, we apply the hybrid version of EQdyna to an SCEC/USGS benchmark problem TPV210, which is the convergence test of the benchmark problem TPV10 (Harris et al., 2009). In TPV10, a normal fault dipping at 60° (30 km long along strike and 15 km wide along dip) is embedded in a homogeneous half-space. Prestresses are depth-dependent and frictional properties are set to result in a subshear rupture. This benchmark problem is motivated by ground motion prediction on Yucca Mountain discussed in the above section. In TPV10, modelers are asked to run simulations at an element size of 100 m on the fault surface. In TPV210, we examine element-size dependence of the solutions by simulating the same problem at a set of element sizes, namely, 200 m, 100 m, 50 m, 25 m, 12.5 m, 6.25 m, and so on. At the time of the benchmark exercise (in November 2009), with OpenMP parallelization on one computing node only, we even had difficulty to perform simulations at the 50-m element size and it was impossible to run the 25-m or smaller element sizes. Recently, with the hybrid MPI/OpenMP version of EQdyna, we successfully performed simulations up to the 25-m element size on a new IBM (iDATA-Plex) Linux system hosted by the Texas A&M Supercomputing Facility. We report our results from the most recent development here.

5.5.1 Element-size Dependence of Solutions

Figure 5.6 compares the rupture arrival time (referred to as "rupture time" hereafter) of the spontaneous rupture propagation on the dipping fault from three simulations with different element sizes: 100 m, 50 m, and 25 m. At the early stage of the rupture propagation (e.g., less than 3 seconds), the contours from the three element sizes essentially overlap in this scale of the figure. Difference in the rupture time contours becomes visible later. In general, faster rupture propagation is associated with finer element size. Particularly, the contours of the 25-m element size are relatively well separated from those of the 100-m and 50-m element sizes, while the contours of the latter two are largely follow each other (with the 50-m element size being a little faster). If we consider the solution from finer element size to be closer to the exact solution, this may indicate that the convergence rate from the 100-m to 50-m element sizes is slower than that from the 50-m to 25-m element sizes.

Figure 5.7 shows time histories of two components of slip velocity at a surface station with a coordinate of (12 km, 0 km) in the coordinate system of Figure 5.6. Right-lateral and normal faulting are positive in the two components, respectively. Given the zero strike-slip initial shear stress in this normal faulting event, nonzero strike-slip component at the surface station is a result of rake rotation due to dynamic stresses. Overall, time histories of slip velocity from the three element sizes match well. In detail, visible difference in the peak slip rate exists. Particularly, finer element size appears to be associated with

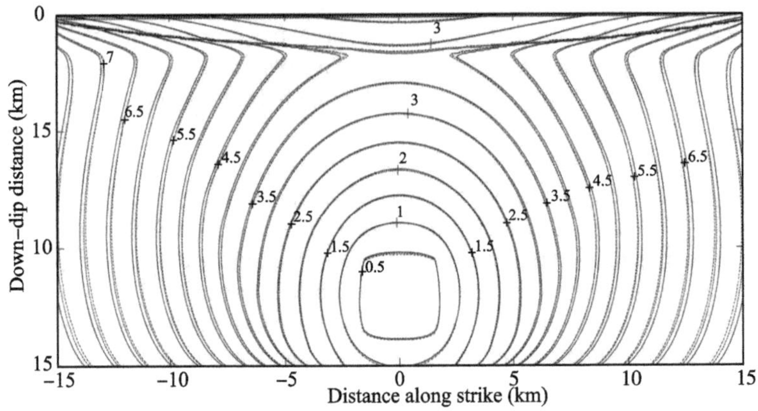

Fig. 5.6 Rupture time comparisons between three simulations with different element sizes on an identical dynamic problem (SCEC benchmark problem TPV210). Dash-dot, dash, and solid lines correspond to 100-m, 50-m, and 25-m element sizes, respectively. Rupture time unit is second.

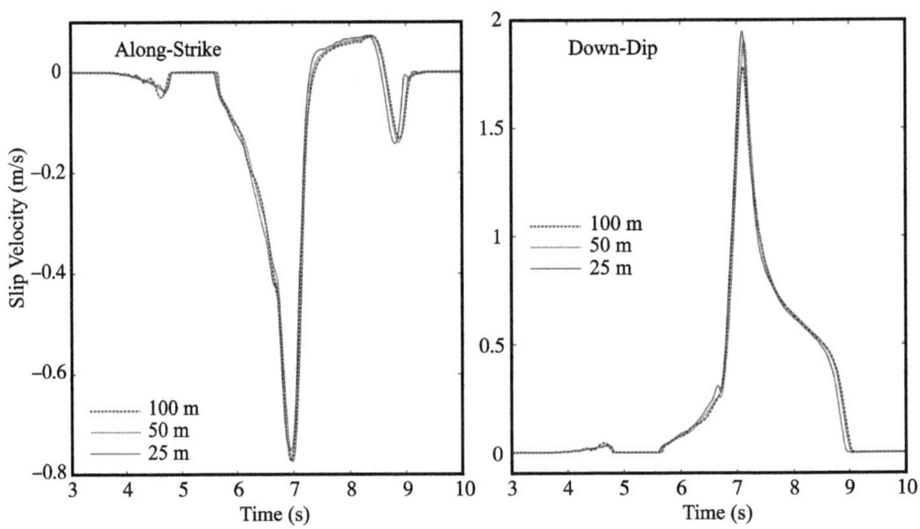

Fig. 5.7 Time histories of along-strike and along down-dip slip velocities at a surface on-fault station with a coordinate (12 km, 0 km) (see Fig. 5.6) from three different simulations with element sizes of 100 m, 50 m, and 25 m, respectively.

higher peak slip rate in the down-dip component, though the opposite seems the case in the along-strike component. This suggests that coarser element size may result in larger amount of rake rotation in normal faulting simulations.

Figure 5.8 illustrates element-size dependence of time histories of ground velocity at an off-fault station. The station is at 300 m depth, 1000 m away from the fault plane on the footwall side along the central cross section that is normal to the fault trace. The station is about the location of the repository site on Yucca Mountain with respect to the Solitario Canyon fault discussed above. Here, the horizontal component is the fault-trace normal component, while the other horizontal component (parallel to the fault trace) has negligible motion. It appears that the waveforms (e.g., locations of wiggles) from the three element sizes match well. However, the amplitudes of wiggles appear to have some dependence on the element size: finer element size generally results in a little larger amplitude.

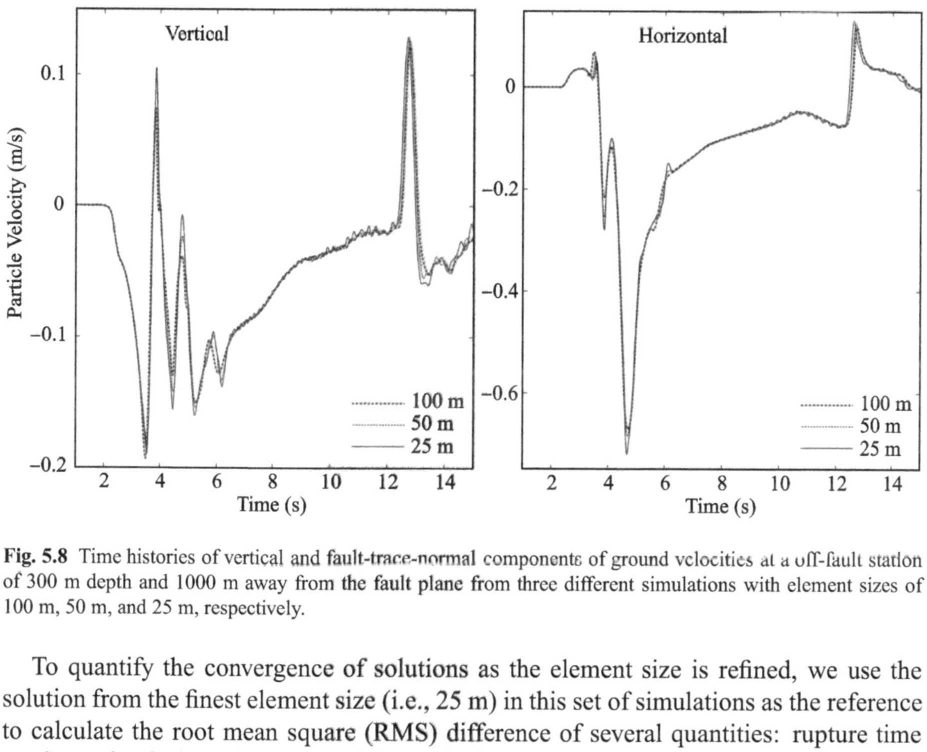

Fig. 5.8 Time histories of vertical and fault-trace-normal components of ground velocities at a off-fault station of 300 m depth and 1000 m away from the fault plane from three different simulations with element sizes of 100 m, 50 m, and 25 m, respectively.

To quantify the convergence of solutions as the element size is refined, we use the solution from the finest element size (i.e., 25 m) in this set of simulations as the reference to calculate the root mean square (RMS) difference of several quantities: rupture time on the entire fault surface (Fig. 5.6), peak slip velocity at the station shown in Figure 5.7, and peak ground motion at the station shown in Figure 5.8. Rupture time difference in a percentage of the rupture time of the reference case at each common node is first calculated and the RMS difference over all common nodes on the fault surface between two element sizes (one being the reference size) is then determined. Peak velocity (both slip and particle) difference in a percentage of the peak velocity of the reference case is first computed and the RMS difference over the two components is then obtained. We also include the solution from an extra element size, 200 m, in this quantification. These

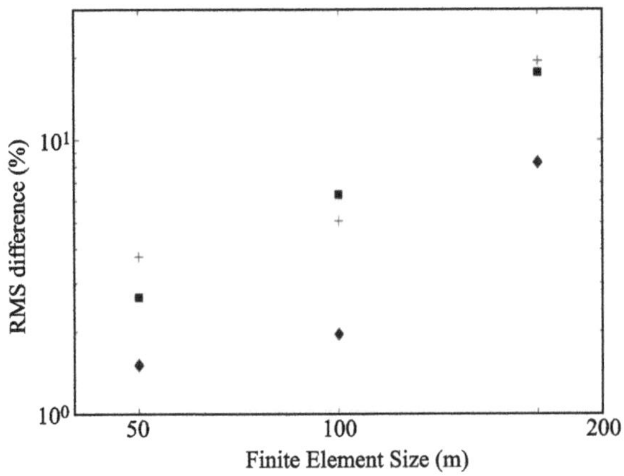

Fig. 5.9 Decrease in RMS differences of rupture time (diamond), peak slip velocity (square), and peak particle velocity (plus) with the reduced element size. RMS differences are relative to the reference solution of the finest element size (25 m). See text for details.

results are summarized in Figure 5.9. From this log-log plot, one can see that the RMS differences in rupture time and peak particle velocity drop more dramatically from the 200-m to 100-m element sizes than those from the 100-m to 50-m element sizes, and the difference in peak slip velocity decreases by following a power law. The power law convergence in the grid size of a finite difference method and a boundary integral method is reported by Day et al. (2005) on an earlier SCEC/USGS benchmark problem TPV3. However, their reference (finest) grid sizes are 50 m and 100 m in the two methods, respectively, and may not be fine enough to resolve the change in the convergence rate observed in our results. Different convergent rates between 200-m versus 100-m and 100-m versus 50-m element sizes in our results may result from the fact that the 200-m element size is too coarse to resolve the cohesive zone at the rupture tip (e.g., Day et al., 2005), while the 100-m element size is fine enough to resolve the cohesive zone. It would be interesting to see how the convergence rate might or might not change with finer element sizes (e.g., 12.5 m, 6.25 m, and so on) using larger cluster systems, as the 25-m element size is the finest element size we can run on the IBM (iDATAPlex) Linux system (EOS) at Texas A&M supercomputing facility. The basic specifications of EOS are given in Table 5.1. In the June 2010 version of the world's TOP500 supercomputing sites, EOS is listed as 420th.

Table 5.1 Basic specifications of EOS cluster system at Texas A&M Supercomputing Facility

Total Nodes	324
Cores/Chip	4
Cores/Node	8
Total Cores	2592
CPU type	Intel's 64-bit 2.8 GHz Nehalem (quad-core X5560) processor
Memory/Node	24 GB

5.5.2 Computational Resource Requirements and Performance Analysis

The required memory (RAM) in EQdyna increases approximately proportionally to the element number in the model. For uniform element size models, the element number increases with the inverse cube of the element size. In EQdyna, we use coarser elements in model regions that are not the focus of a study. Even with this nonuniform element size scheme, the element number and thus the memory requirement still increase dramatically with the reduced element size. Here, we report the results from this set of simulations and also discuss briefly performance (particularly, weak scaling) of the hybrid version of EQdyna at this stage.

Table 5.2 summaries computational resources used on EOS in the above set of simulations with the four different element sizes. When the element size is reduced by half, the element number increases by 4.0 to 4.6 times and memory usage increases by about 4 times. To ensure numerical stability, the simulation time step is also reduced by half. With the same termination time of 15 seconds, the total time step number is doubled. Putting together the spatial and temporal increases, we can see that the model size increases by about 8 to 9 times when the element size is halved. With 8 times increase in the CPU number used between two adjacent element sizes, the closeness in the wall clock time between the 100-m and 50-m element sizes indicates very good scalability (weak scaling) of the hybrid version in the range of tens to hundreds of CPUs usages. However, further work is needed to achieve similar weak scaling to thousands of CPUs, and to tens of thousands of CPUs on larger cluster systems, such as Ranger at the Texas Advanced Computing Center.

Table 5.2 Model sizes and computational resources used in the convergence test of a benchmark problem

Element size (m)	200	100	50	25
Element number	6,166,160	24,651,088	98,985,744	419,554,200
Time step (s)	0.016	0.008	0.004	0.002
Termination Time (s)	15	15	15	15
Memory (GB)	5.9	23.4	94.0	380.0
CPUs	2	16	128	1024
Wall Clock Time (hr)	1.31	2.11	2.38	9.01

5.6 Conclusions

Using dynamic source characterization to incorporate more physics into ground motion prediction is an emerging trend in earthquake ground motion simulations. With rapid development of high performance cluster systems, parallel computing promotes deterministic ground motion simulations in the regional scale (such as southern California) up to several Hz. An explicit finite element dynamic code, EQdyna, has been verified in a community-wide code verification exercise and is under parallelization using hybrid

MPI/OpenMP approaches. It has potential to become a major tool in large-scale deterministic ground motion simulations with dynamic source characterization and parallel computing.

Acknowledgements

The author is grateful to the Texas A&M Supercomputing Facility for use of EOS and its staff for help in hybrid parallel simulations of the SCEC TPV 210 problem presented in this chapter.

References

Andrews, D. J. (1976a), Rupture propagation with finite stress in antiplane strain, J. Geophys. Res., 81, 3575-3582.

Andrews, D. J. (1976b), Rupture velocity of plane strain shear cracks, J. Geophys. Res., 81, 5679-5687.

Andrews, D. J. (1999), Test of two methods for faulting in finite-difference calculations, Bull. Seism. Soc. Am., 89, 931-937.

Andrews, D. J., Thomas C. Hanks, and John W. Whitney (2007), Physical limits on ground motion at Yucca Mountain, Bull. Seismol. Soc. Am., 97, 1771-1792, doi:10.1785/0120070014.

Belytschko, T., J. S. Ong, W. K. Liu, and J. M Kennedy (1984), Hourglass control in linear and nonlinear problems, Comput. Method Appl. Mech. Eng., 43, 251-276.

Burridge, R. (1973), Admissible speeds for plane-strain self-similar shear cracks with friction but lacking cohesion, Geophys. J. Roy. Astron. Soc., 35, 439-455.

Dalguer, L. A., and S. M. Day (2006), Comparison of fault representation methods in finite difference simulations of dynamic rupture, Bull. Seism. Soc. Am., 96, 1764-1778, doi:10.1785/0120060024.

Day, S. M. (1982), Three-dimensional simulation of spontaneous rupture: the effect of nonuniform prestress, Bull. Seism. Soc. Am., 72, 1881-1902.

Day, S. M., L. A. Dalguer, N. Lapusta, and Y. Liu (2005), Comparison of finite difference and boundary integral solutions to three-dimensional spontaneous rupture, J. Geophys. Res., 110, B12307, doi:10.1029/2005JB003813.

Day, S. M., R. Graves, J. Bielak, D. Dreger, S. Larsen, K. B. Olsen, A. Pitarka, and L. Ramirez-Guzman (2008), Model for basin effects on long-period response spectra in Southern California, Earthq. Spectra, 24 (1), 257-277, doi:10.1193/1.2857545.

Dieterich, J. H. (1979), Modeling of rock friction, 1. Experimental results and constitutive equations, J. Geophys. Res., 84, 2169-2175.

Duan, B. (2008a), Effects of low-velocity fault zones on dynamic ruptures with nonelastic off fault response, Geophys. Res. Lett., 35, L04307, doi:10.1029/2008GL033171.

Duan, B. (2008b), Asymmetric off-fault damage generated by bilateral ruptures along a bimaterial interface, Geophys. Res. Lett., 35, L04307, doi:10.1029/2008GL033171.

Duan, B. (2010a), Role of initial stress rotations in rupture dynamics and ground motion: A case study with implications for the Wenchuan earthquake, J. Geophys. Res., 115, B05301, doi:10.1029/2009JB006750.

Duan, B. (2010b), Inelastic response of compliant fault zones to nearby earthquakes, Geophys. Res. Lett., L16303, doi:10.1029/2010GL044150.

Duan, B., and D. D. Oglesby (2006), Heterogeneous fault stresses from previous earthquakes and the effect on dynamics of parallel strike-slip faults, J. Geophys. Res., 111, B05309, doi:10.1029/2005JB004138.

Duan, B., and D. D. Oglesby (2007), Nonuniform prestress from prior earthquakes and the effect on dynamics of branched fault systems, J. Geophys. Res., 112, B05308, doi:10.1029/2006JB004443.

Duan, B., and S. M. Day (2008), Inelastic strain distribution and seismic radiation from rupture of a fault kink, J. Geophys. Res., 113, B12311, doi:10.1029/2008JB005847.

Duan, B., and S. M. Day (2010), Sensitivity study of physical limits on ground motion at Yucca Mountain, Bull. Seismol. Soc. Am., 100(6), 1996-3019, doi: 10.1785/0120090372.

Duan, B., J. kang, and Y.-G. Li (2011), Deformation of compliant fault zones induced by nearby earthquakes: Theoretical investigations in two dimensions, J. Geophys. Res., 116, B03307, doi:10.1029/2010JB007826.

Guatteri, M., P. M. Mai, and G. C. Beroza (2004), A pseudo-dynamic approximation to dynamic rupture models for strong ground motion prediction, Bull. Seism. Soc. Am., 94, 2051-2063.

Harris, R. A., M. Barall, R. Archuleta, et al. (2009), The SCEC/USGS dynamic earthquake rupture code verification exercise, Seism. Res. Lett., 80, 119-126, doi:10.1785/gssrl.80.1.119.

Hughes, T. J. R. (2000), The Finite Element Method: Linear Static and Dynamic Finite Element Analysis, Dover, Mineola, N. Y.

Ida, Y. (1972), Cohesive force across the top of a longitudinal shear crack and Griffith's specific surface energy, J. Geophys. Res., 77, 3796-3805.

Kosloff, D., and G. A. Frazier (1978), Treatment of hourglass patterns in low order finite element codes, Numer. Anal. Methods Geomechan., 2, 57-72.

Liu, Q., J. Chen, S. Li, Y. Li, J. Wang, B. Guo, and S. Qi (2008), Wenchuan *M*s 8.0 earthquake: the crustal velocity structure and stress field from the western-Sichuan seismic array observations, Eos Trans. AGU, 89(53), Fall Meet. Suppl., abs. U22B-01.

Madariaga, R., K. Olsen, and R. Archuleta (1998), Modeling dynamic rupture in a 3D earthquake fault model, Bull. Seism. Soc. Am., 88, 1182-1197.

Olsen, K. B., S. M. Day, J. B. Minster, Y. Cui, A. Chourasia, D. Okaya, P. Maechling, and T. Jordan (2008), TeraShake2: Spontaneous rupture simulations of Mw 7.7 earthquakes on the Southern San Andreas fault, Bull. Seism. Soc. Am., 98 (3), 1162-1185, doi.10.1785/0120070148.

Olsen, K. B., S. M. Day, J. B. Minster, Y. Cui, A. Chourasia, M. Faerman, R. Moore, P. Maechling, and T. Jordan (2006), Strong shaking in Los Angeles expected from southern San Andreas earthquake, Geophys. Res. Lett., 33, L07305,

doi:10.1029/2005GL025472.
Ruina, A. (1983), Slip instability and state variable friction laws, J. Geophys. Res., 88, 10359-10370.
Stepp, J.C., I. Wong, J.Whitney, R. Quittmeyer, N. Abrahamson, G. Toro, R. Youngs, K. Coppersmith, J. Savy, T. Sullivan and Yucca Mountain PSHA project members (2001). Probabilistic seismic hazard analyses for ground motions and fault displacements at Yucca Mountain, Earthquake Spectra **17**, 113-151.
Wu, X., B. Duan, and V. Taylor (2011), Parallel simulations of dynamic earthquake rupture along geometrically complex faults on CMP systems, Journal of Algorithm & Computational Technology, 5(2), 313-340.

Author Informtion

Benchun Duan

Center for Tectonophysics, Department of Geology & Geophysics, Texas A&M University, TX 77843, USA.
Email: bduan@tamu.edu

Chapter 6
Load-Unload Response Ratio and Its New Progress

Xiang-Chu Yin, Yue Liu, Lang-Ping Zhang, and Shuai Yuan

This chapter presents the motivation, basic ideas, fundamental problems of Load-Unload Response Ratio (LURR) and elucidates the earthquake prediction status using LURR. Especially the new progress of LURR, including the evolution law of LURR before strong earthquake and the application of dimensional methods has been described in detail. The results of four methods (experiment, numerical simulation, analytical and the real seismic data) come to a consistent conclusion that at the early stage of seismic period LURR fluctuates around 1, then it rises swiftly and to its peak point (abbreviated to PP). The catastrophic events do not happen at the time of peak point, but lag behind the PP. The evolution law of LURR has great importance to actual earthquake prediction, since we can predict the occurrence time quantitatively (by scale of months) if we can make sure of the time of the PP. Above all, the variation of LURR could depict clearly the seismogenic process, offering more clear ideas and methods to earthquake prediction.

Keywords: Earthquake prediction, LURR, Evolution law of LURR, Peak point of LURR, Dimensional analysis

6.1 Introduction

Strong earthquakes are terrible natural disasters, which usually cause huge casualty and property loss (e.g., more than 200,000 people were killed in Haiti earthquake in 2010 and more than 80,000 people were killed in Wenchuan earthquake in 2008). For this reason, earthquake is regarded as principal one among all natural disasters.

Earthquake is also one of the most complicated natural phenomena. Many aspects of earthquake remain enigmas. But from the viewpoint of mechanics, the physical essence of earthquake is quite clear that is just an abrupt shear rupture in seismic source region accompanied with sudden release of strain energy in it. Consequently, the seismogenic process should be a damage process of the focal media leading to the abrupt shear rupture. In other words, the seismogenic process is one of damage evolutions which finally results in the occurrence of earthquake, so it is mainly a mechanical process, being quite different from that in engineering mechanics.

In the so-called engineering mechanics, the typical problem is to solve the governing equations under appropriate boundary and initial conditions, using analytical, numerical or experimental methods.

For any branch of solid mechanics, the governing equations consist of the equations of motion, the geometric equations and constitutive laws. The equations of motion and the geometric equations are the same for different branches below.
The equations of motion:

$$\begin{aligned}
\rho\frac{\partial^2 u}{\partial t^2} &= \frac{\partial\sigma_x}{\partial x}+\frac{\partial\sigma_{xy}}{\partial y}+\frac{\partial\sigma_{xz}}{\partial z}+X,\\
\rho\frac{\partial^2 v}{\partial t^2} &= \frac{\partial\sigma_{xy}}{\partial x}+\frac{\partial\sigma_{y}}{\partial y}+\frac{\partial\sigma_{yz}}{\partial z}+Y,\\
\rho\frac{\partial^2 w}{\partial t^2} &= \frac{\partial\sigma_{xz}}{\partial x}+\frac{\partial\sigma_{yz}}{\partial y}+\frac{\partial\sigma_{z}}{\partial z}+Z.
\end{aligned}\tag{6.1}$$

The geometric equations (or continuous equations):

$$\begin{aligned}
\varepsilon_x &= \frac{\partial u}{\partial x},\\
\varepsilon_y &= \frac{\partial v}{\partial y},\\
\varepsilon_z &= \frac{\partial w}{\partial z};\\
\varepsilon_{xy} &= \frac{\partial v}{\partial x}+\frac{\partial u}{\partial y},\\
\varepsilon_{yz} &= \frac{\partial w}{\partial y}+\frac{\partial v}{\partial z},\\
\varepsilon_{zx} &= \frac{\partial u}{\partial z}+\frac{\partial w}{\partial x}.
\end{aligned}\tag{6.2}$$

The constitutive laws are different for different materials. It is well known for elastic, elastic-plastic or rheological media, but for the media at the depth of seismic focus, its behavior is not very clear at present. Therefore, concerning the problems related to seismogenic process, the governing equations (e.g. the constitutive relations and damage evolution law of the focal media) and the boundary and initial conditions are difficult to know precisely. Nowadays it is just able to obtain the variations of some physical parameters, such as deformation of crust, and seismicity in terms of field measurements. The problem confronting us now is how to extract the information of damage for the focus media from the measured variations of some geophysical parameters and to predict/forecast the forthcoming earthquake in terms of these measurements.

From the microscopic viewpoint, the damage process for geo-material (rock, or rock-like) has incredible richness in complexity (Meakin, 1991; Bai et al., 1994; Krajcinnovic,

1996; Wei et al., 2000; Xia et al., 2002). In any rock block there must be a large number of disordered defects (cracks, fissures, joints, faults, caves, etc.) with different sizes, shapes and orientations. The damage process involves the nucleation and extension of micro-damages, coalescence between micro-damages and the formation of a main crack that leads to the eventual fracture. It is an irreversible, far-from-equilibrium, nonlinear, multi-scale and multi-physics one, which has been intensively studied for decades, but many fundamental prolems are still open.

From the macroscopic viewpoint, the constitutive relation (stress-strain curve) is a comprehensive description of the mechanical behavior of any materials. A typical stress-strain curve for focal media (rock) is shown in Figure 6.1. For being more universal, in Figure 6.1 the ordinate denotes general load P instead of stress σ and the abscissa is the general response R to load P instead of strain $\boldsymbol{\varepsilon}$. If the load acting on the material increases monotonously, the material will experience the regimes of elastic, damage and failure or destabilization. The most essential characteristic of the elastic regime is its reversibility; i.e., the positive process and the contrary process are reversible. In other words, the loading modulus and the unloading one are equal to each other. Contrary to the elastic regime, the damage one is irreversible, hence the loading response is different from the unloading one, or the loading modulus is different from the unloading one. This difference indicates the deterioration of material due to damage.

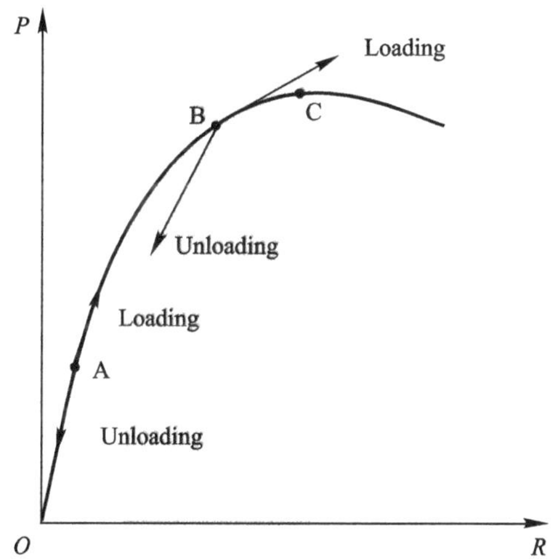

Fig. 6.1 The constitutive relation (stress-strain curve) of geo-material (rock).

In order to measure quantitatively the difference, two parameters are defined bellow. The first one is the response rate X defined as

$$X = \lim_{\Delta P \to 0} \frac{\Delta R}{\Delta P}, \tag{6.3}$$

where ΔP and ΔR denote the increments of load P and response R respectively.

The second one is the Load/Unload Response Ratio (LURR) Y,

$$Y = \frac{X_+}{X_-}, \tag{6.4}$$

where X_+ and X_- refer to response rate under loading and unloading respectively.

It is obvious that LURR should be unity ($Y = 1$) for the elastic regime due to $X_+ = X_-$ and $Y > 1$ for the damage regime due to $X_+ > X_-$. The more severely the material damages, the higher the Y value will be. Therefore, the Yvalue (LURR) could measure quantitatively the damage degree or degree in proximity to failure of media and also could act as a precursor for earthquake prediction/forecasting.

In order to calculate LURR, we have to select specific geophysical parameter as the response. The straightest one is the rigidity or compliance of the examined crust block, but it is very difficult to measure it now. Many scientists have studied the rigidity of a block which contains cracks. According to Oda (1983), this problem could be solved by the fabric tensor, but calculating the fabric tensor of a crust block needs the full information of all the cracks in it, e.g. the shape, size, orientation and status (open or closure) for all cracks (faults). It is almost impossible to get all information of all faults in any rock block in the earth at present so that we have to find another way to clear away such obstacles. It is known that a crack or seismic fault with size a corresponds to an earthquake with specified magnitude and energy release (Kanamori and Anderson, 1975). Therefore, it would be better to define the Y value directly by the released seismic energy in seismology or AE energy in laboratory as follows (Yin et al., 2000; Zhang et al., 2006):

$$Y = \frac{\left(\sum_{i=1}^{N^+} E_i^m\right)_+}{\left(\sum_{i=1}^{N^-} E_i^m\right)_-}, \tag{6.5}$$

where E denotes the radiated seismic energy which can be calculated from magnitude M according to the Gutenberg-Richter formula (Kanamori and Anderson, 1975), the sign "+" means loading and "–" unloading and the parameter m could be selected as $0, 1/3, 1/2, 2/3$ or 1. When $\mathrm{m} = 1, E^m$ is exactly the energy itself; $m = 1/2, E^m$ denotes the Benioff strain; $m = 1/3, 2/3, E^m$ represents the linear scale and area scale of the focal zone respectively; $m = 0, Y$ is equal to N^+/N^-, where N^+ and N^- denote the number of earthquakes occurring during the loading and unloading duration respectively. We adopt $m = 1/2$ in this chapter and it is the most of our works.

In order to predict earthquakes in terms of LURR, another problem should be solved. That is how to load and unload the crustal block with size of hundreds or even thousands of kilometers. One of the measures is the earth tide. Tidal forces exerted by the moon and sun produce continuously varying stresses in the earth's crust. How to calculate the tide-induced stress in the crust has been elucidated in earlier LURR researches (Yin, 1987; Yin and Yin, 1991; Yin, 1993; Yin et al., 1994a,b; Maruyama, 1995; Yin et al., 1995; 2000).

According to the results of rock mechanics (Jaeger and Cook, 1979), the Coulomb criterion was used to judge loading and unloading.

6.2 The Status of Earthquake Prediction Using LURR

After the proposal of the LURR idea and solving some basic problems such as the measurements of loading and unloading the crust block and calculation LURR, the first task for us is to retrospect inspections of historical earthquake cases. The retrospective inspections of hundreds of earthquake cases have been conducted (Yin et al., 1995; Yin et al., 2000). The results from these cases validated the LURR method. For more than 80% examined cases, the Y value appears high value (significantly larger than 1) before the forthcoming strong earthquakes.

Although the case inspections are satisfactory and the laboratory modeling (Shi et al., 1994; Yin et al., 2004a; Yin *et al.,* 2004b; Zhang et al., 2006) and numerical simulation (More et al., 2002; Wang et al., 2004) also proved its validity, LURR theory has to be tested ultimately in real earthquake prediction practice. We have applied LURR to real earthquake prediction practice since 1993. At the early stage, some cases of our prediction were successful and the others failed (Yin et al., 2000). It is exhilarating that the prediction results using LURR have been getting better and better. In 2004–2007, 95% earthquakes with magnitude $M \geqslant 5$ in the mainland of China occurred in the predicted area (Institute of Earthquake Science, China Earthquake Administration, 2008) in terms of LURR except for earthquakes in those areas where data were scarce so that LURR could not be calculated, which are shown in Table 6.1.

It could be concluded from Table 6.1 that from 2004 to 2007, the annual prediction using LURR is pretty good. This result was confirmed by the Institute of Earthquake Science, China Earthquake Administration (Institute of Earthquake Science, China Earthquake Administration, 2008).

Table 6.1 The count of earthquakes with $M_L \geqslant 5$ occurring in the mainland of China in every year from 2004 to 2007 respectively and the count of earthquakes with $M_L \geqslant 5$ occurring in the predicted area of LURR in the same period. Statistics does not include figures for earthquakes occurring in data scarcity regions where the data are unavailable to calculate LURR

Years	Earthquakes with $M_L \geqslant 5$ in every year	Earthquakes with $M_L \geqslant 5$ in the anomaly area of LUR R	Percentage
2004	17	15	88%
2005	13	12	92%
2006	9	9	100%
2007	12	12	100%

Figure 6.2 shows the maps of the LURR anomaly regions in the mainland of China calculated in the ends of 2003, 2004, 2005 and 2006, and also the epicenters distribution of earthquakes with magnitude $M_L \geqslant 5$ occurring in the following years (2004, 2005, 2006 and 2007).

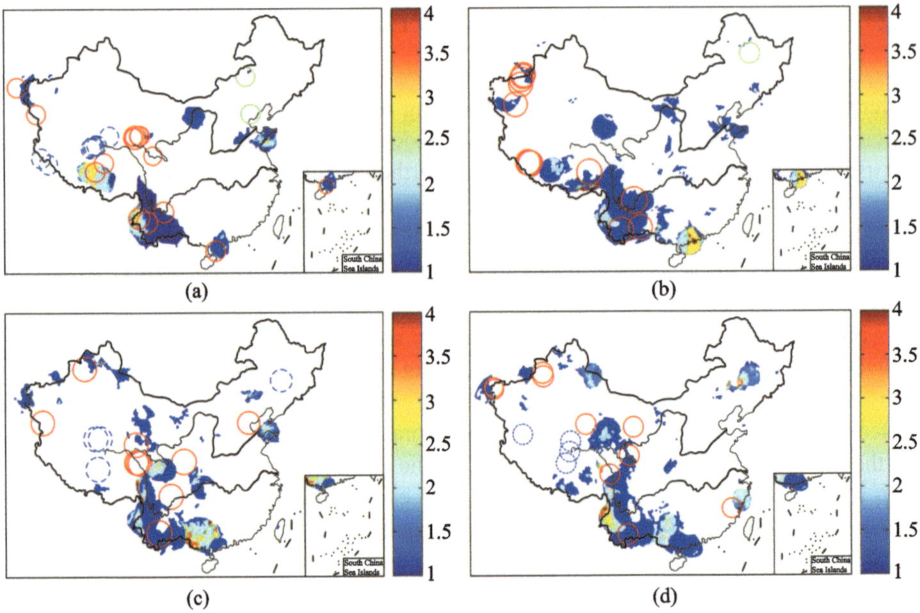

Fig. 6.2 The LURR anomaly regions in the mainland of China at the ends of 2003(a), 2004(b), 2005(c)and 2006(d) and strong earthquakes ($M \geqslant 5$) occurring in the mainland of China in 2004(a), 2005(b), 2006(c) and 2007(d) respectively. The solid line circles denote earthquakes ($M \geqslant 5$) in regions where the data is available to calculate LURR and dash line circles denote earthquakes ($M \geqslant 5$) in data scarcity regions where the data are unavailable to calculate LURR. Among the solid line circles, the earthquakes occurring in the predicted area of LURR are denoted by red thick circles and the earthquakes occurring in the unpredicted area of LURR are denoted by fine green circles.

6.3 Peak Point of the LURR and Its Significance

In recent years, we figured out the evolution laws of LURR before strong earthquakes by a lot of means. The a, b, c, d of Figure 6.3 mean respectively: (a) evolution of LURR vs. time in an acoustic emission experiment for rock sample; (b) evolution of LURR before the October 17, 1989 Loma Prieta Earthquake from actual earthquake data; (c) damage evolution of non-uniform brittle medium simulated by network model and evolution of LURR with time (Liang et al., 1998; Zhang, 2009); (d) the damage evolution of non-uniform brittle medium simulated with Lyakhovsky's model and analytic result of LURR (Lyakhovsky et al., 1997, 2001; Zhang, 2009). The arrows indicate catastrophic events (earthquakes or catastrophic failure of specimen). The results of the four methods come to a consistent conclusion that at the early stage of seismogenic regime, LURR fluctuates around 1, then it rises swiftly and reaches to its PP. The catastrophic events do not happen at the peak point, but the catastrophic events lag behind the PP. The lagged time is denoted by T_2, the time from the beginning of LURR anomaly to the PP is denoted by T_1 and the total abnormal time of LURR is called T,

$$T = T_1 + T_2. \tag{6.6}$$

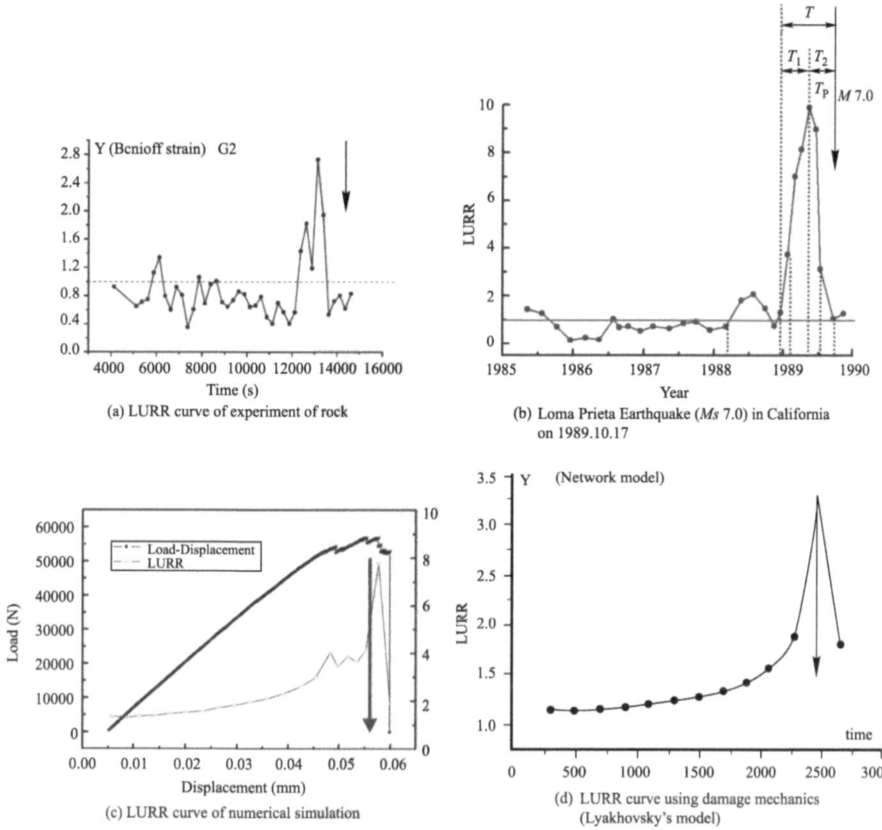

Fig. 6.3 The evolution of LURR before strong earthquake or catastrophic rupture in experiment.

According to research (Zhang , 2006),

$$T = 80(1 - 2.5 \times 10^{-0.09M}), \tag{6.7}$$
$$T_2 = 60(1 - 2.3 \times 10^{-0.08M}), \tag{6.8}$$

where M denotes the earthquake magnitude and T scales with month. It is indicated from Table 6.2 that T_2 is quite a long period, e.g. T_2 is about 14 months for an earthquake with magnitude 6 and T_2 is even more than 2 years for an earthquake with magnitude 7.

The above results have great importance to actual earthquake prediction, since we can predict the occurrence time quantitatively (by scale of months) if we can make sure of the time of the PP. Above all, the variation of LURR could depict clearly the seismogenic course, offering more clear methods and approach to earthquake prediction.

However, it also brings about some complexity. To predict the future earthquake, we should not only consider the value of LURR at one time window, but also study the spatio-temporal evolution of LURR systematically. Take the point T_{f2} in Figure 6.2b as an example, even though the value of LURR is much higher than 1, there is still a relatively long time to the future earthquake, since it is before the peak point, but at point T_{b2} LURR is smaller than T_{f2} (sometimes it is even smaller than 1, which means LURR has returned to "normal"). Actually it is right at the future earthquake because it has passed the PP. So, in order to predict earthquake reliably, we have to trace the variation of LURR, especially discover the emergence of the PP. Then according to formula 6.2, we can calculate T_2 for different magnitude of earthquakes, which are shown in Table 6.2. This Table tells that T_2 is a quite long time for large earthquakes. For example, T_2 could be 36 months (28 ± 8) for earthquakes of magnitude $M8$. It means the earthquake does not happen at the PP (where LURR reaches the highest value), then LURR decreases (sometimes even smaller than 1), and the earthquake happens several years later. This was exactly the case of 2008 Wenchuan Earthquake, although we had discovered anomaly of this area for a long time, but we did not make the short-time prediction successfully due to the misunderstanding that after the elapse of time for 2 years from the PP we thought the anomaly of LURR in that region might be false.

Table 6.2 T, T_1 and T_2 of different magnitudes of earthquakes

Magnitude	T(month)	T_1(month)	T_2(month)
5	9	4	5 ± 2
6	22	8	14 ± 4
7	33	11	22 ± 6
8	42	14	28 ± 8
9	49	15	34 ± 10?

6.4 Earthquake Cases in 2008–2009

Predicted T_2 which is calculated by formula 6.8 (noted by T_{2p}) and actual T_2 (noted by T_{2a}) for the earthquake cases in 2008 and 2009 occurring in the mainland of China are listed in Table 6.3 and Table 6.4.

At the same time, the differentials $\Delta T_2 = T_{2p} - T_{2a}$ are also listed in Tables 6.3 and 6.4. When ΔT_2 is positive, it means T_{2p} is larger than T_{2a}, namely, the actual earthquake is earlier than predicted one and vice versa. From Table 6.3 we can see the biggest ΔT_2 occurred in Yutian and Wuqia earthquakes, both of which occurred in Xinjiang. In Table 6.4, the maximum ΔT_2 happens in the Atushi earthquake, in Xinjiang. So we recognized that T_2 depends on not only the earthquake magnitude but something else (e.g. geological regimes) needed to be involved. Except for these three earthquakes, the differentials ΔT_2 are very small, which means the prediction is pretty good. On the other hand, the differentials ΔT_2 of the 13 earthquakes in 2008–2009 are all positive except the earthquake in Haixi in Qinghai, which means there existed systematic errors in our prediction and which could be reduced or eliminated.

Table 6.3 The contrast between T_{2p} (the predicted T_2) and T_{2a} (the actual T_2) for earthquakes occurring in the mainland of China in 2008

Location	Time	Place	Magnitude(M)	T_{2a}(month)	T_{2p}(month)	ΔT_2(month)
Gaize	2008-01-09	32.5°N, 85.2°E	6.9		Scarce data	
Yutian	2008-03-21	35.6°N, 81.6°E	7.3	8	24	16
Wenchuan	2008-05-12	30.95°N, 100°E	8.0	26	24	−2
Zhongba	2008-08-25	31.0°N, 83.6°E	6.8		Scarce data	
Panzhihua	2008-08-30	26.2°N, 101.9°E	6.1	14	15	1
Wuqia	2008-10-05	39.5°N, 73.9°E	6.8	13	20	7
Dangxiong	2008-10-06	29.8°N, 90.3°E	6.6	18	19	1
Haixi	2008-11-10	37.6°N, 95.9°E	6.3	19	17	−2

Annotation: T_{2a} (month)—actual T_2 ; T_{2p}(month)—predicted T_2;
ΔT_2 (month) = $T_{2p} - T_{2a}$; The special window (radius) is 200 km.

Table 6.4 The contrast between T_{2p} (the predicted T_2) and T_{2a} (the actual T_2) for earthquakes occurring in Chinese mainland in 2009

Earthquakes	Internet	Catalog	Epicenter	Occurrence Time	Peak Point	T_{2a}	T_{2p}	ΔT_2
Chabuchaerxibo	5	5.4	43.3°N, 80.9°E	2009-1-25	2007-11-30	14	5	2
Keping	5.2	5.6	40.7°N, 78.7°E	2009-2-20	2007-10-31	16	7	2
Huichun	5.3	5.6	42.7°N, 130.7°E	2009-4-18				
Aheqi	5.5	5.8	41.3°N, 78.3°E	2009-4-19	2007-10-31	18	10	1
Atushi	5	5.4	40.1°N, 77.4°E	2009-4-22	2007-10-31	18	5	6
Yecheng	5.2	5.6	36.4°N, 77.6°E	2009-5-21	2008-3-31	14	7	0
Yaoan	6	6.3	25.6°N, 101.1°E	2009-7-9	2007-6-30	24	14	2
Nima	5.6	5.9	31.3°N, 86.1°E	2009-7-24				
Qinghai	6.4	6.6	37.6°N, 95.8°E	2009-8-28	2008-1-31	19	18	0

Annotation: T_{2a} (month)—actual T_2; T_{2p} (month)—predicted T_2;
ΔT_2 (month) = $T_{2p} - T_{2a}$; The special window (radius) is 200 km.

6.5 Improving the Prediction of Magnitude M and T_2-Application of Dimensional Method

Integrating with the dimensional method (Buckingham, 1914; Sedov, 1959), LURR could be a hopeful methodology of earthquake prediction. An earthquake prediction should include three parts: prediction of the location, magnitude and occurrence time of the predicted earthquake.

6.5.1 Location

The location of the predicted earthquake should fall into the LURR anomaly region.

6.5.2 Magnitude

The equivalent physical parameter E_s of magnitude M is the radiated energy of an earthquake. They are related by Gutenberg formula ($\log E_s = 4.8 + 1.5M_s$). E_s could be involved with the following parameters: J_{PP}, E_a, γ_r and h.

(i) J is defined as

$$J(t) = \iint\limits_{Y \geqslant 1} Y dxdy = Y_a \cdot A, \tag{6.9}$$

which is used to denote the LURR anomaly region weighted with Y (LURR) and represents the expanse and degree of the seismogenic zone (anomaly region of LURR) during a specific time window [from $(t - t_w)$ to t]. In formula 6.9, A is the area of LURR anomaly

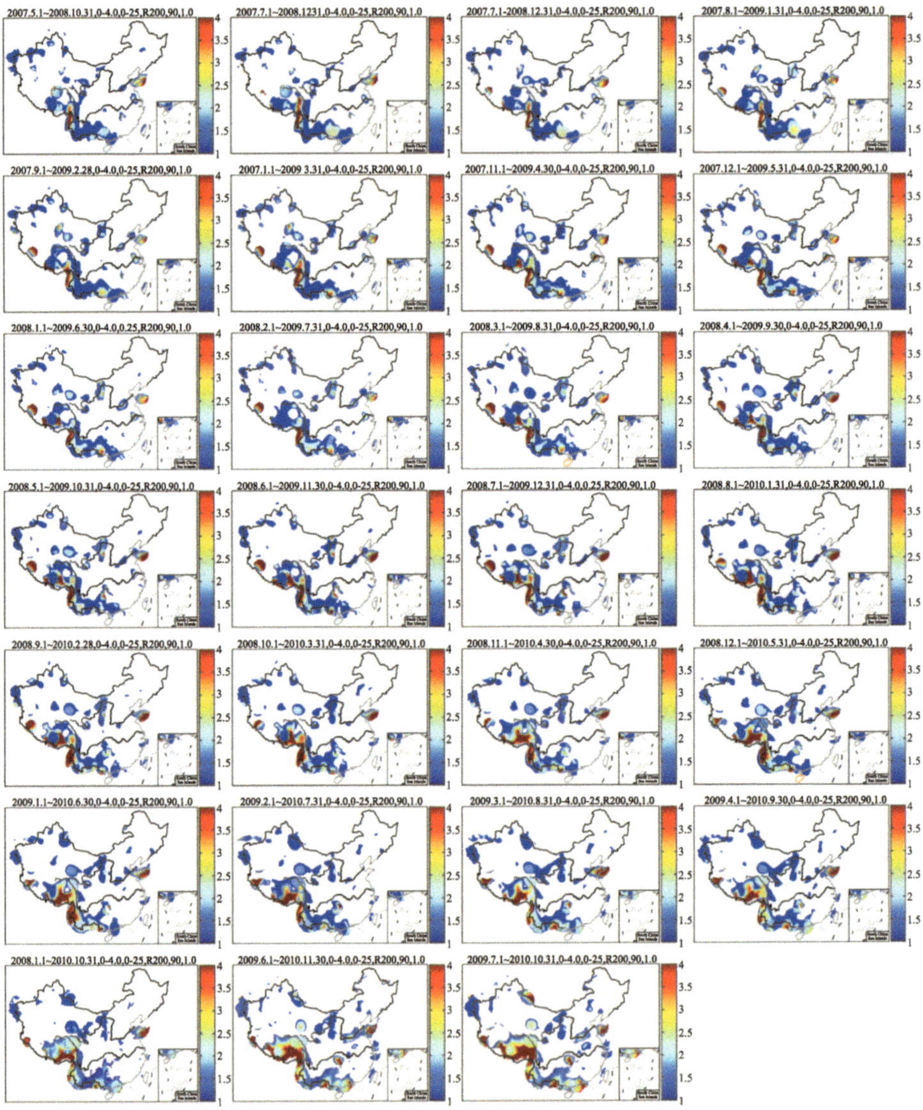

Fig. 6.4 The map of LURR scanning results in the mainland of China in recent years.

region and Y_a is the average value for the whole LURR anomaly region. J_{PP} means the value of J at peak point or the maximum of J. In order to get the value of J_{PP}, the spacial scans of LURR in the mainland of China should be conducted for a serious time window at first (Figure 6.4) and then calculated $J(t)$ according to expression 6.9. As an example, Figure 6.5 shows us the curve of $J(t)$ *for* Kaifeng-Heze region at the border between Henan Province and Shandong Province in China.

(ii) E_a is the sum of radiated energy of all earthquakes occurred in a specific region per year and per area measured during a long time duration. The radiated energy of an earthquake e_r is just a portion of the whole consumed energy of the earthquake e_t. In this context, $e_r = \eta e_t$ *and* η is named efficiency of earthquake in some books. Undoubtedly, η is less than unit and assumed roughly as constant. If the duration is long enough, we can consider that the cumulated deformation energy and the consumed energy should be balanced against each other. In this chapter we use the catalog from 1900 to 2009. That means the duration is 110 years. The distribution of E_a in the mainland of China is shown in Figure 6.6.

(iii) γ_r is the shear strain rate *in situ*. The distribution of γ_r in the mainland of China can be obtained from the measured results of GPS (Shen et al., 2003; Gu et al., 2001; Li, 2004).

(iv) The thickness of the seismogenic body is denoted with h and the volume of the seismogenic body is $A \cdot h$. According to the dimensional formula

$$[E_s] = [E_a]^{\alpha 1} \cdot [J_{PP}]^{\alpha 2} \cdot [h]^{\alpha 3} \cdot [\gamma_r]^{\alpha 4}. \tag{6.10}$$

In which the square brackets [ ] mean dimension, it is obtained that

$$E_s \propto E_a \cdot J_{PP}^{2/3} \cdot h^{2/3} / \gamma_r = E_a \cdot Y_a^{2/3} \cdot A_{PP}^{2/3} \cdot h^{2/3} / \gamma_r .$$

We introduce a nondimensional quantity β,

$$\beta = h / A^{1/2}.$$

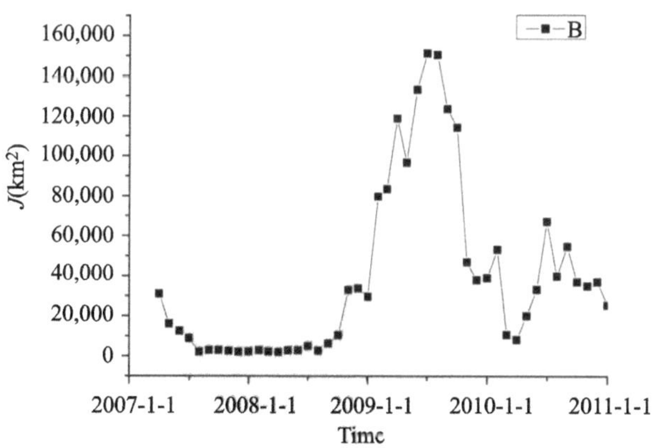

Fig. 6.5 The curve of $J(t)$ *for* Kaifeng-Heze region at the border between Henan Province and Shandong Province in east China.

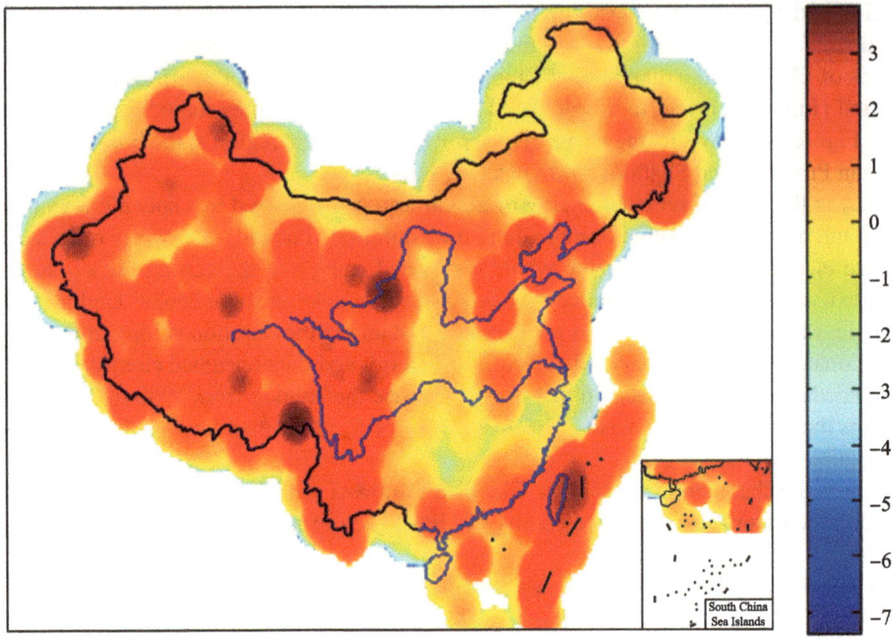

Fig. 6.6 The distribution of E_a in the mainland of China. The color scale indicates the equivalent magnitude for E_a

Then $E_s \propto E_a/\gamma_r \cdot J_{PP}^{2/3} \cdot (\beta \cdot A^{1/2})^{2/3} = \beta^{2/3} \cdot E_a \cdot J_{PP}/\gamma_r$, and $E_s \cdot \gamma_r/E_a \cdot J_{PP} \propto \beta^{2/3}$ $E_s \cdot \gamma_r/(E_a \cdot J_{PP})$ is a nondimensional quantity. According to the tradition of dimensional analysis, it is denoted with π_1.

$$\pi_1 = E_s \cdot \gamma_r/(E_a \cdot J_{PP}), \tag{6.11}$$

and $\pi_1 \propto \beta^{2/3}$. Obviously β is related to magnitude M, So that,

$$\pi_1 = f_1(M) \tag{6.12}$$

Furthermore, we rewrite $\pi_1 = E_s \cdot \gamma_r/(E_a \cdot J_{PP}) = E_s/E_d$, here

$$E_d = E_a \cdot J_{PP}/\gamma_r \tag{6.13}$$

Fitting the data of about 50 earthquake cases in the mainland of China (Figure 6.7), it is obtained that

$$\pi_1 = 1\text{E}-17\text{e}^{3.006Ms} \quad (R^2 = 0.96), \tag{6.14}$$

and then

$$M_s = 5.14\lg E_d - 112.08. \tag{6.15}$$

The magnitude of the predicted future earthquake can be calculated from fomula (6.15) as long as we get the value of E_d.

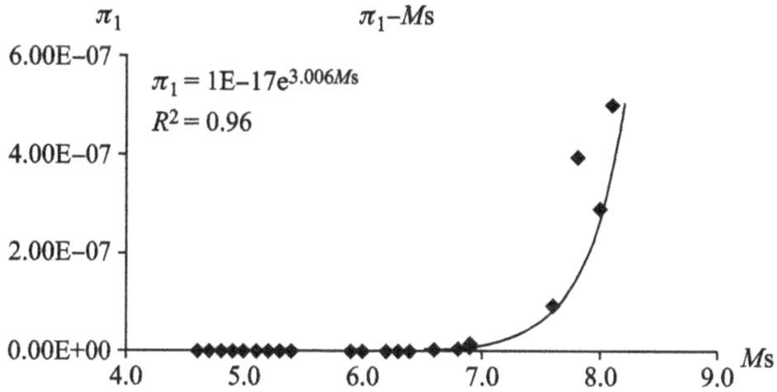

Fig. 6.7 The curve of π_1 *vs.* magnitude *M*.

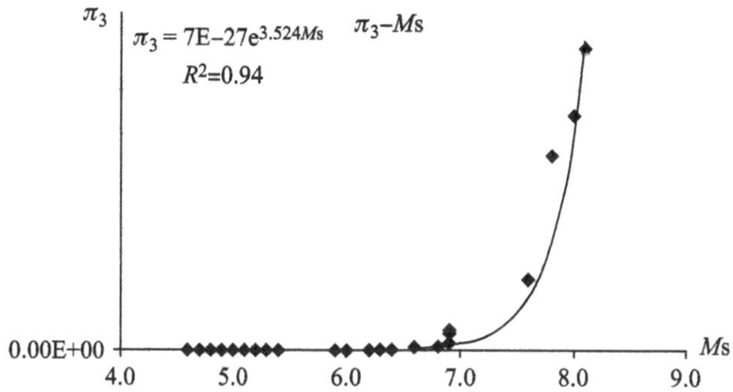

Fig. 6.8 The curve of π_2/π_1 *vs.* magnitude *M*.

6.5.3 Occurrence time (T_2)

Obviously, T_2 times γ_r is also a non-dimensional quantity, which denoted as π_2

$$\pi_2 = T_2 \cdot \gamma_r. \tag{6.16}$$

Of course,

$$\pi_1 \cdot \pi_2 = \pi_3, \tag{6.17}$$

π_3 is another non-dimensional quantity. Fitting the same data of Figure 6.7, it is obtained that

$$\pi_3 = 7E - 27e^{3.524Ms} \quad (R^2 = 0.94), \tag{6.18}$$

and then

$$T_2 = 8.5E_d \cdot 10^{0.03M} \cdot 10^{-30.8}/\gamma_r \cdot \pi_1. \tag{6.19}$$

The unit of T_2 is day and the unit of γ_r is 10^{-9} rad/yr (Figure 6.8).

As examples of this methodology, the magnitude of the forthcoming earthquake in Kaifeng-Heze region will be probably *M*5.7 and with regard to the expansive south-western region (including Tibet, Qinghai and Yunnan) the magnitude of the future earthquake even will be above *M*8.

6.6 Conclusions

Since the proposal of LURR, more than 2 decades have elapsed (Yin et al., 2006). A lot of achievements have been made on LURR in such a long period, but there still exist many problems and also enough improving room, e.g. evolution law of LURR after PP which may be related to its application to short-term earthquake prediction.

Acknowledgements

This research was funded by the National Natural Science Foundation of China (Grant No. 11021262, 19732060 and 40004002) and Informalization Construction Project of Chinese Academy of Sciences during the 11th Five-Year Plan Period (No.INFO-115-B01). The earthquake catalog used in this chapter is provided by CENC (China Earthquake Networks Center), China Earthquake Administration. The calculation in the chapter was conducted partly in Supercomputing Center of Computer Network information Center, Chinese Academy of Sciences.

References

Bai Y.L., Lu C.S., Ke F.J. et al. 1994, Evolution induced catastrophe, Physics Letters A, 185, 196-201.

Buckingham E. 1914, On physically similar systems: illustrations of the use of dimensional analysis, Physical Review, 4, 345-376.

Gu G.H., Shen X.H., Wang M., Zheng G.M., Fang Y., Li P. 2001, General characteristics of the recent horizontal crustal movement in Chinese mainland, Acta Seismologica Sinica, 23 (4): 362-369.

Institute of Earthquake Science, China Earthquake Administration. 2008, The researches on prediction of the strong earthquake tendency for Chinese mainland in 2009, in The Research on Prediction of the Strong Earthquake Tendency in Chinese Mainland for 2009, Seismological Press, Beijing (in Chinese).

Jaeger J.C., Cook N.G.W. 1979, Fundamentals of Rock Mechanics, Chapman and Hall, London

Kanamori H., Anderson D.L. 1975, Theoretical basis of some empirical relation in seismology, Bull. Seism. Soc. Am., 65, 1073-1096.

Krajcinnovic D. 1996, Damage Mechanics, Elsevier, Amsterdam.

Li Y.X., Li Z., Zhang J.H. et al. 2004, Horizontal strain field in the Chinese mainland and its surrounding areas, Chinese Journal of Geophysics, 47 (2): 222-231 (in Chinese).

Liang N.G., Liu H.Q., Wang T.C. 1998, A meso elastoplastic constitutive model for polycrystalline metals based on equivalent slip systems with latent hardening, Science in China, Ser. A 41, 887-896.

Lyakhovsky V., Ben-Zion Y., Agnon A. 1997, Distributed damage faulting and friction, Journal of Geophysical Research, 102, 27635-27649.

Lyakhovsky V., Ben-Zion Y., Agnon A. 2001, Earthquake cycle, fault zones, and seismicity patterns in a rheologically layered lithosphere, Journal of Geophysical Research, 106, 4103-4120.

Maruyama T. 1995, Earthquake prediction in China, *Zisin*, 19, 69-75 (in Japanese).

Meakin P. 1991, Model for material failure and deformation, Science, 252, 226-234.

Mora P., Wang Y., Yin C., Place D., Yin X.C. 2002, Simulation of load-unload response ratio and critical sensitivity in the Lattice Solid Model, Pure and Applied Geophysics, 159, 2525-2536.

Oda M. 1983, A method for evaluating the effect of crack geometry on the mechanical behavior of cracked rock mass, Mech Mater, 2, 163-171.

Sedov L.I. 1959, Similarity and Dimensional Methods in Mechanics, Infosearch Ltd., London.

Shen Z.K., Wang M., Gan W.J., Zhang Z.S. 2003, Contemporary tectonic strain rate field of Chinese Continent and its geodynamic implications, Earth Science Frontiers, Vol.10 suppl., 93-100.

Shi X.J., Xu H.M., Wan Y.Z., Lu Z.G., Chen X.Z. 1994, The characteristic of rock fracture under simulated tide force: laboratory study on the theory of loading and unloading response ratio, Chinese Journal of Geophysics, 37, 437-442.

Wang Y., Mora P., Yin C., Place D. 2004, Statistic tests of load-unload response ratio signals by lattice solid model: implication to tidal triggering and earthquake prediction, Pure and Applied Geophysics, 161, 1829-1839.

Wei Y.J., Xia M.F., Ke F.J., Yin X.C., Bai Y.L. 2000, Evolution induced catastrophe and its predictability, Pure and Applied Geophysics, 157, 1929-1943.

Xia M.F., Wei Y.J., Ke J.F., Bai Y.L. 2002, Critical sensitivity and trans-scale fluctuations in catastrophic rupture, Pure and Applied Geophysics, 159, 2491-2509.

Yin X.C. 1987, A new approach to earthquake prediction, Earthquake Research in China, 3, 1-7 (in Chinese with English abstract).

Yin X.C. 1993, A new approach to earthquake prediction, Russia's "Nature", 1, 21-27 (in Russian).

Yin X.C., Chen X.Z., Song Z.P., Yin C. 1994b, The load-unload response ratio theory and its application to earthquake prediction, Journal of Earthquake Prediction Research, 3, 325-333.

Yin X.C., Chen X.Z., Song Z.P., Yin C. 1995, A new approach to earthquake prediction: the load/unload response ratio (LURR) theory, Pure and Applied Geophysics, 145, 701-715.

Yin X.C., WANG Y.C., Peng K.Y., Bai Y.L., Wang H.T., Yin X.F. 2000, Development of a new approach to earthquake prediction—load/unload response ratio (LURR) theory, Pure and Applied Geophysics, 157, 2365-2383.

Yin X.C., Yin C. 1991, The precursor of instability for nonlinear system and its application to earthquake prediction, Science in China, 34, 977-986.

Yin X.C., Yin C., Chen X.Z. 1994a, The precursor of instability for nonlinear system and its application to earthquake prediction—the load-unload response ratio theory, in Non-linear Dynamics and Predictability of Geophysical Phenomena (eds. Newman

W.I., Gabrelov A., Turcotte D.L.), Geophysical Monograph 83, Iugg Volume 18, 55-60.
Yin X.C., Yu H.Z., Kukshenko V., Xu Z.Y., Wu Z.S., Li M., Peng K.Y., Elizarov S., Li Q. 2004, Load-unload response ratio (LURR), accelerating energy release (AER) and state vector evolution as precursors to failure of rock specimens, Pure and Applied Geophysics, 161 (11-12): 2405-2416.
Yin X.C., Zhang H.H., Yu H.Z., Zhang Y.X., Peng K.Y. 2004, Prediction of seismic tendency of Chinese mainland in 2004 in terms of LURR, in The research on seismic tendency of China in 2005, 282-285 (edited by Center for Analysis and Prediction, CSB), Seismological Press, Beijing (in Chinese).
Yin X.C., Zhang L.P., Zhang, H.H., YIN C., Wang Y.C., Zhang Y.X., Peng K. Y., Wang, H.T., Song Z.P., Yu H.Z., Zhuang J.C. 2006, LURR's Twenty Years ans Its Perspective. Pure and Applied Geophysics 163(11-12), 2317-2341.
Zhang H.H., Yin X.C., Liang N.G., Yu H.Z., Li S.Y., Wang Y.C., Yin C., Kukshenko V., Tomiline N., Elizarovs S. 2006, Acoustic emission experiments of rock failure under load simulating the hypocenter condition, Pure and Applied Geophysics, 163 (11-12), 2389-2406.
Zhang L.P. 2009, Study on damage evolution of heterogeneous brittle media in seismogenic conditions and earthquake prediction, PhD Thesis of The Graduate University of the Chinese Academy of Sciences (in Chinese).

Author Information

Xiang-chu Yin

Institute of Earthquake Science, China Earthquake Administration, Beijing 100036, China.

State Key Laboratory of Nonlinear Mechanics, Institute of Mechanics, Chinese Academy Sciences, Beijing 100190, China.

Institute of Geophysics, China Earthquake Administration, Beijing 100081, China.

E-mail: xcyin@public.bta.net.cn

Yue Liu, Shuai Yuan

State Key Laboratory of Nonlinear Mechanics, Institute of Mechanics, Chinese Academy Sciences, Beijing 100190, China.

Lang-ping Zhang

Institute of Earthquake Science, China Earthquake Administration, Beijing 100036, China.

State Key Laboratory of Nonlinear Mechanics, Institute of Mechanics, Chinese Academy Sciences, Beijing 100190, China.

Chapter 7
Discrete Element Method and Its Applications in Earthquake and Rock Fracture Modeling

Yucang Wang, Sheng Xue, and Jun Xie

This chapter introduces the Esys_Particle, the open source code of Discrete Element Method (DEM), and its evolution history and basic features of the model. We outline the recent developments of the Esys_Particle code, including incorporation of single particle rotation, new contact law, and parameter calibration, parallel algorithm, coupling of thermal and hydrodynamic effects. We discuss the major differences between our model and most current DEMs. Some numerical simulations of rock fracture, earthquake process, fault evolution, hydrothermal effect coupling and a full solid-fluid coupling are presented in this chapter.

Keywords: Discrete element method, the Esys_Particle, earthquake simulation, rock fracture

7.1 Introduction

Earthquake prediction still remains one of the most difficult problems despite extensive research efforts. One of the major reasons is that the nonlinear physics of earthquakes is not well understood yet. It has been recognised that the dynamics processes associated with earthquakes are characterized by a multiplicity of spatial and temporal scales (Rundle et al., 2002). Essentially, an earthquake process is either a slip instability controlled by friction (Brace and Byerlee, 1966), or a rapidly occurring fracture of the fresh rocks in the interior of the earth (Knopoff, 1993). The traditional researches focus on theoretical analyses, observations and laboratory experiments. However, these methods suffer some limitations. For example, the analytical researches are mostly limited to very simple cases (i.e. a single, homogeneous fault) (Aki and Richards, 1980). Currently it is still very difficult to study analytically the whole earthquake occurrence (spatial and temporal distribution of seismicity) with a large number of coupled fault systems in a geologically complex area because a complete theory to predict precisely the failure of brittle solids and capture the complex behaviour of friction phenomenon has not yet come out.

With the present great capacities of supercomputer and massively parallel machines, numerical simulations offer an alterative means of study. In the past decades, some simple conceptual models, such as the spring-block model (Burridge and Knopoff, 1967; Carlson and Langer, 1989; Carlson et al., 1989), cellular automata (Barriere and Turcotte, 1991; Lomnitz-Adler, 1993; Sammis and Smith, 1999), and SOC model (Bak and Tang, 1989), have been proposed to study the general statistical behaviour of the earthquake events in a single fault. Although it may be possible to gain useful insight by ignoring the detailed mechanisms of earthquakes, only a complete physico-based model including the most important ingredients can provide a complete picture. Some of physico-based earthquake models are proposed using different methodologies: finite element method (Bird, 1978; Melosh and Williams, 1989; Tang, 1997; Oglesby and Day, 2001; Li and Liu, 2006; Xing et al., 2007), finite difference (Day, 1982; Madariaga et al., 1997) and boundary integral equation method (Fukuyama and Madariaga, 1995). In modelling the earthquake phenomena, these continuously based models encounter limitations when large-scale slip and opening of a large amount of fractures must be considered since they have an implicit representation of the discontinuities.

An alternative to these continuous approaches is the Discrete Element Method (DEM, Cundall and Strack, 1979), which has been a powerful numerical tool and extensively used in many scientific and engineering problems. The basic idea behind DEM is to represent the material to be modelled as an assemblage of discrete particles interacting with one another. The precise nature of the interaction depends on the scale of interest and the details of the simulation. At each time step, the calculations performed in DEM alternate between integrating equations of motion for each particle, and applying the force-displacement law at each contact, which updates the contact forces based on the relative motions between two particles and their relevant contact stiffness. The unique discrete nature of DEM allows it to study with great simplicity many problems which are highly dynamic with large deformations and a large number of frequently changing contacts.

There are different kinds of DEMs. The un-bonded (cohesionless) DEMs permit no tensile forces transmitted between particles, and are used to model motions of power and behaviour of particulate materials (Morgan et al., 1999a; 1999b; Sheng et al., 2004; Sitharam, 2000). In the bonded (cohesive) DEMs, however, particles are bonded so that tensile forces can be transmitted. It is often used to model wave propagation and fracture of intact materials such as rocks (Mora et al., 1993; 1994; 1998; Place et al., 1999; 2002; Toomey and Bean, 2000; Wang et al., 2000; Chang et al., 2002; Hazzard et al., 2000a, 2000b; Potyondy et al., 1996; 2004; Donze et al., 1997; Hentz et al., 2004a; 2004b).

Currently, two commercial DEM softwares are available: PFC and EDEM. There are plenty of users developed DEM codes. Among them are two open source codes: Esys_Particle and Yade. To our knowledge, Esys_Particle is the earliest bonded DEM to model the earthquake-related phenomena, and has been developed since 1993.

In this chapter, we outline the recent development of the Esys_Particle. In the flowing sections, a brief introduction to the Esys_Particle is first introduced. A recent theoretical development, including the incorporation of single particle rotation, new contact law, parameter calibration, and coupling of thermal and hydrodynamic effects, will follow, and finally some numerical simulations of rock fracture, earthquake process, fault evolution, hydrothermal effect coupling and a full solid-fluid coupling, are presented.

7.2 A Brief Introduction to the Esys_Particle

The ESyS_Particle, previously called Lattice Solid Model or LSMearth, is an open source software developed by the Earth Systems Science Computational Centre (ESSCC), the University of Queensland. Motivated by short-range Molecular Dynamics (MD, Allen et al., 1987; Rapaport, 1995), it is similar to the Discrete Element Model (Cundall et al., 1979), but involves a different computational approach. It was designed to provide a basis to study the physics of rocks and the nonlinear dynamics of earthquakes, specifically to address the computational limitations of existing DEM software. The ESyS_Particle has been recently extended to include single particle rotation and a full set of interactions between particles. The major features that distinguish the ESyS_Particle from existing DEMs are the explicit representation of particle orientations using unit quaternion, complete interactions (six kinds of independent relative movements are transmitted between two 3-D interacting particles) and a new way of decomposing the relative rotations between two rigid bodies so that the torques and forces caused by such relative rotations can be uniquely determined (Wang, et al., 2006; Wang, 2009; Wang and Alonso-Marroquin, 2009a). It has been successfully utilised in the study of physical process such as rock fracture (Mora and Place, 1993; Place et al., 2002), stick-slip friction behaviour (Mora et al., 1994; Place et al., 1999), granular dynamics(Mora et al.,1998;1999), heat-flow paradox (Mora et al., 1998;1999), localization phenomena (Place et al., 2000), Load-Unload Response Ratio theory (Mora et al., 2002b; Wang et al., 2004), Critical Point systems (Mora et al., 2002a), comminution in shear cells, silo flow, rock fragmentation, and fault gouge evolution.

Written using C++, the ESyS-Particle is designed for execution of parallel supercomputers, clusters or multi-core PCs running a Linux-based OS. The simulation engine implements spatial domain decomposition via the Message Passing Interface (MPI). A Python script interface provides flexibility for the users. Before running the code, the initial conditions, physics parameters, integration steps, types of particles (simple or rotational particle), types of loading walls, the contact properties (elastic, frictional, bonded contacts), artificial viscosity, ways of loading (force controlled or displacement controlled) and output fields are specified in the script.

Pre-processing includes a particle generation package, which can generate regular or random-sized particles and bond information if need be. Aggregates or grains, gouges and faults can also be made. Post-processing includes Povray and VTK visualization packages, which can visualize the article and fields (velocity, displacements).

In the past 18 years, the evolution of the Esys_Particle code experienced the following major milestones:

(i) Early stage of Atomic Lattice Solid Model (Mora and Place, 1993; 1994). It was a 2-D Fortran code based on molecular dynamics principles. In the simplified model, there is interaction between particles only in radical direction, and no intrinsic friction between particles exists.

(ii) C++ LSMearth code (Place et al., 2002). It was an objected oriented approach where micro-physics can be easily added or removed. It was extended to 3-D, with script language.

(iii) Paralleled code using MPI (Abe et al., 2004); Incorporation of intrinsic friction between particles (Place and Mora, 1999; Abe et al., 2002, Wang and Mora, 2009); Thermal effect (Abe et al., 2000); Particle rotation and full rigidity between particles (Wang et al., 2006); Python script interface (Latham,2006 finished); Theoretical analysis on particle parameter calibration (Wang and Mora, 2008a); Hydro- effect and Darcy flow (Wang, 2008); Fully solid-fluid coupling based on DEM and Lattice Boltzmann Method (Wang, currently involved).

7.3 Theoretical and Algorithm Development

7.3.1 The Equations of Particle Motion

Particle motion can be decomposed into two completely independent parts, translational motion of the centre of mass and rotation of the centre of mass. The former is governed by the Newtonian equation

$$\boldsymbol{r}(t) = \boldsymbol{f}(t)/M, \tag{7.1}$$

where $\boldsymbol{r}(t)$ and M are position of the particle and the particle mass respectively. $\boldsymbol{f}(t)$ is the total forces acting on the particle, which may include the spring forces by the neighbouring particles, the forces by the walls, viscous force, gravitational force, etc. The equation above can be integrated using the velocity Verlet scheme (Mora and Place, 1994; Place and Mora, 1999; Allen, 1987).

The particle rotation depends on the total applied torque and usually involves two coordinate frames, one is fixed in space, called space-fixed frame, in which Eq. 7.1 is applied, the other is attached to the principal axes of the rotation body, referred to as body-fixed frame. The particle rotation is governed by the Euler's equations (in the body-fixed frame) (Goldstein, 1980)

$$\begin{aligned}
\tau_x^b &= I_{xx}\dot{\omega}_x^b - \omega_y^b\omega_z^b(I_{yy} - I_{zz}),\\
\tau_y^b &= I_{yy}\dot{\omega}_y^b - \omega_z^b\omega_x^b(I_{zz} - I_{xx}),\\
\tau_z^b &= I_{zz}\dot{\omega}_z^b - \omega_x^b\omega_y^b(I_{xx} - I_{yy}),
\end{aligned} \tag{7.2}$$

where τ_x^b, τ_y^b and τ_z^b are components of total torque $\boldsymbol{\tau}^b$ expressed in body-fixed frame, ω_x^b, ω_y^b and ω_z^b are components of angular velocities $\boldsymbol{\omega}^b$ measured in body-fixed frame, and I_{xx}, I_{yy} and I_{zz} are the three principle moments of inertia in body-fixed frame in which the inertia tensor is diagonal. In case of 3-D spheres, $I = I_{xx} = I_{yy} = I_{zz}$.

In our model, the unit quaternion $q = q_0 + q_1 i + q_2 j + q_3 k$ is used to explicitly describe the orientation of each particle (Evans, 1977; Evans and Murad, 1977). The physical meaning of a quaternion is that it represents a one-step rotation around the vector $q_1\hat{i}+$

$q_2\hat{j}+q_3\hat{k}$ with a rotation angle of $2\mathbf{arccos}(q_0)$ (Kuipers, 1998). A quaternion for each particle satisfies the following equation (Evans, 1977; Evans and Murad, 1977),

$$\dot{\boldsymbol{Q}} = \frac{1}{2}\boldsymbol{Q}_0(q)\,\Omega, \tag{7.3}$$

where $\dot{\boldsymbol{Q}} = \begin{pmatrix} \dot{q}_0 \\ \dot{q}_1 \\ \dot{q}_2 \\ \dot{q}_3 \end{pmatrix}$, $\boldsymbol{Q}_0(q) = \begin{pmatrix} q_0 & -q_1 & -q_2 & -q_3 \\ q_1 & q_0 & -q_3 & q_2 \\ q_2 & q_3 & q_0 & -q_1 \\ q_3 & -q_2 & q_1 & q_0 \end{pmatrix}$, $\Omega = \begin{pmatrix} 0 \\ \omega_x^b \\ \omega_y^b \\ \omega_z^b \end{pmatrix}$.

Now, Eqs.7.2 and 7.3 can be solved and algorithm is outlined below. We obtain the quaternion $q(t+dt)$ at the next time step using:

$$q(t+dt) = q(t) + dt\,\dot{q}(t+dt/2) + O(dt^3). \tag{7.4}$$

Hence, the quaternion derivative at mid-step ($\dot{q}(t+dt/2)$) is required. Eq. 7.3 indicates that $q(t+dt/2)$ and $\boldsymbol{\omega}^b(t+dt/2)$ are also required, where the former can be easily calculated using

$$q(t+dt/2) = q(t) + \dot{q}(t)dt/2, \tag{7.5}$$

where $\dot{q}(t)$ again is obtained from Eq. 7.3, and $\boldsymbol{\omega}^b(t)$ can be calculated, using

$$\boldsymbol{\omega}^b(t) = \boldsymbol{\omega}^b(t-dt/2) + I^{-1}\boldsymbol{\tau}^b(t)dt/2 \tag{7.6}$$

and $\boldsymbol{\omega}^b(t+dt/2)$ can be obtained, using

$$\boldsymbol{\omega}^b(t+dt/2) = \boldsymbol{\omega}^b(t-dt/2) + I^{-1}\boldsymbol{\tau}^b(t)dt. \tag{7.7}$$

Detailed algorithms used to solve Eqs.7.2 and 7.3 can be found in Wang et al. (2006) and Wang (2009).

7.3.2 Contact Laws, Particle Interactions and Calculation of Forces and Torques

Three types of interactions exist between contact particles in the current ESyS_Particle model: bonded, solely normal repulsive and cohesionless frictional interaction contacts. Bonded contact interactions between particles allow transmission of tensile forces, which can be used to model behaviour of continuum or intact materials. The breakage of bonds provides an explicit mechanism for microscopic fracture. This is different from the solely normal repulsive and cohesionless frictional interaction contacts, which does not allow tensile forces to be transmitted between particles.

7.3.2.1 Solely normal repulsive interaction

When two particles contact elastically, only the normal force exists when $d < R_1 + R_2$, here d is the distance between the two particles, R_1 and R_2 are radii of two particles.

7.3.2.2 Cohesionless frictional interaction

In the case of the frictional interaction, forces are transmitted in both normal and tangential directions when $d < R_1 + R_2$. The normal force is dealt with exactly the way it is dealt with in case of the elastic interaction. In tangential direction, static-dynamic frictional force is employed in the current ESyS_Particle model (Wang and Mora, 2009).

7.3.2.3 Bonded interaction

For bonded interactions, the three important issues that need to be specified are types of interactions being transmitted between each particle pair, the algorithm to calculate the interactions between bonded particles due to the relative motion and the criterion for a bond to break. These issues are discussed below.

The force-displacement law and interactions transmitted

Theoretically, six independent parameters are required to represent a 3-D particle, thus six types of relative motions exist between two bonded particles (Fig. 7.1). The relationship between interactions and relative displacements between two bonded particles can be written in the linear form

$$\begin{aligned} \boldsymbol{f}_r &= K_r \Delta \boldsymbol{r}, \quad & \boldsymbol{f}_{s1} &= K_{s1} \Delta \boldsymbol{s}_1 \quad & \boldsymbol{f}_{s2} &= K_{s2} \Delta \boldsymbol{s}_2, \\ \boldsymbol{\tau}_t &= K_t \Delta \boldsymbol{\alpha}_t, \quad & \boldsymbol{\tau}_{b1} &= K_{b1} \Delta \boldsymbol{\alpha}_{b1}, \quad & \boldsymbol{\tau}_{b2} &= K_{b2} \Delta \boldsymbol{\alpha}_{b2}, \end{aligned} \tag{7.8}$$

where $\Delta \boldsymbol{r}$, $\Delta \boldsymbol{s}_1 (\Delta \boldsymbol{s}_2)$ are the relative displacements in normal and tangent directions. $\Delta \boldsymbol{\alpha}_t$ and $\Delta \boldsymbol{\alpha}_{b1}$ $(\Delta \boldsymbol{\alpha}_{b2})$ are the relative angular displacements caused by torsion and rolling. $\boldsymbol{f}_r$, $\boldsymbol{f}_{s1}$, $\boldsymbol{f}_{s2}$, $\boldsymbol{\tau}_t$, $\boldsymbol{\tau}_{b1}$ and $\boldsymbol{\tau}_{b2}$ are forces and torques, K_r, K_{s1}, K_{s2}, K_t, K_{b1} and K_{b2} are relevant stiffness. Assuming that the bonds are identical in every direction, $K_s = K_{s1} = K_{s2}$ and $K_b = K_{b1} = K_{b2}$. It should be pointed out that Eq. 7.8 is valid only for infinitesimal

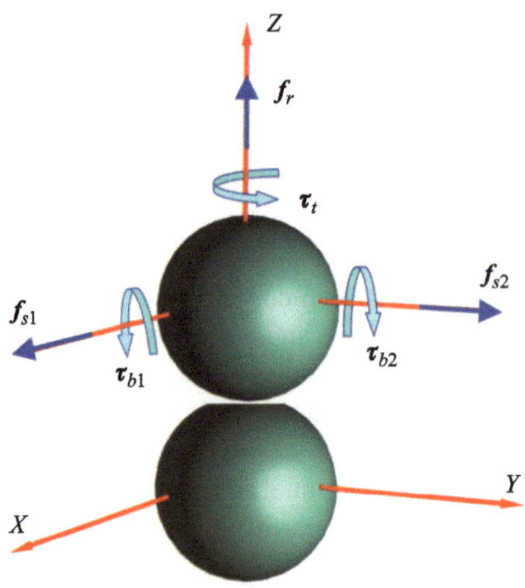

Fig. 7.1 Six types of interactions between bonded particles. $\boldsymbol{f}_r$ is normal force, $\boldsymbol{f}_{s1}$ and $\boldsymbol{f}_{s2}$ are shear forces, $\boldsymbol{\tau}_t$ is twisting torque, $\boldsymbol{\tau}_{b1}$ and $\boldsymbol{\tau}_{b2}$ are bending torque.

small deformation. In case of finite deformation, rotation series are non-commutative, or order-dependent. For example, a rotation of $\Delta\boldsymbol{\alpha}_{b1}$ followed by a rotation of $\Delta\boldsymbol{\alpha}_{b2}$ leads, in principle, to a final orientation which is different from that when we apply $\Delta\boldsymbol{\alpha}_{b2}$ followed by $\Delta\boldsymbol{\alpha}_{b1}$. We will show that the two rolling motions, $\Delta\boldsymbol{\alpha}_{b1}$ and $\Delta\boldsymbol{\alpha}_{b2}$, actually cannot be decoupled. Therefore, in practice, they are treated as one rolling motion, controlled by an orientation angle.

Calculation of interactions due to relative motion

Before solving Eqs. 7.1 to 7.3, forces and torques due to the relative motions between bonded particles need to be calculated based on the position and orientation of each particle. We developed a Finite Deformation Method (FDM) (Wang, 2009) as an alternative to the incremental method (Sakaguchi and Muhlhaus, 2000; Potyondy and Cundall, 2004). In this method the total relative (translational and rotational) displacements between each pair of bonded particles are calculated at each time step. Then, the forces and torques are calculated in terms of those relative displacements. The method is different in essence from the incremental methods, where forces and torques are incremented at each time step based on those of the previous time step.

The FDM scheme requires a complete decomposition of the relative motions between particle pairs. The basic idea of the decomposition is outlined in Figure 7.2. The body-fixed frame of one particle (particle 2) is taken as a reference, and the relative motion between the two particles is decomposed into relative translational motion (without relative rotation) and relative rotation (without translation). The two parts are completely independent. Then, the translation is decomposed into normal and tangent motions. To decompose the relative rotation, a two-step decomposition scheme is developed in which torsion (or twisting) and rolling (or rolling) are fully decoupled and order-independent (Wang, 2009). Note that the relative rotation may contribute to tangent motions.

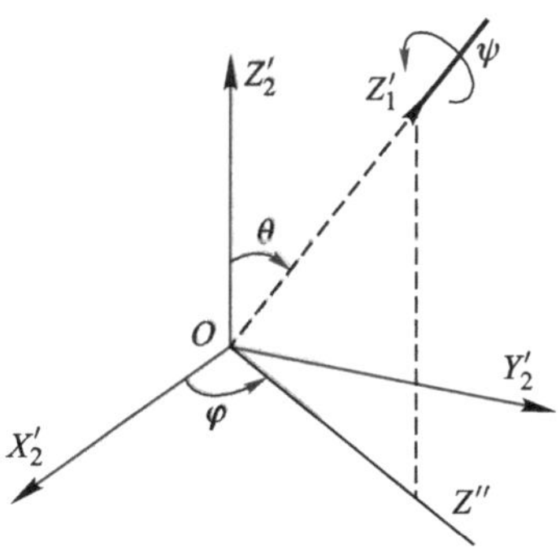

Fig. 7.2 An arbitrary rotation between two rigid bodies or two coordinate systems can be decomposed into two-step rotations, one pure axial rotation of angle ψ around its Z_2' -axis, and one rotation of Z_2' -axis over θ on certain plane controlled by another parameter φ.

Suppose in the space-fixed frame, the initial orientations of particles 1 and 2 at time t=0 are $p^0 = 1+0i+0j+0k$ and $q^0 = 1+0i+0j+0k$. At time t, the current orientations are $p = p_0+p_1i+p_2j+p_3k$ and $q = q_0+q_1i+q_2j+q_3k$. Let $X_2Y_2Z_2$ and $X_1Y_1Z_1$ represent the body-fixed frame of particles 2 and 1. The relative rotation of particle 1 over particle 2, or rotation from $X_2Y_2Z_2$ to $X_1Y_1Z_1$, is represented by quaternion $r_{21}^0 = q^{-1}p = q^*p$.

Suppose $X_2'Y_2'Z_2'$ is an auxiliary frame, obtained by directly rotating $X_2Y_2Z_2$ so that its Z_2' -axis is pointing to particle 1. It is the $X_2'Y_2'Z_2'$ system in which the relative rotation between two particles should be evaluated (Fig. 7.1). Such relative rotation corresponds to a rotation from $X_2'Y_2'Z_2'$ to $X_1'Y_1'Z_1'$ (only the Z_1' -axis is drawn in Fig. 2), and can be decided by quaternion $r_{21} = m^{-1}r_{21}^0 m = r_0 + r_1 i + r_2 j + r_3 k$ (expressed in $X_2'Y_2'Z_2'$ system), where the quaternion m specifies the rotation from $X_2Y_2Z_2$ to $X_2'Y_2'Z_2'$.

Using quaternion algebra, we proved that an arbitrary rotation between two rigid bodies or two coordinate systems cannot be decomposed into three mutually independent rotations around three orthogonal axes (Wang, 2009). However, it can be decomposed into two rotations, one pure axial rotation of angle ψ around its Z_2' -axis, and one rotation of Z_2' -axis over θ on a specific plane controlled by another parameter φ. These two rotations, corresponding to the relative twisting and bending between two bodies in our model, are sequence-independent. Such a two-step rotation is controlled by three independent parameters ψ, θ and φ, which can be decided as follows,

$$\begin{aligned}
\cos\frac{\psi}{2} &= \frac{r_0}{\sqrt{r_0^2 + r_3^2}}, \quad \sin\frac{\psi}{2} = \frac{r_3}{\sqrt{r_0^2 + r_3^2}}, \\
\cos\theta &= r_0^2 - r_1^2 - r_2^2 + r_3^2, \\
\cos\varphi &= \frac{r_1 r_3 + r_0 r_2}{\sqrt{(r_0^2 + r_3^2)(r_1^2 + r_2^2)}}, \quad \sin\varphi = \frac{r_2 r_3 - r_0 r_1}{\sqrt{(r_0^2 + r_3^2)(r_1^2 + r_2^2)}}.
\end{aligned} \tag{7.9}$$

After decomposition is done, torques and forces caused by the relative rotation between two particles are uniquely determined (Wang, 2009).

The criterion for bond breakage

In ESyS_Particle model, a bond breaks if the pure extensional force exceeds the threshold $f_r \geqslant F_{r0}$ (but it does not break under pure compression), or similarly the pure shear force $|f_s| \geqslant F_{s0}$, the pure twisting torque $|\tau_t| \geqslant \Gamma_{t0}$, or the pure bending torque $|\tau_b| \geqslant \Gamma_{b0}$.

When all the interactions exist at the same time, the following empirical criterion is used to judge whether a bond is going to break:

$$\frac{f_r}{F_{r0}} + \frac{|f_s|}{F_{s0}} + \frac{|\tau_t|}{\Gamma_{t0}} + \frac{|\tau_b|}{\Gamma_{b0}} \geqslant 1. \tag{7.10}$$

We set f_r positive under extension and negative under compression so that it is more difficult for a bond to break under compression than under extension, and the effects of normal force on breakage of the bond have been taken into account. As pointed by Wang et al. (2006), Γ_{b0} and Γ_{t0} may be related to F_{r0} and F_{s0}.

7.3.3 Calibration of the Model

We studied analytically how the particle scale stiffnesses are related to the macroscopic elastic constants of materials (Wang and Mora, 2008a). We found that the 2D triangular lattice of equal-sized particles yields isotropic elasticity, and the normal, shear and bending stiffnesses are related to the macro-scopic elastic parameters such as bulk Young's modulus E and Poission's ratio ν:

$$K_r = \frac{\sqrt{3}E}{3(1-\nu)}, \quad K_s = \frac{1-3\nu}{1+\nu}K_r, \quad K_b = \frac{\sqrt{3}(1+\nu)(1-2\nu)ER^2}{36(1-\nu)}. \tag{7.11}$$

However, in 3D case, the closest packing generates anisotropic elasticity, for example, Face-Centered Cubic (FCC, Fig. 7.3) lattice yields cubic elasticity, the simplest case for an orthotropic solid. The normal, shear, bending and twisting stiffnesses are decided using:

$$K_r = \frac{\sqrt{2}ER}{2(1-2\nu)}, \quad K_s = \frac{1-3\nu}{1+\nu}K_r,$$
$$K_b = \frac{\sqrt{2}ER^3}{48(1-\nu)} - \frac{(1-2\nu)R^2}{24(1-\nu)}K_r, \quad K_t - \frac{1-3\nu}{1+\nu}K_b = \frac{(1-2\nu)R^2}{24(1-\nu)}K_s, \tag{7.12}$$

where R is the radius of particles. If the particle stiffnesses are chosen according to Eqs. 7.11 and 7.12, the realistic macroscopic elasticity is guaranteed.

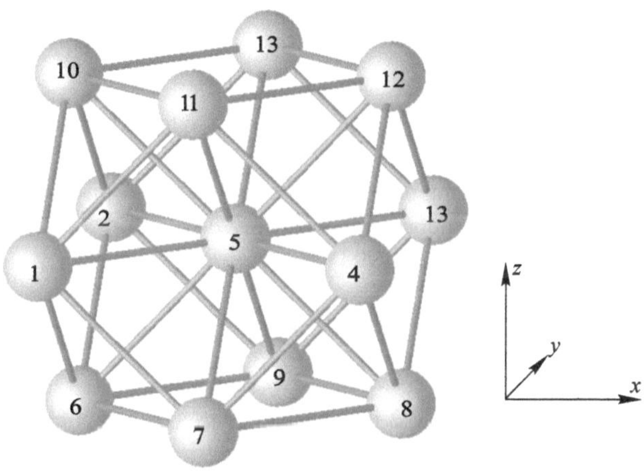

Fig. 7.3 Face-Centred Cubic (FCC) lattice and coordinate system. The centre of the FCC structure is placed on the origin of the coordinates.

7.3.4 Incorporation of Thermal and Hydrodynamic Effects

In the Esys_Particle model, thermal effects have been included. The algorithm is briefly introduced here. The behaviour of temperature in a solid with heat sources can be described by the heat flow equation

$$\frac{\partial T}{\partial t} = k_{ij}\frac{\partial^2 T}{\partial x_i \partial x_j} + \frac{1}{\rho c_p}H(x,t), \tag{7.13}$$

where ρ is the density, c_p is the heat capacity, $H(x,t)$ is the source term (i.e. frictional heat being generated by particles rubbing together) and k_{ij} is the thermal diffusivity tensor. Eq. 7.13 can be solved numerically using the following algorithm. First the source term heat is added to the particles and then the heat is extrapolated in time by an explicit finite difference method (Abe et al., 2000).

$$T'^{i}(t) = T^{i}(t) + \frac{1}{\rho c_p} H^{i}(x,t), \tag{7.14}$$

$$T^{i}(t+\Delta t) = T'^{i}(t) + \xi k \Delta t \sum_{j} \frac{T^{j}(t) - T^{i}(t)}{d_{ij}^{2}}, \tag{7.15}$$

where $T^i(t)$ and $T^i(t+\Delta t)$ are the temperature of particle i at time t and $t+\Delta t$, d_{ij} is the distance between touching particles i and j, and ξ is a factor depending on the lattice geometry, and $\xi = 1.5$ in the 2-D case and $\xi = 2$ in the 3-D case. The summation in Eq. 7.15 goes through the neighbouring particles of particle i.

The pore flow can be regulated similarly. First it is assumed that the voids inside particle are much smaller than the particle sizes; therefore the porosity is just an average concept for each particle. There is an average and uniform pore pressure p_i for each particle i, For two contacted particles i and j, the fluid exchange is determined by Darcy's Law

$$\Delta V_f = C(p_i - p_j)\Delta t, \tag{7.16}$$

where C is the conductance of the link, which is related to local permeability and geometry of the material.

The hydro-mechanical coupling can be implemented based on the Biot's linear poreelastic theory. According to this theory, the constitutive equations of a porous medium can be written as (Detournay and Cheng, 1993)

$$\varepsilon = -(P - \alpha p)/K_m, \tag{7.17}$$

$$\varsigma = -\alpha(P - p/B)/K_m, \tag{7.18}$$

where p is pore pressure, $P = -\sigma_{kk}/3$ is the mean or total mechanical pressure (isotropic compressive stress), $\varepsilon = \varepsilon_{kk} = \Delta V/V$ is the volumetric strain (positive for extension), $\varsigma = V_f/V$ is the variation of fluid content (positive corresponds to a "gain" fluid), α is Biot coefficient, B is the Skempton pore pressure coefficient and K_m is the drained bulk modulus of the material. V and V_f are the volume of the material and fluid respectively. From Eq. 7.18, the following equation is obtained

$$p = BP + K'\varsigma, \tag{7.19}$$

where $K' = BK_m/\alpha$.

The pore pressure for particle i is updated according to

$$p_i(t+\Delta t) = p_i(t) + B\Delta P_i + K'\Delta\varsigma_i, \tag{7.20}$$

where $\Delta P_i = P_i(t+\Delta t) - P_i(t)$ and $\Delta\varsigma_i = \varsigma_i(t+\Delta t) - \varsigma_i(t) = C\Delta t \sum_j (p_j - p_i)/V$.

The summation j goes though all the neighbouring particles of particle i.

7.3.5 Parallel Algorithm

DEM simulations require a significant number of particles and a huge amount of computational resources. In order to increase the computational capabilities, a parallel version of the Esys Particle has been implemented using MPI as underlying communication library (Abe et al., 2004). The parallel process structure follows a modified master-worker model (Fig. 7.4). A master process provides high level control and external communication such as a user interface or I/O facility. The local computational work is done in the worker processes. Direct communication between worker processes is also permitted whenever possible. Benchmarks using large models with several millions of particles have shown that the parallel implementation of the Esys_Particle can achieve a high parallel efficiency of about 80% for a large number of processors on different computer architectures.

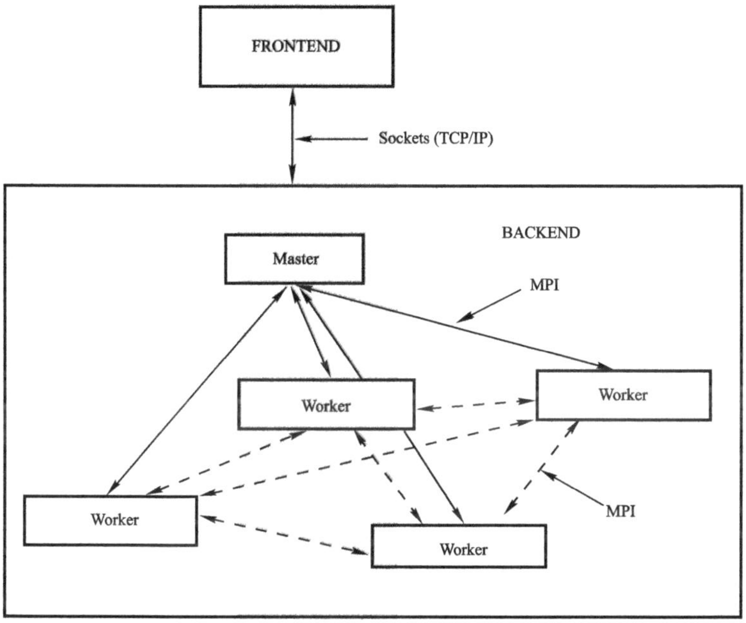

Fig. 7.4 Process structure of the parallel Esys_Particle. The communication between the parallel processes is carried on using MPI (from Abe et al., 2004).

7.4 Some Numerical Results Obtained by Using the Esys_Particle

7.4.1 Earthquakes

Earthquake faults

Using the early version of the Esys_Particle model, Mora and Place (1993; 1994) studied the stick-slip frictional instability (Fig. 7.5). The numerical experiments involve 2-D blocks with rough surfaces rubbing against one another at a constant rate. Slip pulses are observed when two interlocking asperities push against one another, and propagate approximately with the speed of the Rayleigh wave. Particle trajectories during slip have motions normal to the fault, similar to those observed in stick-slip experiments. They demonstrate that stick-slip frictional behaviour responsible for earthquakes can be modelled even using simple particle interaction.

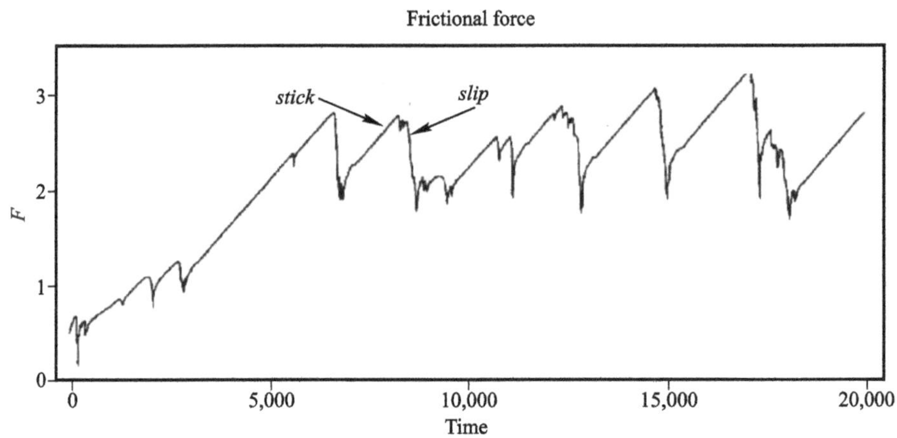

Fig. 7.5 The frictional force as a function of time between two elastic-brittle blocks with rough surfaces rubbing against one another. The sawtooth shapes are stick-slip cycles (from Mora and Place, 1994). In time scale, 20,000 units of time=100,000 time step in modeling.

Shane et al. (2006) studied the macroscopic friction of a 3-D fault using the Esys_Particle model (Fig. 7.6). In the mode, two rectangular elastic blocks of bonded particles, with a rough fault plane and separated by a region of randomly sized un-bonded gouge particles, are sheared in opposite directions by normally-loaded driving plates. It is found that the gouge particles in the 3-D models undergo significant out-of-plane motion during shear, and the 3-D models also exhibit a higher mean macroscopic friction than the 2-D models. These simulations demonstrate the important influence of the third dimension on both the microscopic and macroscopic dynamics of shear zones.

In a similar 3-D fault simulation, Abe et al. (Abe et al., 2006) modelled the temporal evolution and spatial distribution of slip events. A very simple friction law without any rate dependency and no spatial heterogeneity in the intrinsic coefficient of friction was used in the model. Slip events with a wide range of event sizes were produced, and in some of the larger events highly complex slip patterns were observed. Investigation of the temporal evolution and spatial distribution of slip during each event shows a high degree of variability between the events, indicating that the presence of geometric roughness is sufficient to cause complex stick-slip dynamics in a fault model, and variation in the intrinsic constitutive parameters or complex friction laws are not necessary.

In the recent simulations of fault gouge evolution, Abe and Mair (Abe and Mair, 2009) investigated the effects of grain fragment shape on macroscopic friction using the 3-D

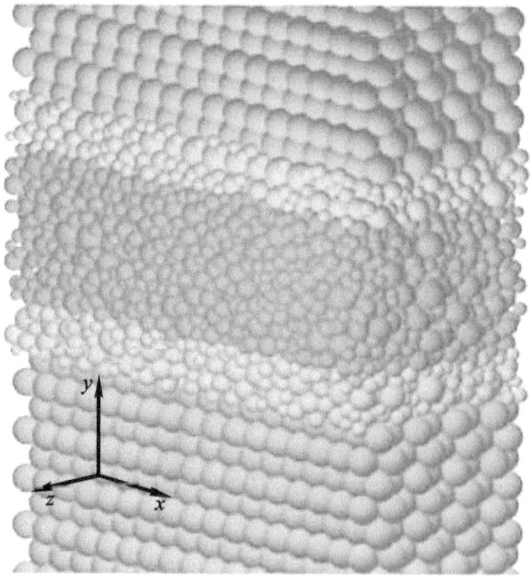

Fig. 7.6 3-D fault model setup and the initial gouge-region particle configurations. Un-bonded gouge particles are seen in the middle (from Latham et al., 2006).

Esys_Particle model. In the model, they introduced aggregate grains composed of particle clusters (cluster-simulations), where bond strength is much higher inside particle clusters than between the clusters, ensuring that broken daughter fragments are predominantly pseudo-angular rather than spherical which is common when the aggregate grains consist of bonded single particles (non-cluster simulations, Abe and Mair, 2005). They noticed that cluster-simulations yield realistic frictional strength (0.6), but non-cluster simulations have much lower friction levels. This means that gouge fragment angularity is crucial in controlling macroscopic friction. It also suggests limitations of using ideal spherical indestructible particles and necessity of the implementing more complicated grain shapes in simulations of fault gouge behaviour.

Heat paradox

A long-standing paradox in earthquake studies is the low heat flow observed around the San Andrews fault, compared with the theoretical value computed using the value of rock friction measured in laboratory experiments. Using the Esys_Particle model, Peter and Place (1998; 1999) simulated the heat generation in transform faults with and without fault gouge. The experiments consisted of a narrow fault gouge placed between two elastic books. They found that without the presence of a fault gouge, the fault is strong, and the heat generation and stress drops are correspondingly high, in accordance with the theoretical prediction. When a fault gouge was present, the fault is weak, and the heat generation as well as stress drops are low, in quantitative agreement with observational constraints. An initially strong gouge may evolve into a weak state after a long time. The simulation results suggest that the fault gouge plays a crucial role in the dynamics of fault slip and cannot be ignored. The further investigations revealed that the mechanism for low fault strength and heat with the presence of a fault gouge is rolling and jostling of gouge grains during slip. As this dynamical mechanism operates during seismic and aseismic slips, it provides an explanation for the lack of a heat flow anomaly in both the seismic and creeping parts of the San Andreas fault.

Localisation

Place and Mora (2000a) used the Esys_Particel model to study fault gouge evolution in time. Experiments consist of a weak and heterogeneous fault zone placed between two elastic blocks being subjected to shear and normal stress. The gouge region constitutes a zone of weakly bonded unbreakable grains of various shapes and sizes. Grains are composed of several particles linked to one another by strong bonds, and bonds between surface particles of different grains were allowed to break after a prescribed separation was exceeded. It is found that during the numerical experiments, a fault gouge layer forms with localisation of shear into bands (Fig. 7.7), and decreases of the gouge layer strength correlate with decreases in gouge layer thickness. After a large displacement, the

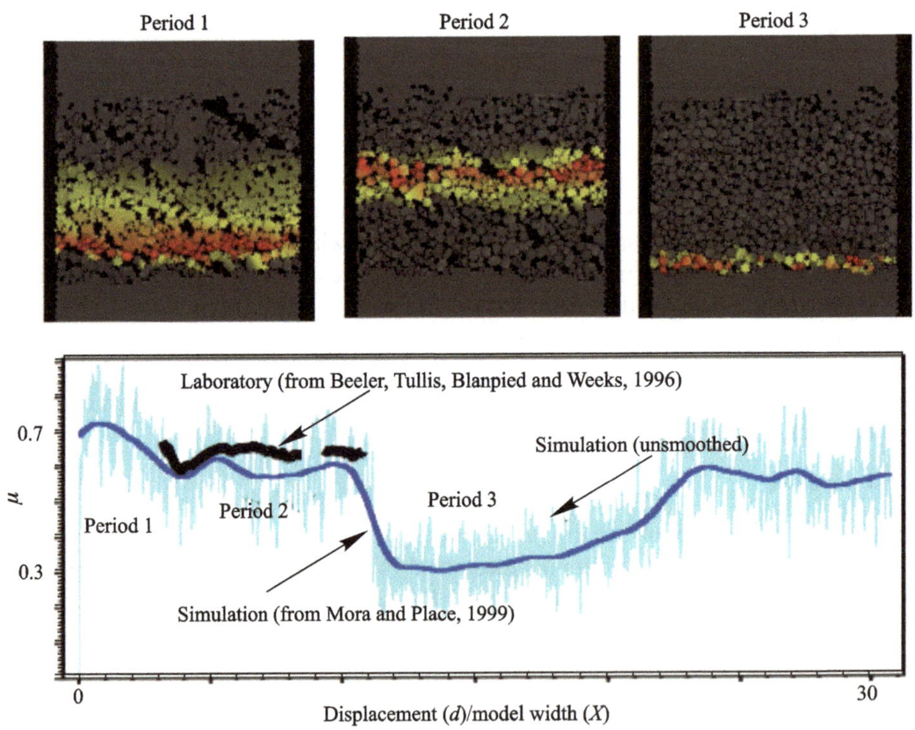

Fig. 7.7 Localisation and fault strength. Top: Snapshots showing how shear strain in a fault gouge layer localises into a narrow band. The lighter grey depicts the zone of active shear. Bottom: Fault friction in the numerical experiment compared to that in a long displacement laboratory experiment (from Place and Mora, 2000).

strength of the fault decreases due to the rolling-type mechanism, and a re-organisation of the model fault gouge is observed with slip becoming highly localised in a very narrow basal shear zone near the base of the gouge layer.

Stress correlation

It has been proposed that power-law time-to-failure of cumulative Benioff strain and an evolution in size-frequency statistics in the lead-up to large earthquakes are evidence that the crust behaves as a Critical Point (CP) system. Such a CP system predicts that stress correlation lengths should grow in the lead-up to large events through the action of small to moderate ruptures and drop sharply once a large event occurs. Mora and Place (2002) tested this idea by using the Esys_Particle model. They first defined the scalar stress correlation function as the function of scalar variable r and the vector stress correlation function, which defines the correlation between deviatoric stress values separated by vector r. Then they studied the evolution of correlation lengths in discontinuous elasto-dynamic systems subjected to shear and compression. In the case of an elasto-dynamic 2-D numerical model of a granular system subjected to shear, stress correlation evolution is built up in the lead-up to the largest event and is accompanied by an increasing rate of moderate-sized events and power-law acceleration of Benioff strain release. The stress correlation length dramatically drops after a large global rupture event (Fig. 7.8). In the case of an intact sample system subjected to compression, when the block is intact and undergoing breakdown, there is no clear evolution in the stress correlation function. Once a fracture system is well developed, there is evidence for CP-like evolution of the correlation function. When the block has almost failed entirely along a dominant fracture, there is no longer an obvious CP-like evolution of the correlation function. These results suggest that well developed systems of interacting faults may behave as CP systems but regions with few interacting faults will not. The catastrophic failure of such granular systems may be predictable in some cases if it is possible to observe the evolution of stress correlation length.

The Load-Unload Response Ratio

The Load-Unload Response Ratio (LURR, Yin et al., 1995; 2000) method is an intermediate-term earthquake prediction approach that has shown considerable promise. It is defined as the ratio of a specified energy release measure during loading and unloading where loading and unloading periods are determined from the earth tide induced perturbations in the Coulomb Failure Stress on optimally oriented faults. In the lead-up to large earthquakes, high LURR values are frequently observed a few months or years prior to the event.

As the first step towards studying the underlying physical mechanism for the LURR observations, the Esys_Particle model has been used to simulate a 2-D elastic-brittle system being subjected to uniaxial compression. A sinusoidal stress perturbation is added to the gradual compressional loading to simulate loading and unloading cycles. In each case, fractures develop and seismic energy is radiated within the model as the system is compressed until the sample fails catastrophically, and LURR is calculated. The results show that LURR values become high and then drop prior to the main event, and remain low thereafter (Fig.7.9), similar to those that have often been observed in earthquake pre-

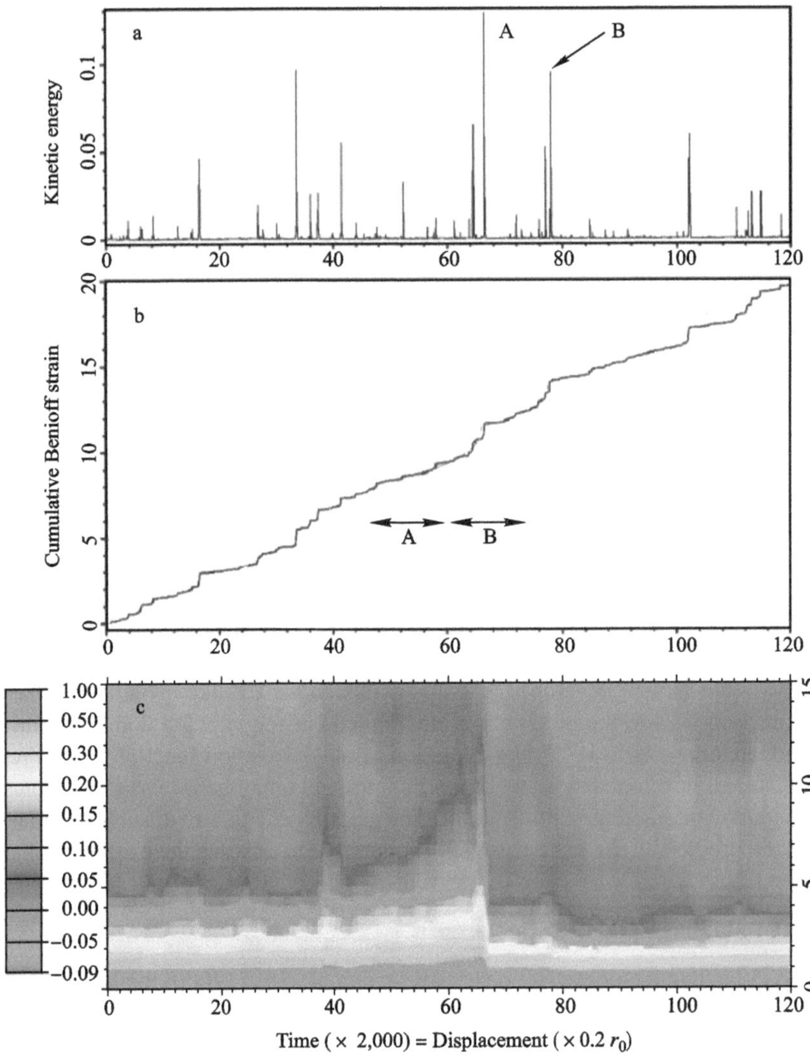

Fig. 7.8 Stress correlation length evolution. Top: Kinetic energy; Middle: Cumulative Benioff strain. Bottom: Scalar stress correlation function (from Mora and Place, 2002).

diction practice. Statistical tests of LURR values in shearing fault model show the similar results (Wang et al., 2004).

The results suggest that LURR provides a good predictor for catastrophic failure in elastic-brittle systems, and provides encouragement for the prospects of earthquake prediction using LURR and the use of advanced numerical models to probe the physics of earthquakes.

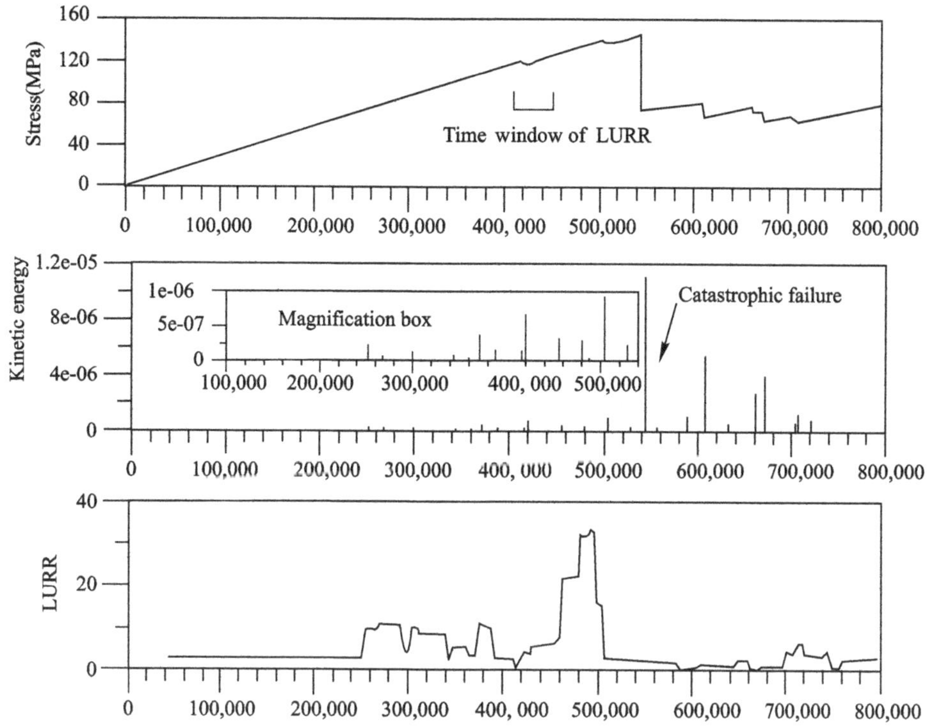

Fig. 7.9 Variation of LURR values in the strain-controlled experiment. From top to bottom, stress measured on the rigid driving plates, kinetic energy within the model, LURR values (form Mora, Wang et al., 2002).

7.4.2 Rock fracture

Uni-axial tests of material fracture were conducted using the early version of the Esys_Particle (Mora and Place, 1993; 1994; Donze et al., 1994). In these simple models, conjugates faults are reproduced. However, there are two issues of the early model. The first one is that the brittle behaviour is not well reproduced, a relatively large residual stress is observed after the peak stress in the case of the uni-axial tests without confining pressure (see Fig. 15 in Place and Mora, 2002), this is not in good agreement with the laboratory tests, where the stress quickly drops to zero with abrupt fracture of the rock specimen. The second one is that some laboratory tests, such as the realistic wing crack extension which is widely observed in uni-axial compression of brittle material with a pre-existed crack, can not be reproduced using the early model. The reason is that the early version of the model was too simple, and some important physical mechanisms are missing. We found that only when normal, shear and bending stiffness exist and single particle rotation is permitted, it is possible to reproduce the laboratory tests mentioned above, and particle rotations and rolling resistance play a significant role and cannot be neglected while modelling such phenomenon (Wang and Mora, 2008b). Due to these reasons, single particle rotation and a full set of interaction are implemented in the new model. Fig. 7.10 shows 2-D simulation of brittle fracture under uni-axial compression for different aspect ratios (3 in Fig. 7.10a, 2 in Fig. 7.10b and 1 in Fig. 7.10c), which are similar to laboratory tests (see the photos from the right of each picture, Andreev, 1995). Fig. 7.11 presents fracture patterns of a 3-D brittle rock-like material subjected to uni-axial compression. When the main faults are formed, two intact cores can be clearly observed with small fragile parts shattering away (Fig. 7.11, left). Fig. 7.12 is

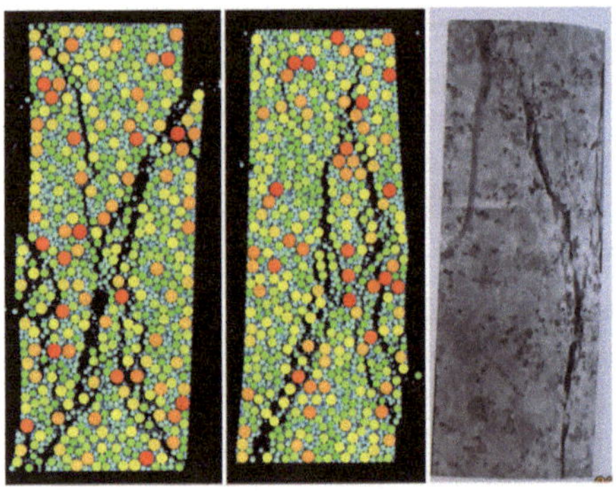

(a)

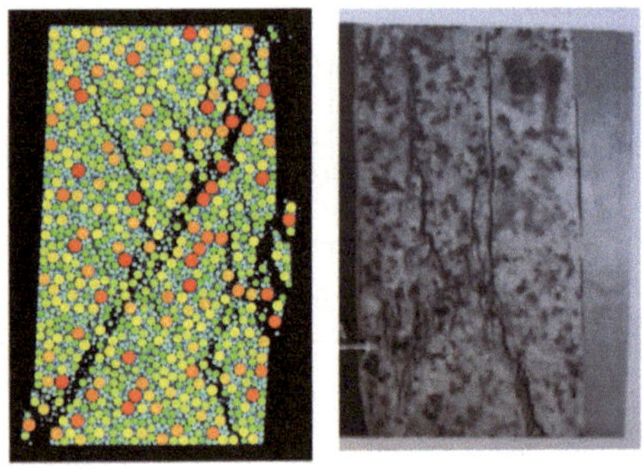

(b)

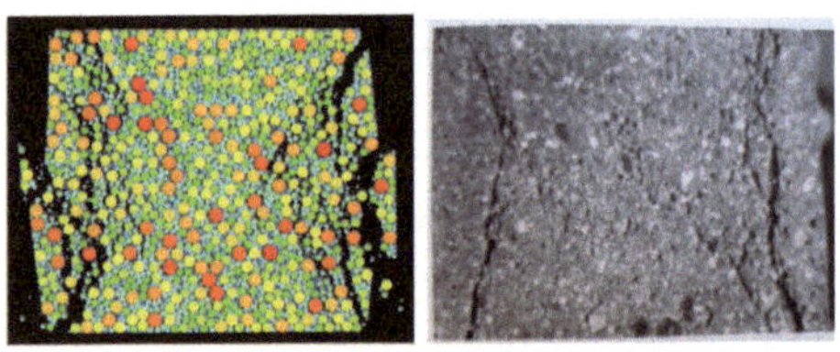

(c)

Fig. 7.10 2-D simulation of brittle fracture under uni-axial compression for different aspect ratios (3 in Fig.7.10a, 2 in Fig.7.10b and 1 in Fig.7.10c). The photos from the right of each picture are laboratory tests (Andreev, 1995).

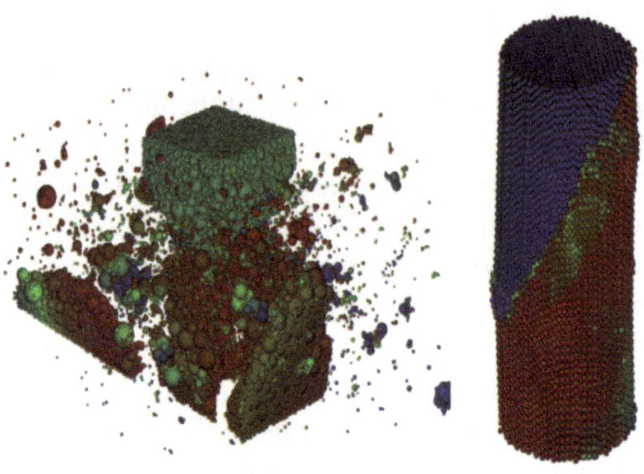

Fig. 7.11 Fractures of 3-D brittle rock-like material under slow uni-axial compression. Left: cubic shape; Right: cylinder shape.

the force versus time step curve for the left of Fig. 7.11. Brittle behaviour is clearly seen when the force suddenly drops to a value close to zero, corresponding to the abrupt brittle fracture of the sample. Fig. 7.13 shows the simulated wing crack extensions in the 2-D and 3-D cases. Small blue particles in the right plot are the centres of the removed bonds (pre-existing fault), while the bigger red ones are the centres of the fractured bonds, representing the modelled events. These simulations are very similar to the laboratory tests (Wang and Mora, 2009). Fig. 7.14 gives the simulations of brittle crack extension

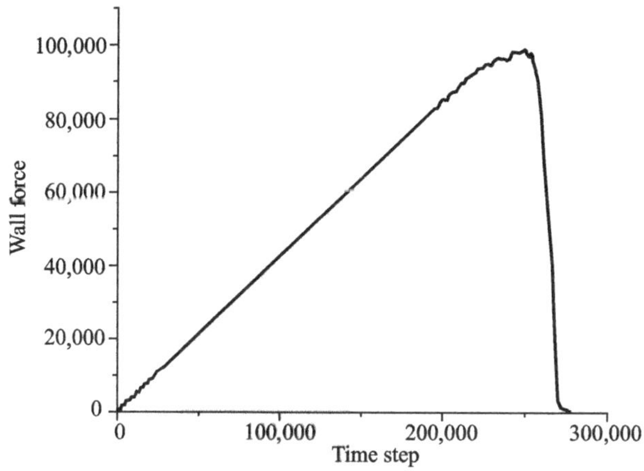

Fig. 7.12 Force versus time step curve for the left of Fig. 7.11. Brittle behaviour is clearly seen when the force suddenly drops to a value close to zero, corresponding to the abrupt brittle fracture of the sample.

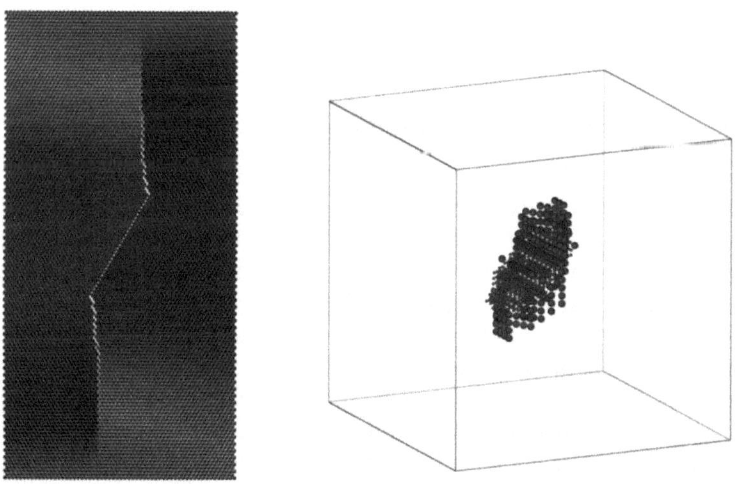

Fig. 7.13 Simulated wing crack extensions, Left: 2-D; Right: 3-D, Small blue particles are the centres of the removed bonds (pre-existing fault). The bigger red particles are the centres of the fractured bonds, representing the modelled fracturing events.

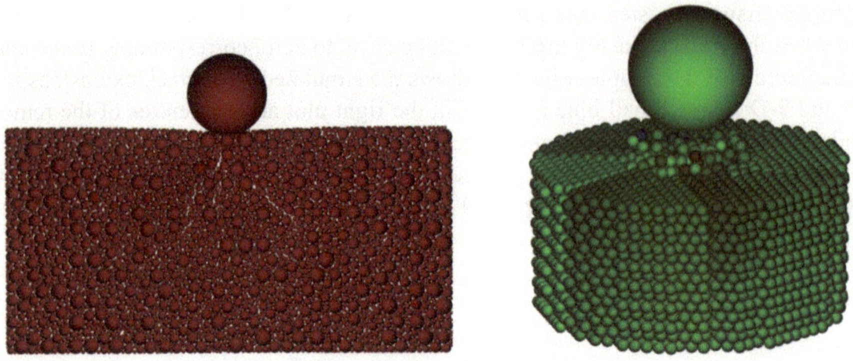

Fig. 7.14 Simulation of brittle crack extension of rock-like materials under impact by a hard ball, Left: 2-D, Right: 3-D.

of rock-like materials under impact by a hard ball. The cracks induced by the impact are clearly seen.

7.5 Coupling of Multiple Physics

Different physical processes are always involved in the earthquake-related phenomena. These processes, such as thermal- and hydro-effects, coupled with mechanical process, play important roles in the evolution and occurrence of earthquakes. In the latest version of the Esys_Particle model, these coupling mechanisms are incorporated.

7.5.1 Thermal-Mechanical Coupling

The Esys_Particle has been extended to include the generation and transport of heat and two of these temperature-related effects: heat transfer and thermal expansion (Abe et al., 2000). When two un-bonded particles are moving past each other, heat is generated due to friction. Heat transfer is implemented by adding the frictional heat to the heat flow equation as the source term and solving the equation. Numerically, two steps are employed: (1) the heat produced by interparticle friction is added to the particles and (2) subsequently the heat is extrapolated in time by an explicit finite difference method. Thermal expansion is implemented by using a linear relation between the temperature and the radius of a particle. The preliminary simulations show that temperature-related effects can have a major influence on the dynamics of single slip events and can affect the statistical distribution of events sizes, but thermal expansion has only minor influence on the dynamics of fault rupture (the size of the main events and the distribution of small to medium events) although the time of the event is influenced by the thermal expansion.

7.5.2 Hydro-Mechanical Coupling

Underground fluid may exist near the earthquake focus region, and its influences need to be evaluated. Generally, change of mechanical pressures will cause a change in pore pressure. On the other hand, fluid pressure gives rise to extra forces on the solids, which in turn affects the movement of the particles. In the Esys_Particle code, a two-way coupling between hydro- and mechanical-processes is implemented using Darcy's Law and Biot poroelasticity theory, as briefly introduced in Section 3.4 (Wang 2008).

Fig. 7.15 shows snapshots from a 2-D simulation of hydraulic fracture. A small and constant confining pressure is applied on the four boundaries. The image on the left shows the initial state of the simulation, the hole in the middle of the sample represents the location where liquid is to be injected through increasing pressure. The middle image of Fig. 7.15 shows the appearance of some cracks, and once fluid flows into the cracks it will cause them to propagate as shown on the right hand side of Fig. 7.15. It reproduces the most basic features of hydraulic fracture.

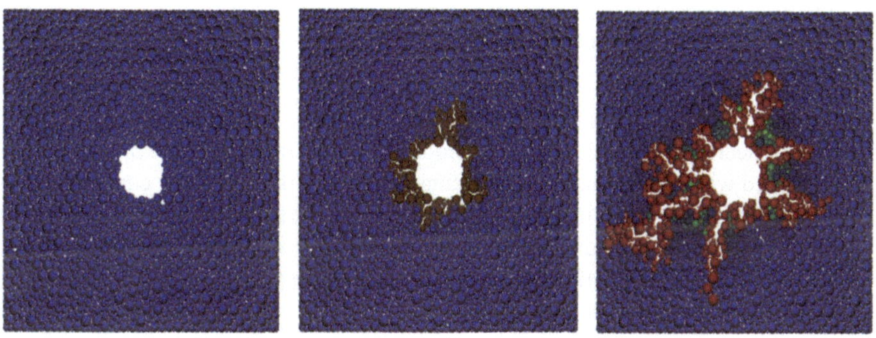

Fig. 7.15 2-D simulation of hydraulic fractures. The colours represent pore pressure (blue for low and red for high).

7.5.3 Full Solid-Fluid Coupling

Currently we are developing a fully coupled code for the direct simulation of particle-fluid system based on DEM and Lattice Boltzmann Method (LBM, Chen and Doolen, 1998). The Esys_Particle code models the movement of solid particles, while Lattice Boltzmann Method models fluid flow. The two codes are integrated into one. The interaction between solid and fluid is treated in a two-way fashion: fluid exerts a hydrodynamics force to particles and particles in return provide moving boundaries and transfer momentum to fluid. Fig. 7.16 shows the preliminary simulation results of several particles moving in inertial flow.

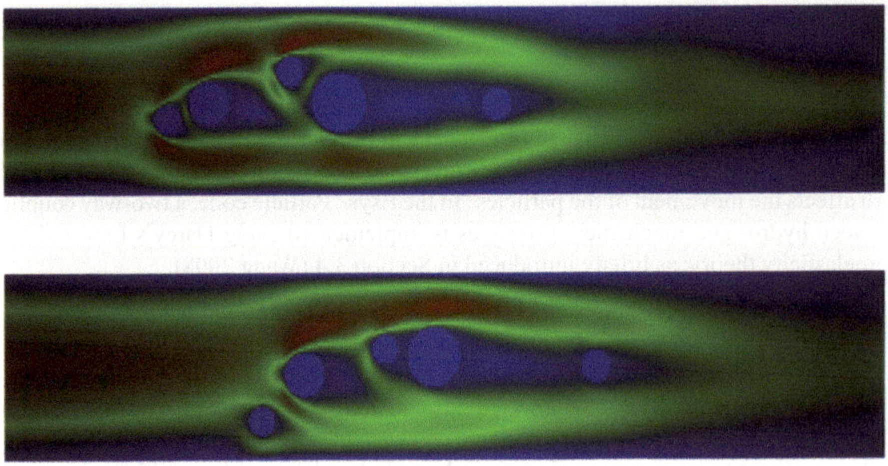

Fig. 7.16 Several particles moving in inertial flow. The colours represent fluid velocity (blue for low and red for high).

7.6 Discussion and Conclusions

The Discrete Element Method (DEM) has been a powerful numerical tool in many scientific and engineering applications. When modelling fault mechanics and brittle rock fracture associated with earthquakes, one advantage of DEM is that highly complex systems can be modelled using basic methodologies without any assumptions on the constitutive behaviour of the materials and any predisposition to where and how cracks may occur and propagate. Owing to its discrete nature, DEM is extremely suitable for modelling large deformation and dynamics phenomena.

Different types of DEMs have appeared in the past three decades. The Esys_Particle code is one of them, and has been developed since 1993. We outlined the major development of the Esys_Particle, including new particle interactions, single particle rotation, parameter calibration and parallel algorithm. Compared with the other DEMs, the Esys_Particle is different and advantageous in the following aspects:

Firstly the Esys_Particle has a unique and *explicit* representation of 3-D particle orientation using unit quaternion. There are no *explicit* rotational degrees of freedom for particles in most existing DEMs. By *explicit*, we mean the exact orientation of the particles are uniquely determined once the explicit variable is given. In the most DEMs, three angular velocities around three orthogonal axes are used to *implicitly* represent orientations. In the 2-D case this does not have any problem. However, in the 3-D case, one can not extract the exact orientation for each particle by simply integrating from the three angular velocities, because even if three angles are obtained by integration, the exact orientation of a particle still can not be uniquely determined, since finite rotations in 3-D are order-dependent. In other words, for the same three angles, different sequences of rotations will result in different final orientations. This important physical principle about finite rotations in 3-D is ignored by the other DEMs designers.

Secondly, when dealing with particle rotations, the Esys_Particle is numerically more stable than other DEMs, because in other DEMs the incremental method is adopted to update particle angular velocities. This requires a very small time step since only infinitesimal rotations are order-independent in 3-D cases. Therefore errors in the incremental method increase faster than in the algorithm used in the Esys_Particle when the time step increases (Wang, 2009).

Thirdly, in nearly all of the existing DEMs, shear forces and torques are computed in an incremental fashion (Potyondy and Cundall 2004). This means that three incremental angles from time $t - \Delta t$ to t are computed using three angular velocities, and then three incremental torques are calculated and added to the torques at time $t - \Delta t$ to get the torques at time t. The orders of the three incremental rotations are ignored, which is theoretically permitted only in case of infinitesimal rotations. It should be pointed out that when torsion and rolling exist at the same time, rolling changes the axis of torsion, which is not decoupled in the incremental method. Using the new decomposition technique, rolling and torsion are completely decoupled and uniquely determined in the Esys_Particle, so do the forces and torques between particles.

In summary, in dealing with particle rotations and calculating torques, the finite deformation scheme used in the Esys_Particle respects the physical principle and is numerically stable, while the incremental method used in other DEMs ignores the certain physical principle and is only accurate when time step and rotation are very small. Other differences between the Esys_Particle and other DEMs include criterion to judge breakage of a bond, stick-slip friction and theoretical studies on how to choose model parameters.

The applicability of the ESyS_Particle is illustrated through several numerical simulations. Most qualitative features of rock fracture observed in laboratory tests are well reproduced, including fracture of brittle materials under compression, wing crack extension in both 2-D and 3-D, crushing of aggregates and fracture caused by dynamic impact.

Besides the simulations above, the potential applications of the Esys_Particle may include:

Hydraulic fracturing in mining engineering and geo-thermal energy extraction;
Induced seismicity by reservoir and mining activities;
Tsunami source generation;
Effect of underground fluid on the earthquake generation;
Dynamic triggering of earthquakes;
Landslide and slop stability ...

The future development of the ESyS_Particle will include the coupling with other physical processes. Currently, the model has been extended to include heat transfer, thermal expansion, friction-generated heat, pore flow (Darcy flow) and full coupling between mechanical and pore effects. A fully coupled solid-fluid code is developing, and will be used to model outbursts of coal and gas in underground mines, in which gas flow and desorption play crucial roles. A fully Mechanical-Hydro-Thermal-Chemical coupled code will be in great demand in some new engineering fields such as nuclear waste disposal and CO_2 capture and sequestration, geo-thermal energy extraction.

Finally, it should be pointed out that DEM has some disadvantages. The first is that it is very time consuming since it always involves a large number of particles and time steps. The rapid development of supercomputers and parallel techniques help to improve this situation. The second is that it is difficult to choose the input parameters. For regular lattices of equal-sized spheres it is possible to derive relations between the microscopic parameters (particle scale stiffnesses) and macroscopic elastic properties (Wang et al., 2008a). Choice of micro-physical fracture parameters is more difficult, requiring empirical calibration against laboratory results.

Acknowledgements

Funding support is gratefully acknowledged, given by the Australian Computational Earth Systems Simulator Major National Research Facility, the University of Queensland and SGI. The ACcESS MNRF is funded by the Australian Commonwealth Government and participating institutions (Univ. of Queensland, Monash U, Melbourne U., VPAC, RMIT) and the Victorian State Government.

References

Abe, S. and Mair, K., 2005, Grain Fracture in 3D Numerical Simulations of Granular Shear, Geophys. Res. Lett. 32, L05305, doi:10.1029/2004GL022123.

Abe, S., Dieterich, J., Place D. and Mora, P., 2002, Simulation of the Influence of Rate and State dependent Friction on the Macroscopic Behaviour of Complex Fault Zones with the Lattice Solid Model, Pure Appl. Geophys., 159(9), pp. 1967-1983.

Abe, S., Latham, S., and Mora, P., 2006, Dynamics Rupture in a 3-D Particle-Based Simulation of a Rough Planar Fault, Pure Appl. Geophys.,163, 9, 1881-1892.

Abe, S., Mair, K., 2009, Effects of gouge fragment shape on fault friction: New 3D modelling results , Geophys. Res. Lett.36, L23302, 4 PP., doi:10.1029/2009GL040684.

Abe, S., Mora, P. and Place, D., 2000, Extension of the Lattice Solid Model to Incorporate Temperature Related Effects, Pure Appl. Geophys., 157(11/12), pp.1867-1887.

Abe, S., Place, D., and Mora, P., 2004, A Parallel Implementation of the Lattice Solid Model for the Simulation of Rock Mechanics and Earthquake Dynamics, Pure Appl. Geophys., 161(11/12), pp. 2265-2277.

Aki, K., Richards. P.G., 1980, Quantitative Seismology: Theory and Methods, vol. 1, Freeman and Company, USA.

Allen, M.P., Tildesley, D.J., 1987, Computer simulation of liquids. Oxford Science Press, Oxford.

Andreev, G.E., 1995, Brittle failure of rock materials: test result and constitutive models. Rotterdam AA Balkema.

Bak, P., and Tang, C., 1989, Earthquakes as a Self-organized Critical Phenomena, J. Geophys. Rev. 94, 15,635-15,637.

Barriere, B. and Turcotte, D. L., 1991, A Scale-invariant Cellular Automata Model for Distributed Seismicity, Geophys. Res. Lett. 18, 2011-2014.

Bird, P., 1978, Finite-element modeling of lithosphere deformation: the zagros collision orogeny. Tectonophysics 50, 307-336.

Brace, W.F. and Byerlee, J.D., 1966, Stick-slip as a Mechanism for Earthquakes, Science 153, 990-992.

Burridge, R., and Knopoff, L., 1967, Model and Theoretical Seismicity, B.S.S.A. 57, 341-371.

Carlson, J.M., and Langer, J.S., 1989, Mechanical Model of an Earthquake Faults, Phys. Rev. A 40, 6470-6484.

Carlson, J.M., Langer, J. S., and Shaw, B. E., 1989, Dynamics of Earthquake Faults, Rev. Mod. Phys. 66, 657-670.

Chang, S,H., Yun, K,J., Lee, C.I., 2002, Modeling of fracture and damage in rock by the bonded-particle model. Geosystem Engrg 5:113-120.

Chen, S. and Doolen, G. 1998, Lattice Boltzmann method for fluid flows. Anu. Rev. Fluid. Mech. 30, 329-364.

Cundall, P. A. and Strack, O. D. L., 1979, A Discrete Element Model for Granular Assemblies, Geotechnique 29, 47-65.

Day, S. M., 1982, Three-dimensional Finite Difference Simulation of Faults Dynamics: Rectangular Faults with Fixed Rupture Velocity, B.S.S.A. 72, 705-727.

Detournay, E , Cheng, A. H. D., 1993. Fundamentals of poroelasticity (Chapter 5), in J.A. Hudson (ed), Comprehensive rock engineering, Vol. 11, 113-169, Oxford, Pergaman Press.Majer, E., 2006.

Donze, F., Mora, P., Magnier, S. A., 1994, Numerical simulation of faults and shear zones, Geophys. J. Int., 116, 52-56.

Donze, F.V., Bouchez, J., Magnier, S.A. 1997, Modeling fractures in rock blasting. Int J Rock Mech Min Sci 34:1153-1163.

Evans, D,J., Murad, S., 1977, Singularity free algorithm for molecular dynamic simulation of rigid polyatomice. Molecular Phys 34:327-331.

Evans, D.J. 1977, On the representation of orientation space. Molecular Phys 34:317-325.

Fukuyama, E. and Madariaga, R. 1995, Integral Equation Method for Plane Crack with Arbitrary Shape in 3 -D Elastic Medium, B.S.S.A. 85, 614-628.

Goldstein, H. 1980, Classical Mechanics. 2nd edn, Addison-Wesley.

Hazzard, J.F., Collins, D.S., Pettitt, W.S., Young, R.P. 2000a, Simulation of Unstable Fault Slip in Granite Using a Bonded-Particle Model. Pure Appl. Geophys., 159 :221 - 245.

Hazzard, J.F., Young, R,P. 2000b, Simulation acoustic emissions in bonded-particle models of rock. Int. J. Rock Mech Min Sci 37:867-872.

Hentz S., Donze, F.V., Daudeville, L. 2004b, Discrete Element modeling of concrete submitted to dynamics loading at high strain rates Comput Struct 82:2509-2524.

Hentz, S., Daudeville, L., Donze, F.V. 2004a, Identification and validation of a discrete element model for concrete. J Eng Mech 130:709-719.

Knopoff, L. 1993, Self-organization and the Development of Pattern: Implications for Earthquake Prediction, Proceedings of the American Philosophical Society 137, 3.

Kuipers, J.B. 1998, Quaternion and rotation sequences. Princeton University Press, Princeton, New Jersey.

Latham, S., Abe, S., and Mora, P., 2006, Parallel 3D Simulation of a Fault Gouge

Using the Lattice Solid Model, Pure Appl. Geophys., 163 (9), pp. 1949-1964, doi:10.1007/s00024-006-0106-2.

Li, Q., Liu, M., 2006. Geometrical impact of the San Andreas Fault on stress and seismicity in California. Geophys. Res. Lett. 33, L08302, doi:10.1029/2005GL025661.

Lomnitz-Adler, J. 1993, Automation Models of Seismic Fracture: Constraints Imposed by Magnitude Frequency Relation, J. Geophys. Rev. 98, 17,745-17,756.

Madariaga, R., Olsen, K.B., Archuleta, R.J., 1997, 3 -D Finite-difference Simulation of a Spontaneous Rupture, Seismol. Res. Lett. 68, 312.

Melosh, H.J., Williams, C.A., 1989, Mechanics of Granben Formation in Crustal Rocks: A Finite Element Analysis. J. Geophys. Res. 94 (B10), 13961-13972.

Mora P., Place, D., 2002a, Stress Correlation Function Evolution in Lattice Solid Elastodynamic Model of Shear and Fracture Zones and Earthquake Prediction, Pure Appl. Geophys., 159:2413-2427.

Mora, P., Place, D. 1993, A lattice solid model for the nonlinear dynamics of earthquakes. Int J Mod Phys C4:1059-1074.

Mora, P., Place, D. 1994, Simulation of the Frictional Stick-Slip Instability, Pure Appl. Geophys., 143:61-87.

Mora, P., Place, D. 1998, Numerical simulation of earthquake faults with gouge: towards a comprehensive explanation for the heat flow paradox. J G R 103:21067-21089.

Mora, P., Place, D. 1999, The weakness of earthquake faults. G R L 26:123-126.

Mora, P., Wang, Y.C., Yin, C., Place, D. and Yin, X.C., 2002b, Simulation of the Load-Unload Response Ratio and Critical Sensitivity in the Lattice Solid Model, Pure Appl. Geophys., 159, 2525-2536.

Morgan JK 1999, Numerical simulations of granular shear zones using the distinct element method: II. The effect of particle size distribution and interparticle friction on mechanical behavior J G R B104:2721-2732.

Morgan JK, Boettcher M.S. 1999, Numerical simulations of granular shear zones using the distinct element method: I. Shear zone kinematics and micromechanics of localization. J G R B104:2703-2719.

Oglesby, D.D., Day, S.M., 2001. Fault geometry and the dynamics of the 1999 Chi–Chi earthquake. Bull. Seismol. Soc. Am. 91, 1099-1111.

Place D., Lombard F, Mora P., Abe S. 2002, Simulation of the Micro-Physics of Rocks Using LSMearth, Pure Appl. Geophys., 159:1911-1932.

Place, D., Mora, P. 1999, A lattice solid model to simulate the physics of rocks and earthquakes: incorporation of friction. J Comp Phys 150:332-372.

Place, D., Mora, P. 2000, Numerical Simulation of Localisation Phenomena in a Fault Zone, Pure Appl. Geophys., 157:1821-1845.

Potyondy D. and Cundall P. 2004, A bonded-particle model for rock. Int J Rock Mech Min Sci 41:1329-1364.

Potyondy D., Cundall P., Lee C.A. 1996, Modelling rock using bonded assemblies of circular particles. Rock Mechanics, Hassani and Mitri (eds) Balkema,Rotterdam.

Rapaport D.C. 1995, The Art of Molecular Dynamic Simulation. Cambridge University Press.

Rundle, JB, PB Rundle, W Klein, J Martins, KF Tiampo, A Donnellan and LH Kellogg, 2002, GEM plate boundary simulations for the Plate Boundary Observatory: Understanding the physics of earthquakes on complex fault systems, Pure Appl. Geophys., 159, 2357-2381.

Sakaguchi H, Muhlhaus H. 2000, Hybrid Modeling of Coupled Pore Fluid-Solid Deformation Problems, Pure Appl. Geophys., 157:1889-1904.

Sammis, C. G. and Smith, S.W. 1999, Seismic Cycles and the Evolution of Stress Correlation in Cellular Automation Models of Finite Faults Networks, Pure Appl. Geophys., 155, 307-334.

Sheng Y, Lawrence CJ, Briscoe BJ, Thornton C, 2004, Numerical studies of uniaxial powder compaction process by 3D DEM. Engrg Comp 21:304-317.

Sitharam TG 2000, Numerical simulation of particulate materials using discrete element modeling. Current Science, 78:876-886.

Tang, C. A. 1997, Numerical Simulation of Rock Failure and Associated Seismicity, Int. J. Rock Mech. Min. Sci. 34, 249-262.

Toomey A, Bean C.J. 2000, Numerical simulation of seismic waves using a discrete particle scheme. G J I 141:595- 604.

Wang, Y.C., 2008, Discrete element simulation of hydraulic fracture and geothermal induced seismicity. Proceedings of the Sir Mark Oliphant International Frontiers of Sciences and Technology , Australian Geothermal energy Conference, Geoscience Australia, Gurgenci and Rudd (editors), 129-133.

Wang, Y.C., 2009, A new algorithm to model the dynamics of 3-D bonded rigid bodies with rotations, Acta Geotechnica, 4,117-127.

Wang, Y.C., Abe, S., Latham, S., Mora, P., 2006, Implementation of particle-scale rotation in the 3D Lattice Solid Model, Pure Appl. Geophys., 163, 1769-1785.

Wang, Y.C., Alonso-Marroquin, F., 2008, DEM simulation of rock fragmentation and size distribution under different loading conditions. Proceedings of the 1st Southern Hemisphere International Rock Mechanics Symposium (SHIRMS 2008), Potvin, Carter, Dyskin and Jeffrey (editors). V2,149-156.

Wang, Y.C., Alonso-Marroquin, F., 2009a, A new Discrete Element Model: Particle rotation and parameter calibration, Granular Matter, 11,331-343.

Wang, Y.C., Alonso-Marroquin, F., 2009b, Calibration of DEM simulation: Unconfined compressive test and Brazilian tensile test. Powders and Grains 2009 Conference, 13-17 July, Colorado School of Mines.

Wang, Y.C., Alonso-Marroquin, F., 2009c, DEM simulation of comminution: Energy consumption and size distribution. Powders and Grains 2009 Conference, 13-17 July, Colorado School of Mines.

Wang, Y.C., Mora, P., 2008a, Elastic properties of regular lattices, J Mech Phys Solids, 56,3459-3474.

Wang, Y.C., Mora, P., 2008b, Modeling wing crack extension: implications to the ingredients of Discrete Element Model, Pure Appl. Geophys., 165, 609-620.

Wang, Y.C., Mora, P., 2009, ESyS_Particle: A new 3-D discrete element model with single particle rotation. Advances in Geocomputing, Springer, eds by HL Xing, 183-228.

Wang, Y.C., Mora, P., Yin, C., Place, D., 2004, Statistical test of the Load-Unload Response Ratio (LURR) signals using the Lattice Solid Model (LSM): implication to tidal triggering and earthquake prediction, Pure Appl. Geophys., 161, 1829-1839.

Wang, Y.C., Yin, X.C., Ke, F.J., Xia, M.F., Peng, K.Y., 2000, Numerical Simulation of rock failure and earthquake process on mesoscopic scale, Pure Appl. Geophys., 157, 1905-1928.
Xing, H.L., Makinouchi, A, Mora, P., 2007, Finite element modeling of interacting fault systems, Physics of the Earth and Planetary Interiors 163, 106-121.
Yin, X.C., Chen, X.Z., Song, Z.P., and Yin, C. 1995, A New Approach to Earthquake Prediction: The Load/Unload Response Ratio (LURR) theory, Pure Appl. Geophys., 145, 701-715.
Yin, X.C., Wang, Y.C., Peng, K.Y., and Bai, Y.L. 2000, Development of a New Approachto Earthquake Prediction: Load/Unload Response Ratio (LURR) Theory, Pure Appl. Geophys., 157, 2365-2383.

Author Information

Yucang Wang, Sheng Xue, Jun Xie

CSIRO Earth Science and Resource Engineering
PO Box 883, Kenmore, QLD 4069, Brisbane, Australia
Email: yucang.wang@csiro.au

Bei Fragen zur Produktsicherheit wenden Sie sich bitte an:
If you have any questions regarding product safety,
please contact:

Walter de Gruyter GmbH
Genthiner Straße 13
10785 Berlin
productsafety@degruyterbrill.com